Statistics Applied with the R Commander

Franz Kronthaler

Statistics Applied with the R Commander

Data Analysis Is (Not) an Art

Springer

Franz Kronthaler
Hochschule Graubünden FHGR
Chur, Switzerland

ISBN 978-3-662-69106-9 ISBN 978-3-662-69107-6 (eBook)
https://doi.org/10.1007/978-3-662-69107-6

This book is a translation of the original German edition "Statistik angewandt mit dem R Commander," 2nd edition, by Franz Kronthaler, published by Springer-Verlag GmbH, DE in 2021. The translation was done with the help of an artificial intelligence machine translation tool. A subsequent human revision was done primarily in terms of content, so that the book will read stylistically differently from a conventional translation. Springer Nature works continuously to further the development of tools for the production of books and on the related technologies to support the authors

This Springer imprint is published by the registered company Springer-Verlag GmbH, DE, part of Springer Nature.
The registered company address is: Heidelberger Platz 3, 14197 Berlin, Germany

A Note for the Reader

It is often observed that many students have great respect for the subject of statistics and that many people, although they may need to learn statistics, are not well trained in thinking about statistical correlations and statements. One reason for this is probably that statistics are often taught in a highly mathematical and somewhat incomprehensible way for many. More than five years ago, I therefore tried to write a simple, understandable statistics book that can be used across disciplines and that helps to apply statistics and train statistical thinking. My goal was to provide all people with a simple and intuitive access to statistics. The many feedbacks from colleagues, female and male students show that the original idea and the concept have proven themselves. The book, and I am very pleased about this, is now recommended and used at many universities and in various departments. However, a book can always be improved. This is to be done with this new edition, without shaking the proven concept. For example, in recent years there has been intense discussion in statistics about the term "statistical significance". In addition, there are also minor errors that need to be corrected.

But let's go back again, why statistics at all. A simple reason is that we are confronted daily with statistical numbers or correlations. When we open a newspaper, electronically or traditionally, we automatically read about correlations and these need to be understood. For example, I recently read in a daily newspaper that young people drive cars that emit a lot of CO_2. The headline of the article was something like "Young people don't care about the climate when it comes to cars". This is about a relationship between age and the environmental harm of the driven car. Is this really a relationship between age and environmentally harmful cars? Or is there perhaps a third variable, the available money, behind this relationship? It's worth thinking about.

The ability to apply statistics and analyze data is becoming increasingly important. This applies universally to all disciplines, businesses, public administrations, and people. If we observe the behavior of individuals, companies, and other things and evaluate these observations well, then we can also make informed decisions. The better our education, the more sound our decisions are, the better hopefully the world we live in. Ultimately, this is the motive why I started writing this book more than five years ago. It is intended to help analyze data, apply statistics, and think in statistical contexts. It shows that data analysis is not as complicated as many believe. The book provides easy access to the analysis of a dataset. At the same time, it enables better assessment of data analyses and statements. Who doesn't know the phrase: Don't trust any statistics you didn't fake yourself.

The special feature of the book is that it discusses the methods of statistics step by step using a dataset. This makes it understandable how the methods of statistics build on each other and how more and more information can be drawn from a data set over time. The focus is on the contents of statistics that are needed to analyze a data set. Contents that are rarely used in data analysis are omitted. This keeps the book slim.

The second feature of the book is the focus on application. The book is not written mathematically. From experience, I know that mathematics often deters readers from learning and applying statistics. For this reason, the book focuses on the concepts and ideas of statistics and their application. You will be enabled to use statistical methods. I do not believe that statistics can be taught entirely without mathematics. However, it is possible to reduce the mathematics to the essentials and incorporate it in such a way that the application of statistics, rather than the mathematics, is in the foreground. The reader of the book should not even notice anymore that mathematical concepts are being used to generate knowledge. With enjoyment, it should be learned what benefit there is in using statistical methods.

The third feature is the easy replaceability of the data set. It is readily possible to use a different data set and work through the book with it. The analysis of a data set requires a systematic approach. This systematic approach is reflected in the structure of the book.

All three features together enable you to systematically analyze a data set without much effort.

Data analysis is fun!

New Features and Additions

Writing a textbook is always "work in progress", i.e., one can always do something better or differently, or there are developments that should be integrated. Therefore, it makes sense to reconsider the content and its presentation after some time. In addition, one always receives suggestions that can contribute to the improvement of the book. So far, it appears that the original concept of the first edition has proven itself, and therefore there is no need to revise the book conceptually. However, there are numerous contents that are worth integrating.

In statistics, there has been an intensive discussion about the term "statistical significance" and its use for some time. An important addition is therefore the consideration of this discussion and the inclusion of the resulting recommendations. In particular, the previous statements are supplemented by the effect size.

I pondered a bit longer whether price and quantity indices should be integrated. Ultimately, due to the importance of the topic, for example in considerations of inflation, I decided to integrate this topic into the chapter "Ratios".

A new chapter on the analysis of variance has also been introduced. This allows the reader to test not only for group differences in two groups, but also in multiple groups. The great advantage, however, is that this discusses an important concept of statistics, the concept of how to divide variance in an explained and an unexplained part.

A need of students is always the ability to directly apply what they have learned using applications and tasks. The new edition naturally addresses this, the number of applications is further increased. In particular, real data and data sources are integrated, so that the previous limitation to exclusively simulated data is eliminated. In addition, the students learn to find real data.

In addition, minor errors were corrected and some concepts of statistics are discussed in more detail. Furthermore, the new edition was used to switch to the latest version of the R Commanders.

Learning Features of the Book The book discusses statistics using a simulated real-life example. After reading the book, the reader should be able to systematically analyze a data set and filter out the information contained in the data set. To achieve this, the book uses the following additional features in addition to the discussion on the specific example.

Calculating by Hand and Calculating with R Examples are calculated both by hand (for a manageable number of observations) and with the R Commander (on a complete data set). This facilitates the understanding of the ideas and concepts of statistical methods, while we also learn to apply them to a larger data set with the R Commander. With the latter, we also gain access to the command language of R, so we learn to use R even without a graphical interface. In addition, the R Commander can be used to check tasks calculated by hand.

Freak-Knowledge The Freak-Knowledge section serves to address knowledge and concepts that go beyond the contents of the book. It brings up interesting aspects that generate additional knowledge, without fully discussing them. The section is also intended to whet the appetite for more.

Checkpoints At the end of each chapter, the key points are briefly summarized in bullet points. This provides the reader with an overview of what they should particularly retain from each chapter. Additionally, they are encouraged to reflect on these points again.

Applications & Solutions In addition to the checkpoints, exercises are provided at the end of each chapter to deepen the discussed knowledge. Understanding the concepts of statistics is promoted by manual calculation, practicing with the R Commander serves to further deepen the analysis of a dataset. The solutions can be found at the end of the book. New in this book is also the integration of tasks with real data. This data is provided in manageable quantities in the applications. However, it is always possible and recommended to refer back to the original source and work with updated data. This enables access to real data.

Additional Data Sets Also at the end of the book, another data set is provided, which allows the reader to apply what they have learned to another example. Furthermore, lecturers and students from other disciplines can find additional data sets at http://www.statistik-kronthaler.ch, which enable them to learn statistics based on specific subject-related application areas.

Acknowledgments

A project like this cannot be realized alone. It relies on support and inputs. With this, I would like to thank everyone who has contributed to the success of the project. Some are particularly noteworthy. First and foremost, my students. They have the patience, sometimes they have to, to listen to me, and support me with valuable questions and advice. But also my family, who stand by me with advice and action. Special thanks also go to the Springer team, Mrs. Denkert, Mr. Heine, Mrs. Herrmann, Mrs. Ruhmann and Mrs. Stricker, who believed in the project from the beginning and opened up the opportunity for me to publish the book. Mrs. Ruhmann deserves special thanks for this new edition.

Contents

List of Figures

List of Tables

Part I

Basic Knowledge and Tools to Apply Statistics

1 Statistics is Fun

Statistics is fun and interesting. Interesting because with statistics we discover relationships and understand how individuals and objects, e.g., companies, behave. Fun, simply because it's fun to discover things. At the same time, statistics is important. If we can observe and evaluate how individuals, companies, or other objects behave, then we understand many things much better. In this chapter, we clarify why statistics is interesting and relevant, while also clarifying important statistical terms.

1.1 Why Statistics?

Why statistics? This question is asked by many students of business economics, information science, psychology, sociology, tourism, among others. Let the statisticians deal with the statistics. Why should we learn statistics? If this question is answered, then we have already won. We are motivated to deal with statistics and statistics is already half learned. Statistics is important for all of us for a number of good reasons.

Firstly, a good part of the knowledge in the world is based on data and has accordingly arisen through data analysis. We can divide knowledge into three types, experiential knowledge, theoretical knowledge, and empirical knowledge. Experiential knowledge is knowledge that we acquire over the course of life based on our own experiences, e.g., I burn my fingers on a hot stove! If we think about it for a moment, we immediately realize that the smallest part of our knowledge is based on such experiences. Theoretical knowledge is, colloquially speaking, knowledge that is logically derived from observations. This brings us to the data. Theoretical knowledge is based on observations. But that's not all. It is checked and improved with the help of new observations and data, which brings us to empirical knowledge. For example, not too long ago it was generally accepted knowledge that the number of children born per woman in richer

F. Kronthaler, *Statistics Applied with the R Commander*,
https://doi.org/10.1007/978-3-662-69107-6_1

countries is lower than in poorer countries. This knowledge arose from the observation that the birth rate decreases as countries become richer. Newer observations show that this still applies today, but with the small difference that from a certain degree of wealth, the number of children per woman increases again.

The second argument is simply that data is sometimes used to create opinion and mood, at worst to manipulate people. If we understand how knowledge or supposed knowledge is generated from data, then we are better able to recognize attempts of manipulation.

And finally, data analysis helps to make good and correct decisions. Hal Varian, chief economist at Google, said some time ago, loosely translated, the following: "The coolest job is the job of a statistician. [...] The ability to extract and communicate information and knowledge from data will be incredibly important in the coming decades [...]". Companies like Google, but also other large and small companies, process data to get to know their customers and employees, to open up new markets or to check the effectiveness and image of their products.

There are plenty of good reasons to deal with statistics. But the best reason for us is that learning the techniques of data analysis is not as complicated as some might fear. It just needs a lecturer who explains the concepts of statistics simply, "a lecturer on your side".

1.2 Checkpoints

- *Knowledge is based on experiences, theories, and data, with the largest part of our knowledge based on data.*
- *Data is sometimes used to manipulate us.*
- *Sound data analysis helps us make good decisions.*

1.3 Data

In statistics, we deal with the analysis of data. So first, we need to clarify what data is and how a data set is structured. A data set is a table that contains information about individuals, companies, or other objects. The information is the data. The data set contains three pieces of information: 1) about whom we have information, 2) about what we have information, and 3) the information itself. Let's imagine we are studying ten people and want to know their gender, age, and income, and also how much they like to eat cheeseburgers. We can display this information in a table.

Table 1.1 shows us a first data set. This data set is structured as follows: The first column shows us about whom or which objects we have information. The No. 1 can stand for Mr. Tchaikovsky, the No. 2 for Mrs. Segantini, the No. 3 for Mrs. Müller, etc. Column 1 technically gives us the objects of study.

Table 1.1 Data set persons, gender, age, income, burger

Person No.	Gender	Age	Income	Cheeseburger
1	Male	55	10,000	Very much
2	Female	23	2800	Very much
3	Female	16	800	Less
4	Male	45	7000	Not much
5	Male	33	8500	Like
6	Female	65	15,000	Less
7	Male	80	2000	Less
8	Female	14	0	Not much
9	Female	30	5200	Very much
10	Female	45	9500	Not much

In the first line, we can read what information we have. In this data set, this is information about gender, age, income, and how much the people like to eat cheeseburgers. In a statistical sense, these are the variables. They describe the characteristics of the objects of study.

The cells from column A and row 2 (short A2) to column E and row 11 (E11) provide us with information about the individual persons and variables. Person No. 1, Mr. Tchaikovsky, is male, 55 years old, has an income of 10,000 francs, and really likes cheeseburgers. Person No. 10 is female, 45 years old, has an income of 9500 francs, and does not like cheeseburgers. In this way, we can read each row and get information about each individual person. But we can also calculate how the people behave overall with regard to the individual variables. For example, it can be quickly seen that the data set contains 4 men and 6 women, and that 3 out of 10 people (30% of the subjects) really like cheeseburgers, etc.

A data set is usually organized in the way shown above. The rows contain the objects. The columns contain the variables. The intersections between columns and rows provide the information about the objects.

In addition, there is another detail that we need to know. As a rule, a data set contains only numbers in addition to the variable names, no other words. The reason is simple, it is easier to calculate with numbers. Exceptions are possible, for example, if we deliberately collect words or texts and want or need to work with them. For example, many questionnaires have a question where the respondents can provide further suggestions for the questionnaire.

Table 1.2 shows us the same data set, only now all information is given in numbers. The information that was presented as words has been converted into numbers. Technically speaking, we have encoded the verbal information, i.e., backed it up with numbers.

Table 1.2 Coded data set persons, gender, age, income, burger

Person No	Gender	Age	Income	Cheeseburger
1	0	55	10,000	1
2	1	23	2800	1
3	1	16	800	3
4	0	45	7000	4
5	0	33	8500	2
6	1	65	15000	3
7	0	80	2000	3
8	1	14	0	4
9	1	30	5200	1
10	1	45	9500	4

Table 1.3 Legend for the data set persons, gender, age, income, burger

Variable name	Variable description	Values	Missing values	Scale
Person Nr	Running index of the interviewed persons			
Gender	Gender of the interviewed persons	0 = male 1 = female	n. d.	Nominal
Age	Age of the interviewed persons	In years	n. d.	Metric
Income	Income of the interviewed persons	In CHF	n. d.	Metric
Cheeseburger	How much do the interviewed persons like eating cheeseburgers?	1 = very much 2 = somewhat 3 = not really 4 = not at all	n. d.	Ordinal

Source: Own survey 2013

The encoding requires creating a legend in which we note the meaning of the numbers. The legend should contain the following information: variable name, variable description, values, missing values, scale, and source.

Table 1.3 shows us the legend for our data set. The variable name is the short designation used for the variable. The variable description precisely describes the meaning of the variable. The idea is simply that we must still know, perhaps in a year, what, for example, the variable cheeseburger actually means. The values column tells us how the data is defined, e.g., that for the variable gender, 0 stands for a man and 1 for a woman. Missing values are marked in this legend with the letters n. d., where n. d. stands for not defined. We do not need to define missing values, as our data set contains information about all individuals and all variables. However, it is often the case that not all individuals have information for all variables. We then help ourselves by either leaving the cell in the data set empty or filling it with a defined value that cannot occur with the corresponding variables, usually 8, 88, 888, …, 9, 99, 999, …. We can annotate the defined value with a note indicating why the value is missing, e.g., 8, 88, 888, …, can stand for "no information available" or 9, 99, 999, …, for "response refused". In the first case, the person interviewed was not able to give an answer. In the second case, the person interviewed, for whatever reasons, refused to answer. The scale column informs us about the scale level of the data. This information is of great importance when selecting statistical methods. We will discuss this in the next section. We also need information about the source of the data. This information helps us assess the reliability of the data. In our case, we collected the data ourselves, the source is our own survey from 2013. If we have worked reliably, we know that the data set is of good quality. But if we know that the data come from a less trustworthy source, we do not have to discard them. However, we must take into account the uncertainty regarding data quality in the data analysis.

Perhaps you have also noticed in the legend that in the second row only variable name and the variable description contain an information. The cells for values, missing values, and scale are empty. Here we deal with the running index of the observed persons, accordingly the values are a consecutive number, we have no missing values, as all observed persons are included in the data set and we do not need a scale, as this column is not used for calculations.

1.4 Checkpoints

- *A data set is organized in columns and rows. The objects of study are contained in the rows, the variables in the columns.*
- *A data set is usually constructed with the help of numbers. This makes it easier to calculate.*
- *The legend of the data set should contain the variable name, the variable description, the definition of the values, and information about missing values and the scale.*

1.5 Scales—Lifelong Important in Data Analysis

We are rapidly approaching data analysis. Before we start, we need to deal with the scale level of variables. The scale level of the data essentially determines which statistical methods are applicable. We will see this in each of the following chapters, and we will also repeatedly find this in our further statistical life. It is worth investing time here and memorizing the types of scales and their properties.

In our legend, we have already learned about the three scale levels. Data can be nominal, ordinal, and metrically scaled.

Nominal means that the variable only allows a distinction between the persons and objects of study. In our example, the variable gender distinguishes between man and woman, a man is a man, a woman is a woman and a man is not equal to a woman. Technically speaking, in this case $0 = 0$, $1 = 1$ and $0 \neq 1$. Further examples are the religion with the characteristics Catholic, Reformed, Russian Orthodox, Buddhist, etc., the corporate form of companies with the charateristics limited company, public limited company, ..., the blood group of persons or the breed of dogs. In these examples, we also see that there are nominal variables with two or more than two characteristics. For a nominal variable with two characteristics, e.g. the variable gender with the characteristics man and woman, we also speak of a dichotomous variable or a dummy variable.

For an ordinal variable, in addition to the possibility of distinction, we also have a ranking, such as in the form of better-worse, friendlier-unfriendlier, etc. In our example, the variable cheeseburger indicates how much people like to eat cheeseburgers. We have a ranking from very much to somewhat, not really to not at all. We know whether a person likes a cheeseburger very much, somewhat, not really or not at all. However, we do not know the distance between our characteristics. Technically speaking, in our example $1 = 1$, $2 = 2$, $3 = 3$, $4 = 4$ and $1 > 2 > 3 > 4$. The distances between the characteristics 1, 2, 3 and 4 are not defined. A classic example of this are grades. With grades, we only know whether we have achieved a very good, good, satisfactory, sufficient, poor or unsatisfactory grade. The additional learning effort we need to go from a sufficient to a satisfactory grade or from a good grade to a very good grade varies from grade to grade, from teacher to teacher and from exam to exam. Further examples of ordinal scaled variables are satisfaction with a product with the characteristics very satisfied, satisfied, less satisfied, dissatisfied or self-assessment of the risk of injury in a sport with the characteristics high, medium, low.

In the case of metric variables, in addition to distinction and ranking, the distance between two characteristics is also clearly defined. In our example, age and income are metric. A person who is 21 years old is older than one who is 20 years old, and the difference between the two values is exactly 1 year. We know exactly how much time passes from the 20th to the 21st or from the 35th to the 36th birthday. Technically speaking, $X > Y > Z$ and $X - Y$ or $Y - Z$ is clearly defined. Furthermore, we have theoretically infinite intermediate values between two values. Let's look again at the example from the 20th to the 21st year of life. In between are 365 days or 8760 hours or 525,600 minutes, etc. Other examples of metric data are temperature, sales, distance between two places, etc. (Table 1.4).

In applied data analysis, the question often arises whether we can treat ordinal data in the same way as metric data. We will understand this later. This question can be answered with actually not, but …. Actually, metric is metric and ordinal is ordinal. Nevertheless, ordinal data is often treated like metric data. This is justifiable when an ordinal variable has enough characteristics, about seven (e.g., from 1 to 7), more is better, and the variable is normally distributed. What normally distributed means, we will clarify later. More

Table 1.4 Scales of measurement

Scale level	Properties	Examples
Nominal	Distinction $0 = 0$, $1 = 1$ and $0 \neq 1$	Gender: male, female Hair color: black, brown… Blood group: A, B, AB and 0
Ordinal	Distinction $1 = 1, 2 = 2, 3 = 3, 4 = 4$ and $1 \neq 2, \ldots$ Ranking $1 > 2 > 3 > 4$	Grades: very good, good, satisfactory... Rating: very much, somewhat, not much, not at all
Metric	Distinction $X = X$, $Y = Y$, $Z = Z$ and $X \neq Y, \ldots$ Ranking $X > Y > Z$ Distance $X - Y$ or $Y - Z$ is defined	Age: 5 years, 10 years, 20 years… Pocket money: CHF 10, CHF 20, CHF 25, … Weight: 50 kg, 65 kg, 67 kg, …

elegant is the creation of a quasi-metric variable. Instead of asking, "How much do you like eating cheeseburgers?" and having this question answered on a scale from 1 to 7, we can ask the same question and instead have a cross placed on a line.

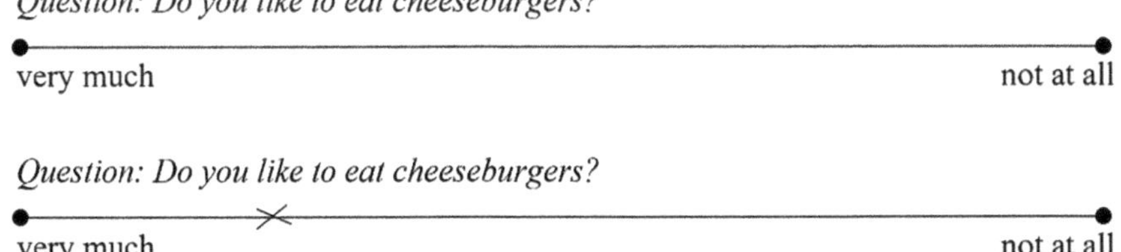

We then measure the distance between the beginning of the line and the cross as accurately as possible. A small distance in this case means that someone likes eating cheeseburgers very much. The closer the distance comes to the maximum possible distance, the less the person likes eating cheeseburgers. If we try this a few times, i.e., simulate for several people, we quickly realize that we get very many different values and thus have created a quasi-metric variable.

As long as we analyze data, the scale levels will not leave us alone!

1.6 Checkpoints

- *Nominal variables allow a distinction of the objects of study.*
- *Ordinal variables include a ranking in addition to a distinction.*
- *Metric variables allow a precise distance determination in addition to the distinction and the ranking.*

- *Ordinal data can be treated like metric data if enough characteristics are possible and the data is normally distributed. A more elegant approach is the creation of a quasi-metric variable.*
- *The scale level plays a decisive role in determining which statistical methods we can apply.*

1.7 Software: Excel, SPSS, or "R"

Data analysis today fortunately no longer happens by hand. We have various computer programs available to assist us, e.g. Excel, SPSS or "R". Which of these programs should we use?

Excel is Microsoft's spreadsheet program. It is not pure statistical software, but it has a variety of functions that can be used for data analysis. The advantages of Excel are that it can be operated via a graphical interface, that we usually know it and that it is installed on our private computer. In addition, it is usually also available in the company we work. The disadvantage is that not all statistical applications are available.

The statistical program SPSS **S**tatistical **P**ackage for the **S**ocial **S**ciences is a professional statistical software. Accordingly, it has most of the known applications of statistics. Like Excel, SPSS works via a graphical interface and no programming skills are necessary. The big disadvantage of SPSS is the price. Also, we cannot easily access it in a company where SPSS is not available.

A free and also professional statistical software is available with the program "R". "R" is an open source software, many people are involved in its development, and like SPSS, it allows the application of most of the known methods of statistics. The great advantage of "R" lies in its free availability, the wide range of statistical methods and the ever-growing mass of people who use this program and offer help on the Internet. However, a disadvantage is that the use of "R" requires some programming skills. This is particularly the case with complex statistical applications.

Which software should we use? For various reasons, R and Excel are good solutions. We already know Excel, Excel is available on our computer and on the company computer, and Excel has a variety of statistical methods. In addition, many organizations provide data as Excel files. Excel allows us to learn statistics without having to resort to new software (this book is therefore also available in an Excel version).

Those who are not afraid to learn a new software will be well served with R and the R Commander. R is free, available as open source software and we can find numerous aids on the Internet. In addition, R is becoming the standard in data analysis. Anyone who wants to deal more professionally with data analysis in the future should consider switching to R at some point, so why not start with it? We do this in this book with Chap. 2. We show how to install R and the R Commander and how to work with this software.

1.8 Case Studies—The Best Way to Learn

Data analysis is best learned using a concrete example. In the book, we use a central data set for this purpose. The used data set contains 10 variables and 100 observations. It is designed to be as realistic as possible and to cover most of the questions that arise when analyzing a data set. The data set is available on the internet platform.

www.statistik-kronthaler.ch.

By using the central data set throughout the entire book, we see how the methods of statistics build on each other and can be used for the systematic analysis of a data set. At one point or another in the book, the data set will not be sufficient and we will supplement it with additional data sets. These data sets are also available at www.statistik-kronthaler.ch.

1.9 Case Study: Company Growth

Let's imagine we want to write a study paper, bachelor's thesis, master's thesis or our doctoral thesis on the growth of young enterprises. For this purpose, we collect information about the enterprises and are now faced with the task of analyzing the data set. Specifically, we have the following data set with information on growth and other characteristics of companies and founders (Table 1.5).

In the table, we see the data available to us for the first twelve companies surveyed. We use this data in the book when we calculate by hand. When we analyze the data using the R Commander, we refer to the complete data set with a total of 100 companies. Larger data sets of 100 and more observed objects are the norm in practice. But we will also occasionally have data sets with fewer observed objects. Now all we need is the legend so we can work with the data (Table 1.6).

1.10 Application

1.1. Name three reasons why statistics is useful.

1.2. What information is contained in a data set?

1.3. What scale levels are important for statistical data analysis and what properties do they have?

1.4. What scale level does the variable sector have with the characteristics industrial companies and service companies, the variable self-assessment with the characteristics very experienced, experienced, little experienced and not experienced, and the variable turnover measured in francs?

1.5. What scale level does the variable education have with the characteristics secondary school, high school, university of applied sciences, university, PhD?

Table 1.5 Data set young enterprises

Enterprise	Growth	Expectation	Marketing	Innovation	Sector	Motive	Age	Experience	Self-Assessment	Education	Sex
1	5	2	29	3	0	3	32	7	5	1	0
2	8	1	30	5	1	2	26	8	4	2	1
3	18	3	16	6	0	2	36	10	5	1	0
4	10	2	22	5	1	1	43	7	3	3	1
5	7	1	9	9	1	2	43	6	3	1	1
6	12	2	14	8	1	3	48	10	4	4	0
7	16	2	26	0	1	2	19	11	5	3	0
8	2	1	26	4	1	1	33	4	2	1	0
9	4	1	28	8	1	2	42	0	1	4	1
10	10	1	31	0	1	3	27	4	2	4	1
11	7	1	6	3	1	2	30	8	4	1	1
12	9	1	30	0	1	2	23	12	5	1	1

Table 1.6 Legend for the data set young enterprises

Variable name	Variable description	Values	Missing values	Scale
Enterprise	Running index for the surveyed enterprises			
Growth	Average growth rate of sales in the last five years	In percent	n. d.	Metric
Expectation	Expectation about future development	1 = better than before 2 = no change 3 = worse than before	n. d.	Ordinal
Marketing	Expenditure for marketing in the last five years	In percent of turnover	n. d.	Metric
Innovation	Expenditure on innovationi in the last five years	In percent of turnover	n. d.	Metric
Sector	Sector in which the company operates	0 = Industry 1 = Service	n. d.	Nominal
Motive	Founder's motive for founding	1 = Unemployment 2 = Implementing an idea 3 = Higher income	n. d.	Nominal
Age	Age of the company founders at the time of founding	In years	n. d.	Metric
Experience	Industry professional experience of the company founders	In years	n. d.	Metric
Self-assessment	Self-assessment of the industry professional experience of the company founders	1 = very experienced 2 = experienced 3 = neither 4 = little experienced 5 = not experienced	n. d.	Ordinal
Education	Highest level of education of the company founder	1 = Secondary school 2 = High school 3 = University of applied sciences 4 = University	n. d.	Ordinal
Sex	Sex of the company founder	0 = Man 1 = Woman	n. d.	Nominal

Source: Own survey 2021

1.6. Variables can often be measured ordinally or metrically. Consider how the education level of individuals can possibly be measured metrically.
1.7. The hotel satisfaction of guests should be measured quasi-metrically. For this, create eight lines with a length of 10 cm, measure a hypothetical value for each guest and create a small data set with the variables guest, age, gender and hotel satisfaction.
1.8. Why are scale levels important for statistical data analysis?
1.9. Why do we need a legend for our data?
1.10. Open our data set data_growth and familiarize yourself with it. Answer the following questions: How large is the number of observations? How many metric, ordinal and nominal variables are contained in the data set? How are the metric, ordinal and nominal data measured?
1.11. How trustworthy is our data set data_growth?
1.12. Name a trustworthy data source for your country, your continent, and the world?
1.13. Think of three questions that are likely to be analyzed with data. Search for the appropriate data sources for this.

2 R and R Commander: A Brief Introduction

2.1 Interesting Facts about R and the R Commander

Before we get to the installation and description of R and the R Commander, I would like to provide some information about R.

The development of R began in the early 1990s. It was initiated by George Ross Ihaka and Robert Gentleman[1]. Ihaka and Gentleman attempted to combine the advantages of the programming languages S and Scheme with the development of R. The project has been so successful that R has developed into a standard in data analysis within a few years. R is intensively used in science and business for data analysis and forecasting purposes.

R is a system driven by a command language. To perform a statistical analysis, a command is usually entered and then executed. At the beginning of the development of R, there was no graphical interface with which data could be analyzed and graphics could be generated. This changed, among other things, with the R Commander developed by John Fox.[2] The R Commander is a graphical interface that allows R to be operated without a command language, via a drop-down menu, and to perform statistical analyses. With the R Commander, it is thus possible to use R in introductory statistics without having to learn a complex command language. At the same time, as we will see, the R Commander enables the intuitive learning of R's command language.

R is a dynamic system that is constantly being developed. This development is being driven by the R Core Development Team (see www.r-project.org).

---

[1] Ihaka, R. & R. Gentleman (1996). R: A Language for Data Analysis and Graphics, Journal of Computational and Graphical Statistics 5(3), 299–314.

[2] Fox, J. (2005). The R Commander: A Basic Statistics Graphical User Interface to R, Journal of Statistical Software 14(9), 1–42.

F. Kronthaler, *Statistics Applied with the R Commander*,
https://doi.org/10.1007/978-3-662-69107-6_2

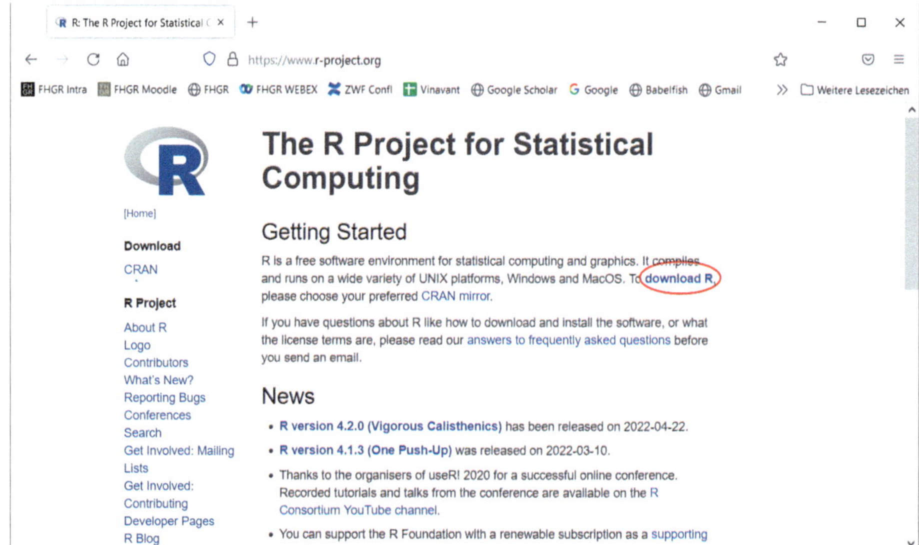

Fig. 2.1 The R project for statistical data analysis and graphics

2.2 The Installation of R and the R Commander

We can obtain the R software and installation instructions on the website www.r-project.org (Fig. 2.1). The website is the central source for R. It is worth taking a closer look at the website. Here we find, in addition to the installation files, a wealth of information about R, its origins, the developers, manuals, a journal, and, for example, the FAQs section (frequently asked questions about R).

To install R on our computer, we click on the link "download R" and we are taken to a list of countries. In this list, we select our home country and click on the corresponding link, for Switzerland for example on https://stat.ethz.ch/CRAN/, for Germany for example on https://cran.uni-muenster.de/.

We then have the opportunity to save the installation files of R for Linux, Mac, or Windows on our computer. We select the program version we need and click on the corresponding link. In the following, we will only discuss the installation of the Windows version. Mac or Linux users are on their own at this point. When we click on the link to the Windows version, the window for R for Windows opens (Fig. 2.2).

In this window, we click on "install R for the first time" and then save the version that is suggested: "Download R 4.0.5 for Windows".[3] We can now save the R installation file on our computer in a folder of our choice. Once we have done this successfully, we can begin to install R.

[3]At the time this book was written, the current version available was version 4.0.5. This may have changed. In general, I recommend always installing the latest version.

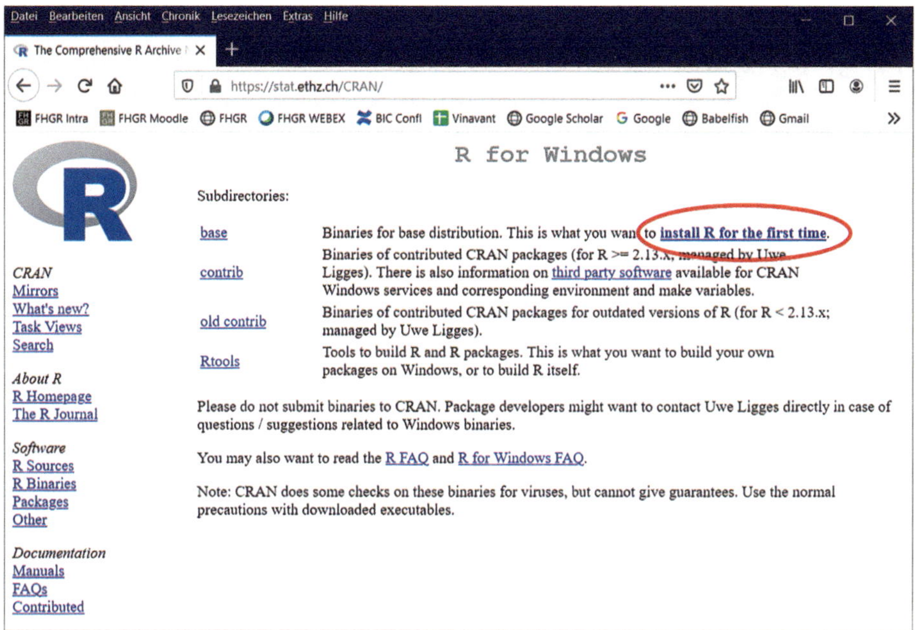

Fig. 2.2 Initial installation of R

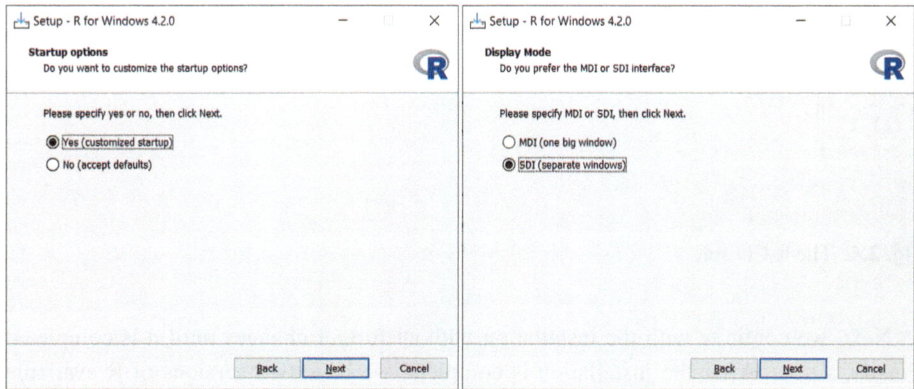

Fig. 2.3 Selection of SDI (Single Document Interface) during installation

We start the installation of R by double-clicking on the file R- 4.0.5-win.exe, the setup assistant appears, which helps us with the installation of R. For optimal use of the R Commander, a slight adjustment of the standard installation settings is helpful. We therefore confirm the installation instructions with next until we reach customize startup options. Here we mark yes and continue. In the following instruction, we select SDI (Single Document Interface) (Fig. 2.3). The Single Document Interface mode is the preferred operating mode of the R Commander.

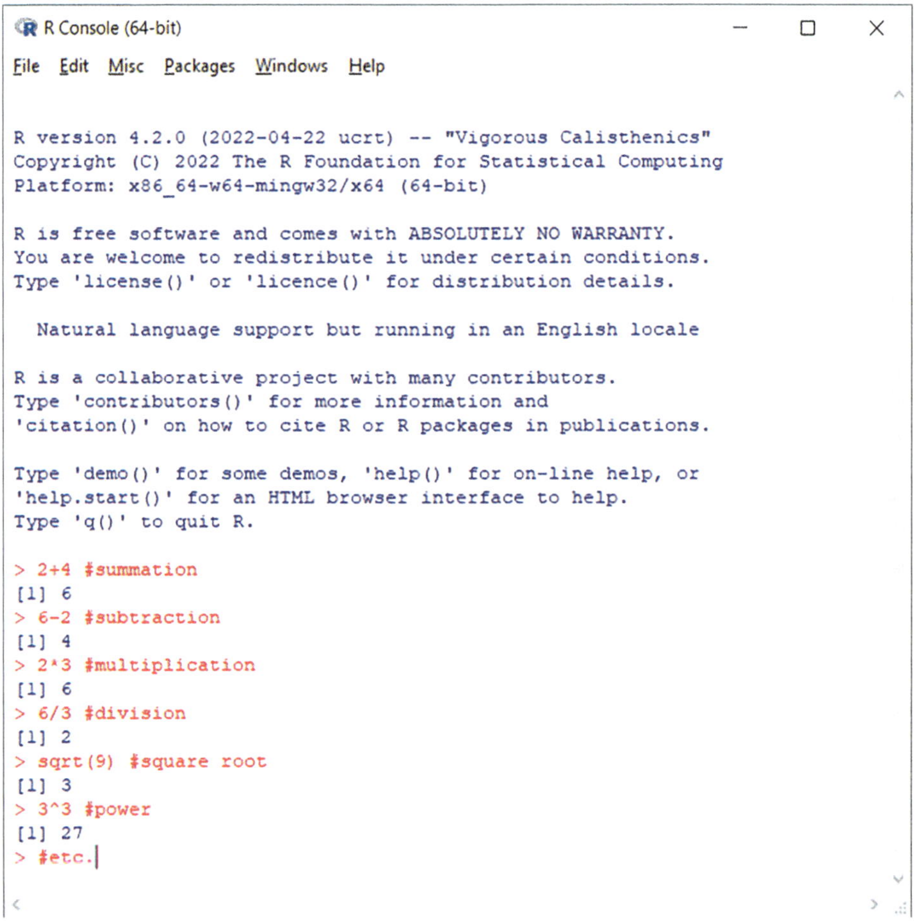

```
R Console (64-bit)
File Edit Misc Packages Windows Help

R version 4.2.0 (2022-04-22 ucrt) -- "Vigorous Calisthenics"
Copyright (C) 2022 The R Foundation for Statistical Computing
Platform: x86_64-w64-mingw32/x64 (64-bit)

R is free software and comes with ABSOLUTELY NO WARRANTY.
You are welcome to redistribute it under certain conditions.
Type 'license()' or 'licence()' for distribution details.

  Natural language support but running in an English locale

R is a collaborative project with many contributors.
Type 'contributors()' for more information and
'citation()' on how to cite R or R packages in publications.

Type 'demo()' for some demos, 'help()' for on-line help, or
'help.start()' for an HTML browser interface to help.
Type 'q()' to quit R.

> 2+4 #summation
[1] 6
> 6-2 #subtraction
[1] 4
> 2*3 #multiplication
[1] 6
> 6/3 #division
[1] 2
> sqrt(9) #square root
[1] 3
> 3^3 #power
[1] 27
> #etc.
```

Fig. 2.4 The R Console

Next, we continue with the installation without further changes until it is completed. Finally, a note. After the installation is complete, we have two versions of R available, the 32-bit version and the 64-bit version. Since computers today are usually designed for 64-bit, we can normally use the 64-bit version.

After R has been successfully installed, we can start R, for example, via the R icon on the desktop and open the R Console (Fig. 2.4). If we have knowledge of the command language, we can already work with R and perform statistical analyses. But we can also use R as a calculator. For example, if we type 2 + 4 behind the red arrow and confirm the instruction with the enter key, we get the result of the calculation. In Fig. 2.4 there are further examples for basic arithmetic operations.

In addition, to have the graphical interface, the R Commander, available, we need to install it separately. The installation is done via the R Console. In this, we select the

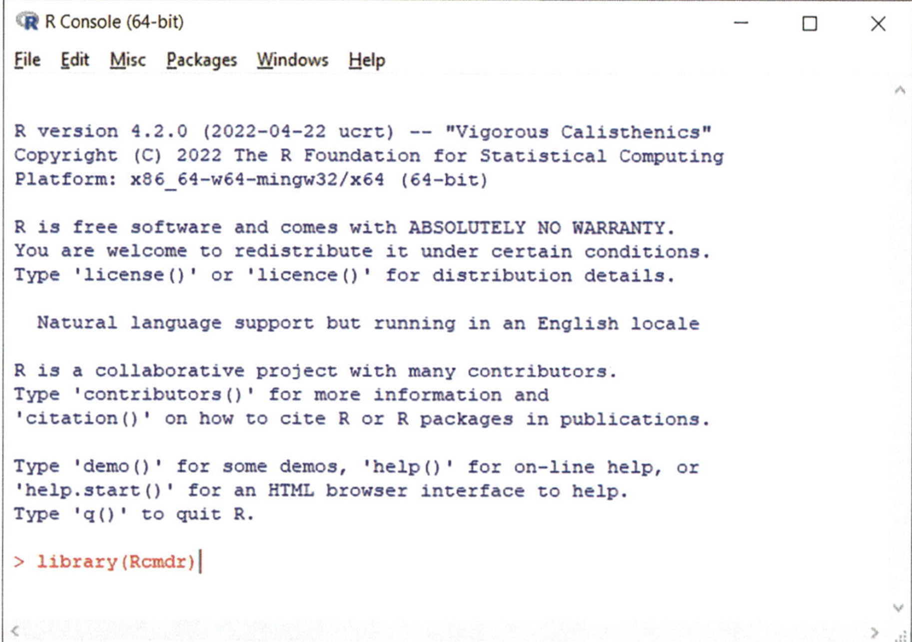

Fig. 2.5 Starting the R Commander

menu item Packages and then click on the command "Install Packages". A list of countries opens, from which we select our home country and confirm with OK. Then a somewhat longer list opens. This list contains the packages available for R. Each of these packages contains a statistical application. This gives us a first impression of the statistical possibilities available to us in R. We install the R Commander with the Rcmdr package. We select Rcmdr from the list, mark the package and confirm again with OK. The R Commander is installed.

In order for us to access the R Commander, we must first close and restart the R Console. In the R Console, we then enter the command library(Rcmdr) (see Fig. 2.5).

We confirm the command with the enter key and the R Commander opens.[4] It will look approximately like in Fig. 2.6 shown.

If we work with the R Commander, library(Rcmdr) is the only command we need to remember to perform statistical analyses with R, usually we will work with the available menus and the commands contained therein.

[4] It may happen that the first time you call up the R Commander, a message appears to install additional packages. We do this, after confirmation it happens automatically.

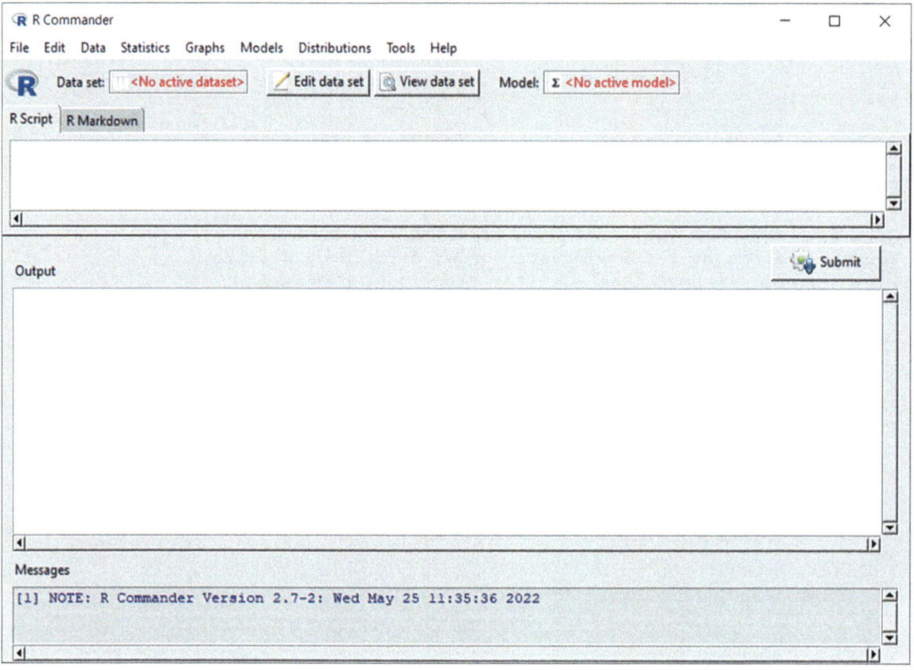

Fig. 2.6 The R Commander

2.3 The R Commander and its Functions

Let's take a look at the R Commander and its functions (Fig. 2.7).

At the very top of the R Commander is the menu bar, where the existing commands are organized. The menu items will only be briefly described here. We will get to know them in more detail throughout the book.

- File: In the File menu, we find commands for changing the working directory and saving our work results.
- Edit: Here we find commands that we may need to change the contents of the command window and the result window.
- Data: The Data menu allows us to load data into memory, import it, and change the variables in the data set as well as create new variables.
- Statistics: Statistics is the central menu for performing statistical analyses. We will mainly move in this menu during data analysis.
- Graphs: With the help of the Graphs menu, we can use the R Commander to create figures and analyze our data graphically.

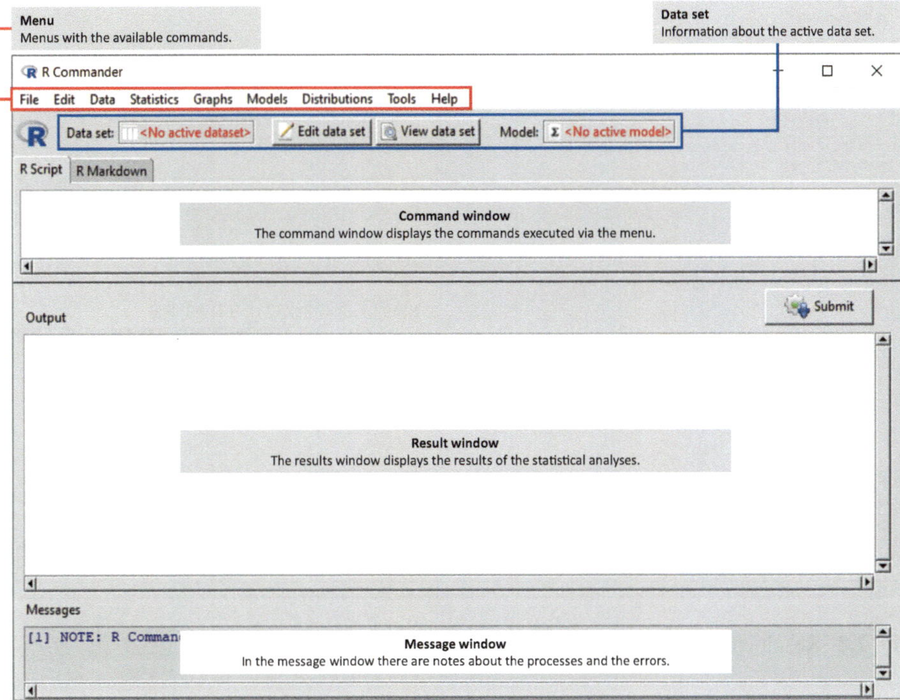

Fig. 2.7 Functions of the R Commander

- Models: Here are, among other things, the commands with which statistical models can be evaluated.
- Distributions: The Distributions menu contains the most important probability distributions. We can use these, for example, to see how these distributions look graphically.
- Tools: We mainly use the Tools menu to load packages with statistical applications that are not yet installed.
- Help: As always, the Help menu provides information about the software and offers numerous aids for problems with the R Commander.

Below the menu bar are buttons with information about the active data set, the possibility to edit the active data set, and view it. Currently, the first button says No current dataset. This is because we have not yet loaded any data into memory. As soon as we have loaded data, the name of the active data set is displayed there. We can also use the button to switch between data sets. If we have loaded more than one data set, the button allows us to change the active data set by clicking on it.

Next to the button with the active data set is the Edit data set and View data set button. We always need the latter when we want to take a look at our data. We should always do

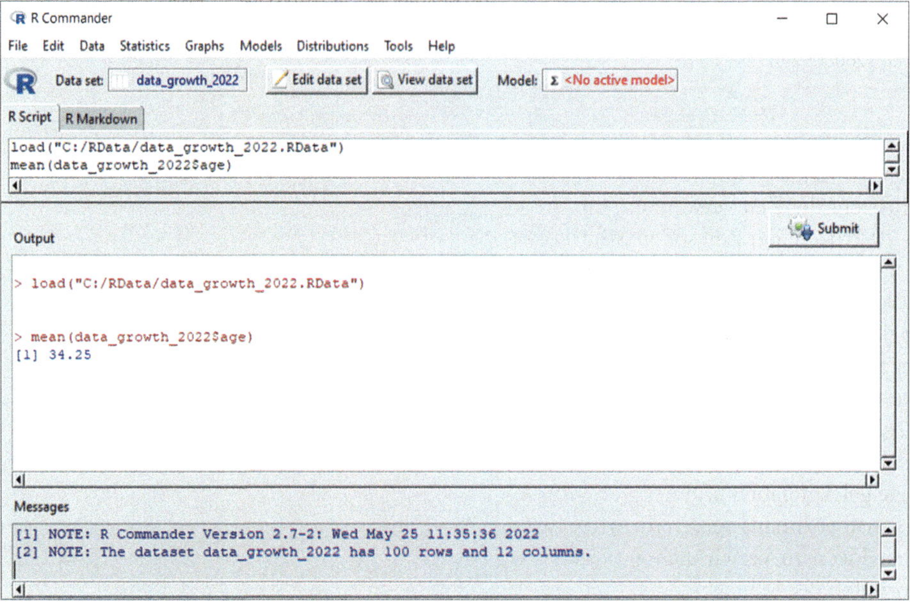

Fig. 2.8 Result window of the R Commander

this after loading the data matrix into memory to see if the data has been read correctly. In addition it is recommended to take a look at the data from time to time during our data analysis. We will rarely use the Edit data set button. To edit our data set, we better use the Data menu.

Other central elements of the R Commander are the command window, the result window, and the message window.

In the command window (named R Script), the commands we perform via the menus are displayed. The command window allows us to easily learn the command language of R. Furthermore, once we know the command language or parts of it, we can directly enter statistical instructions in the command window and have them executed. To execute, we place the cursor behind the relevant command line and click on the Submit button. Alternatively, we can use the keyboard shortcut Ctrl r (the letter r for run).

In the result window, we find the results of our statistical analyses (see Fig. 2.8).

The commands are listed in red behind the arrow, the results of the statistical analyses in blue. In Fig. 2.8, for example, an average value was calculated with the command mean(data_growth_2022$age). Graphics are not displayed in the result window, but in a new window.

The message window gives us hints about the statistical procedures, e.g., how many rows and columns the data set has (see Fig. 2.8) and thus about the number of variables and observations. We also find a message here if something does not work. Error messages are usually listed in red, warnings in green, other messages in blue.

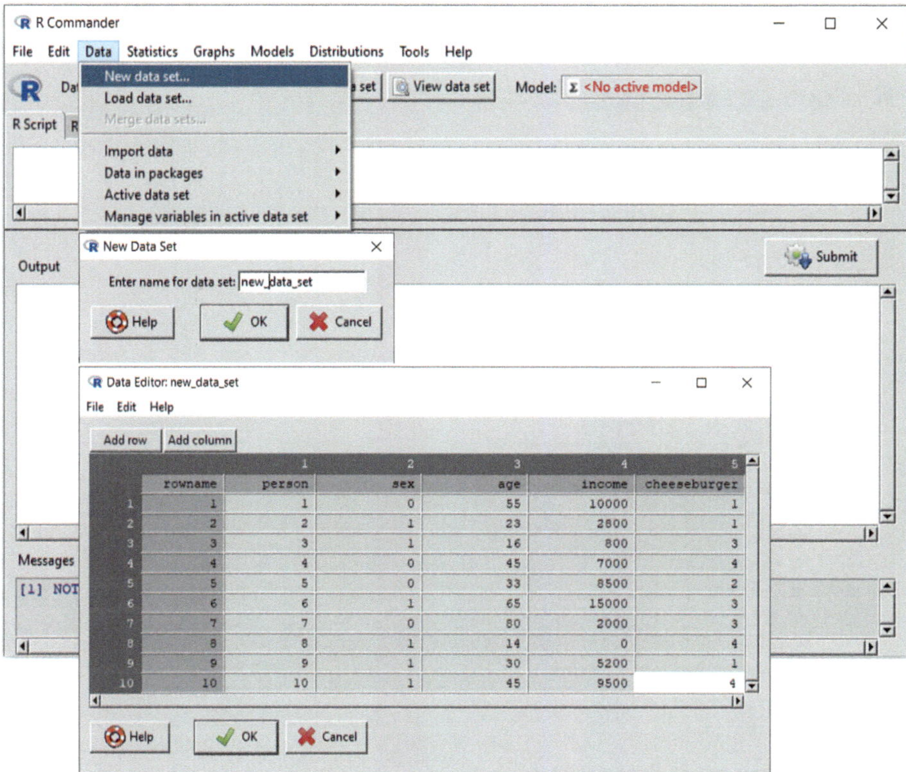

Fig. 2.9 Entering data with the R Commander

We can now start working with the R Commander. Before we use it in the rest of the book and for analysis, it should be shown how data can be read into memory.

2.4 Reading Data with the R Commander

In order to work with the R Commander, we first need to read in data. For this, several options are available in the data menu. In the following, we will discuss direct data entry, importing an Excel file (many organizations provide data in Excel format), and reading data in R format. Other options would include importing a text file or reading SPSS data.

First, the R Commander offers the possibility to enter data directly. We select the command New data set in the data menu. A window appears where we can give the data set a name (Fig. 2.9). We confirm with OK and the data editor opens with two columns and two rows. With the Add row and Add column buttons, we can add columns and rows. Then we can name the variables and enter the values. For this, we click on a cell and enter the text or number.

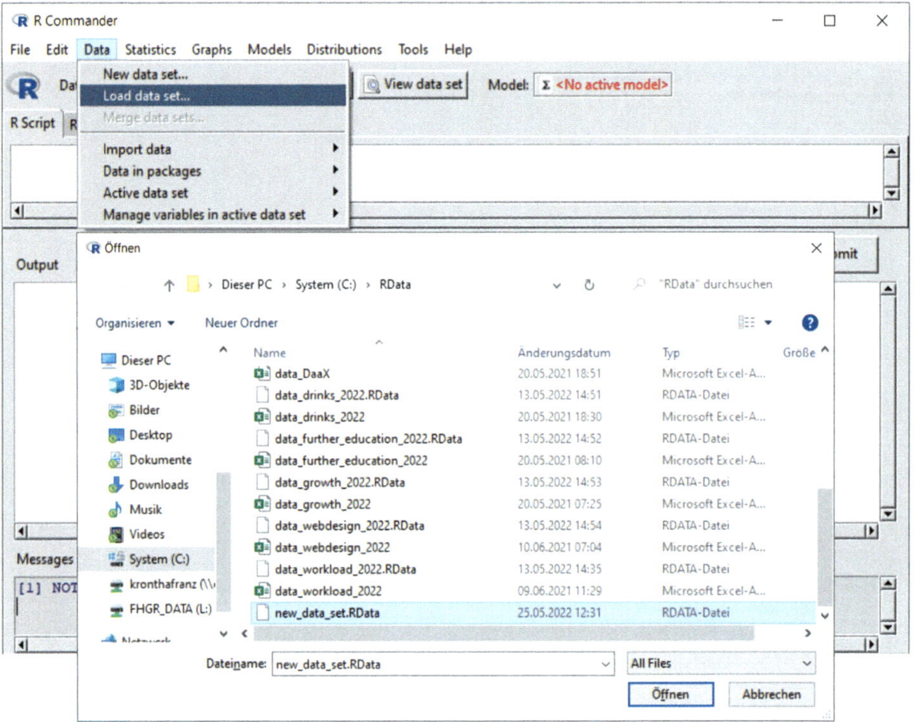

Fig. 2.10 Loading data in R data format

We move between the cells best with the arrow keys on the computer. When we are finished, we confirm with OK and the data set is available. In Fig. 2.9. we have reproduced the data set from Chap. 1.

To prevent the data from being lost after closing the R Commander, we need to save it. For this, we can use, for example, the command Save active data set in the data menu. The R Commander offers us to save the data as R data in RData format in our working directory.

Reading R data in R data format is very simple. We select the command Load data set in the data menu. Then we go to the corresponding directory and the available data is displayed. We select the data and confirm with Open. The data is loaded (Fig. 2.10).

If we want to import Excel data, we again use the data menu. But this time we select the command Import data and the subcommand from Excel file (Fig. 2.11) A small window opens, with which we can name the data set, it is best to choose the name of the Excel file, so we stay consistent. It also asks whether the variable name is noted in the first line of the Excel worksheet and whether the row name is contained in the first column. We click on the former and confirm with OK. Then we select the Excel file and

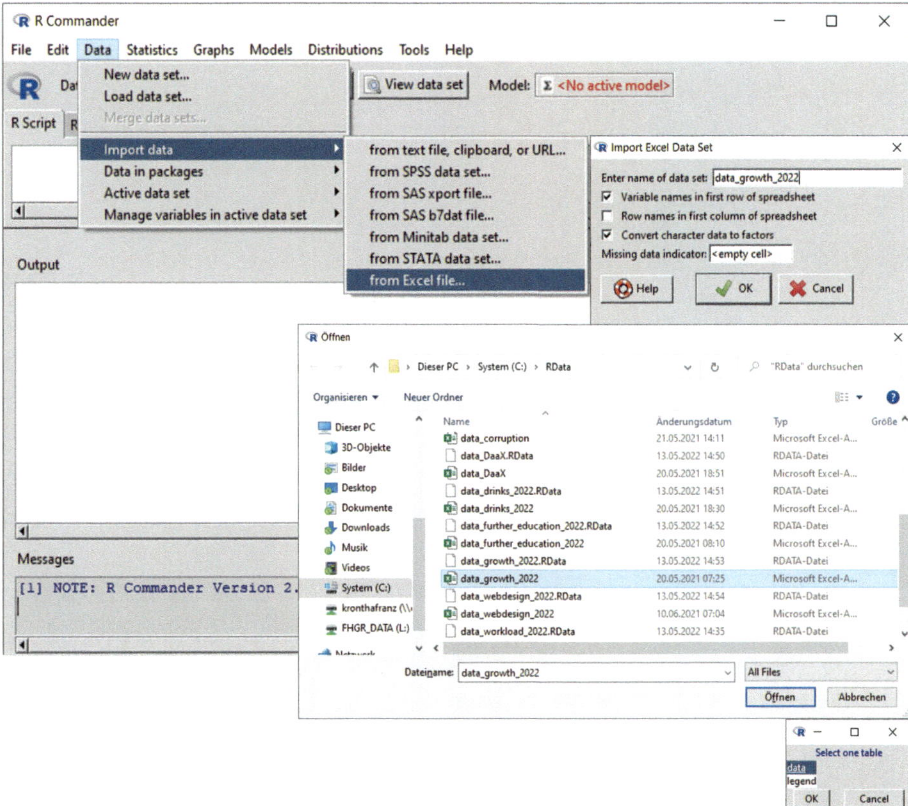

Fig. 2.11 Importing Excel data with the R Commander

confirm this with Open. A small window opens, with which we can select the worksheet. We take our data and confirm again with Ok. The data is now in the memory.

Before we close the R Commander, after importing the Excel data, we should save the data, as described above with enter data. The next time we want to use the data, we do not need to import it again, but can simply read the R data.

After entering, reading, or importing data, we should click on the Data set button and evaluate whether the data is correctly present. An error may have crept in.

If there are problems working with R or the R Commander, I would like to refer again to the help available on the internet. In addition to the website www.r-project.org, there are also numerous internet forums where R users discuss problems and offer help.

We can now use R and the R Commander. Let's start analyzing data.

Part II
Describe, Nothing but Describe

Describing People or Objects or Simply Descriptive Statistics
In descriptive statistics, we report on people and objects that we observe. For example, as customers of McDonald's or Burger King, we want to know how many grams of meat the cheeseburgers we buy contain. In such a case, we observe the cheeseburgers, that is, we weigh the meat in each cheeseburger, then we gather the information and evaluate it for all observed cheeseburgers.

In our case study on growth, we can gather information about the surveyed newly established enterprises using descriptive statistics. After the analysis with descriptive statistics, we understand how the observed companies behave. We have various tools at our disposal to describe our observations. In Chap. 3 we deal with the average, in Chap. 4 with the deviation from the average, and in Chap. 5 we examine how groups of observations behave and how we can graphically display the data. Subsequently, in Chap. 6 we consider the relationship between variables, and in Chap. 7 we show how new knowledge can be generated from existing variables.

3 Averages: How People and Objects Behave in General

3.1 Averages—Why We Need Them

We use average values to analyze how people and objects behave on average, e.g., how much meat the cheeseburgers contain on average. We have different average values available for this: the arithmetic mean, the median, the mode, and the geometric mean. These values are also called measures of central tendency. We will look at their meaning and application shortly. This brings us finally to data analysis.

3.2 The (Arithmetic) Mean

The arithmetic mean, often just called mean, is the measure of central tendency most commonly used. It is easy to calculate. We add up all the values of a variable and then divide them by the number of observations:

$$\overline{x} = \frac{\sum x_i}{n}$$

Σ	is the Greek symbol for sum. Whenever we see this symbol, we add up the values,
$\overline{x}$	is the (arithmetic) mean,
x_i	are the observed values and
n	is the number of observations.

It is important that the arithmetic mean requires metric data, i.e., we should only calculate it if our scale level is metric.

F. Kronthaler, *Statistics Applied with the R Commander*,
https://doi.org/10.1007/978-3-662-69107-6_3

Let's take our case study growth and recall the associated data (see Table 1.5). We want to know how the first six companies have grown on average. The data for this can be found under the variable growth. The growth rates are:

$$x_1 = 5$$
$$x_2 = 8$$
$$x_3 = 18$$
$$x_4 = 10$$
$$x_5 = 7$$
$$x_6 = 12$$

The sum of the growth rates is equal to:

$$\sum x_i = x_1 + x_2 + x_3 + x_4 + x_5 + x_6 = 5 + 8 + 18 + 10 + 7 + 12 = 60$$

If we now divide the sum by our six companies, we get the average growth rate:

$$\overline{x} = \frac{\sum x_i}{n} = \frac{60}{6} = 10$$

Since the growth rate was measured in percent (see Table 1.6), the average growth rate is 10%. So, our six companies have grown by an average of 10% over the last 5 years. We are also interested, for example, in how old our founders are on average. This can be calculated as follows:

$$\overline{x} = \frac{\sum x_i}{n} = \frac{32 + 26 + 36 + 43 + 43 + 48}{6} = 38$$

Our founders are on average 38 years old. They spend on average 20% of their revenue on marketing, another 6% of their revenue they spend on average for innovaton, and on average they have 8 years of industry professional experience.

Bingo, with that we already know a lot about our enterprises and their founders.

Freak Knowledge

We have denoted the arithmetic mean with $\overline{x}$ and the number of observations with the lower case letter n. Strictly speaking, we only denote them this way when analyzing a sample. Here we are not analyzing all objects or people that are of interest, but only a subset. If we have observed the entire population, i.e., all people or objects that interest us, the arithmetic mean is denoted by the Greek letter μ and the number of observations by the upper case letter N.

Sometimes it happens that the data available have already been processed and are only available in categories. For example, the growth rates of our 100 companies might be available as follows (Table 3.1).

In such a case, we have the growth rates only for groups of companies, not for individual companies. If we still want to calculate the arithmetic mean, we use the following formula:

$$\overline{x} \approx \frac{\sum \overline{x}_j \times n_j}{n}$$

$\overline{x}_j$	are the mean values of the respective classes,
n_j	the number of companies in the respective classes and
n	the total number of observed companies.

For class 1 with three companies, the mean value of the growth rate is −7.5, for class 2 with eight companies −2.5, for class 3 with 27 companies 2.5, etc. Thus, the mean can be calculated as follows:

$$\begin{aligned}\bar{x} &\approx \frac{\sum \bar{x}_j \times n_j}{n}\\ &\approx \frac{-7.5 \times 3 - 2.5 \times 8 + 2.5 \times 27 + 7.5 \times 34 + 12.5 \times 23 + 17.5 \times 4 + 22.5 \times 1}{100} \approx 6.6\end{aligned}$$

The average growth rate of the companies over the last 5 years is therefore approximately 6.6%. The wavy lines $\approx$ instead of the equals sign $=$ mean "approximately equal". Since we no longer have the individual information for the companies, but only bundled values for several companies, we can only calculate the arithmetic mean approximately.

Table 3.1 Growth rate in classes of 5% steps

Class	Growth rate from over … to … [in percent]	Number of enterprises
1	−10 to −5	3
2	−5 to 0	8
3	0 to 5	27
4	5 to 10	34
5	10 to 15	23
6	15 to 20	4
7	20 to 25	1

3.3 The Median

The median is also an average value, just defined a bit differently. To determine the median, we ask ourselves, which value divides our observations into 50% of the smaller and 50% of the larger observations. For this, we first have to sort the values in ascending order. For our six companies, we then have the following order:

$$\begin{aligned} x_1 &= 5 \\ x_5 &= 7 \\ x_2 &= 8 \\ x_4 &= 10 \\ x_6 &= 12 \\ x_3 &= 18 \end{aligned}$$

The median in our example is then the value that divides our six companies by growth rate into the 50% of companies with the lower growth rate and the 50% of companies with the higher growth rate. In our case, it lies between the growth rates of 8% and 10%, thus at 9%. We can represent this well on a line (see Fig. 3.1).

The three companies 1, 5 and 2 have a lower growth rate than 9%, companies 4, 6 and 3 have a higher growth rate than 9%.

If we have an odd number of observations instead of an even number, then the median is exactly the value in the middle. If we add the seventh observation to our example from above and sort them in ascending order, the following order results:

$$\begin{aligned} x_1 &= 5 \\ x_5 &= 7 \\ x_2 &= 8 \\ x_4 &= 10 \\ x_6 &= 12 \\ x_7 &= 16 \\ x_3 &= 18 \end{aligned}$$

With seven observations, we have an odd number of observations. If we count from the top, then the fourth value with the characteristic 10 is the value in the middle, i.e., about

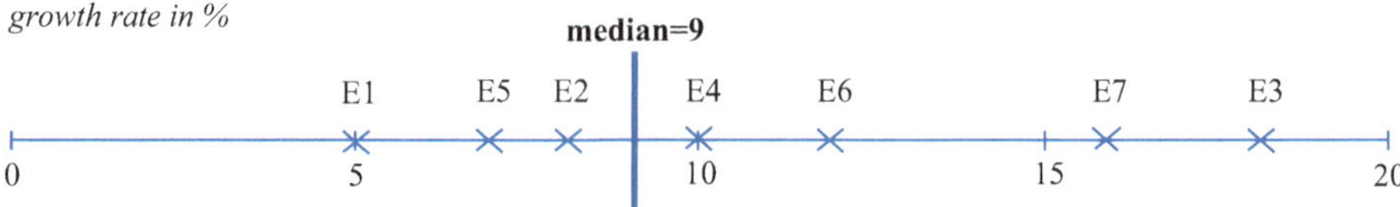

Fig. 3.1 The median for the company's growth rate

50% of the companies have a growth rate lower than 10% and about 50% of the companies have a growth rate higher than 10%. We cannot divide the company in the middle, so there is a slight blur, but we can neglect this.

The median can be used for both metric data as above, and for ordinal data. For example, we might be interested in how the company founders assess their industry experience. We have collected this on an ordinal scale from 1 (very experienced) to 5 (not experienced). The evaluation of our first six companies results in a median of 4. Thus, we can make the statement that the characteristic of being less experienced divides our company founders in the middle.

3.4 The Mode

The mode or most common value is the third average value that usually interests us in data analysis. Basically, the mode can be calculated for metric data, ordinal data, and nominal data. It is particularly interesting when values occur more frequently. When we look at our data, this is especially the case with ordinal and nominal data. For ordinal data, the mode can complement the median, for nominal data it is the only average value available. In our data set growth, the variables industry, motive, and gender are nominal and the variable self-assessment is ordinal scaled. If we determine the mode for these variables for the first six companies, it turns out that the variable industry most frequently shows 1. Four newly founded companies are service companies, two are industrial companies. The most frequently mentioned founding motive is 2. Most founders want to realize an idea. For gender, 0 and 1 appear three times each, i.e., 3 companies were founded by a woman and 3 by a man. For the variable self-assessment, all three characteristics occur equally often. The last two examples show that we can have more than one mode value.

With the help of the averages, we have already found out quite a lot.

3.5 The Geometric Mean and Growth Rates

What remains is the geometric mean. It is used in particular for calculating average growth rates over time. There are plenty of examples for this. We want to know how the population has grown on average in recent years, we are interested in the average interest rate of an investment, how high the average profit growth of a company is, or what the inflation rate was in recent years. To answer all these questions, we need the geometric mean:

$$\bar{x}_g = \sqrt[n]{x_1 \times x_2 \times \ldots \times x_n}$$

$\bar{x}_g$	is the geometric mean or the average growth factor,
x_i	are the relative changes from one year to the next, or the growth factors, and
n	the number of growth factors.

To calculate average growth rates, we first need the growth factors, i.e. the relative changes from one year to the next. Depending on the available numbers, there are two ways to calculate these changes. If we have the absolute numbers, we can simply divide the absolute numbers using the following formula:

$$GF_t = \frac{Y_t}{Y_{t-1}}$$

GF_t	is the growth factor at time t (t for "time" or "point in time"),
Y_t	is the value at time t (t can refer to years, months, days, etc.), and
Y_{t-1}	is the value at time t−1.

If we only have the growth rates available, we can use the following formula to calculate the growth factors:

$$GF_t = 1 + \frac{GR_{t,\%}}{100}$$

GF_t	is again the growth factor at time t and
$GR_{t\%}$,	is the growth rate in percent from time t−1 to time t.

Here we can also provide the formula for the growth rate. This is always needed. The formula is:

$$GR_{t,\%} = \frac{Y_t - Y_{t-1}}{Y_{t-1}} \times 100\%$$

Let's best look at all of this together in an example. Our data set contains the average company growth from 2007 to 2012. We first need to calculate this. For example, we know the sales figures of our first company from 2007 to 2012 (Table 3.2). Using the formulas mentioned above (and minor rounding up and down), we arrive at the annual growth rates and the growth factors.

The growth factors are identical, regardless of whether they are calculated with the first or the second formula. Now we can calculate the average growth factor using the formula for the geometric mean. This is:

$$\bar{x}_g = \sqrt[5]{1.15 \times 1.10 \times 0.98 \times 1.05 \times 0.98} = 1.05$$

Table 3.2 Turnover of the first company in the data set from 2007 to 2012

Year	Turnover in CHF	$GR_{t,\%} = \frac{Y_t - Y_{t-1}}{Y_{t-1}} \times 100\%$	$GF_t = 1 + \frac{GR_{t,\%}}{100}$	$GF_t = \frac{Y_t}{Y_{t-1}}$
2007	100,000			
2008	115,000	15%	1.15	1.15
2009	126,500	10%	1.10	1.10
2010	123,970	−2%	0.98	0.98
2011	130,169	5%	1.05	1.05
2012	127,566	−2%	0.98	0.98

If we now subtract 1 from the average growth factor and multiply by 100%, we get the average growth rate over all years.

$$\text{Ø}GR = \left(\overline{x}_g - 1\right) \times 100\% = (1.05 - 1) \times 100\% = 5\%$$

$\overline{x}_g$ is, as mentioned above, the average growth factor and
$\text{Ø}GR$ is the average growth rate.

The first company we observed, therefore, grew on average by 5% between 2007 and 2012. We now know how the average growth rates of the young companies were calculated in our data set data_growth.

Even better, we can also use the average growth factor for forecasting purposes, i.e., if we want to know what the company's turnover is in 2015, we just need to take the last known number and multiply it by the average growth factor for each additional year.

$$\begin{aligned} Turnover_{2015} &= Turnover_{2012} \times 1.05 \times 1.05 \times 1.05 = Turnover_{2012} \times 1.05^3 \\ &= 147{,}674\ CHF \end{aligned}$$

Of course, this forecast only works well if the average growth factor does not change. When we make a forecast, we naturally have to make an assumption. The assumption in our case is that the average growth rate of the past will continue in the future. If this assumption is good, then our forecast is also good. If the assumption is bad, then our forecast is not very useful.

3.6 Which Average Value Should We Use and What else Do We Need to Know?

First and foremost, the scale level of the variable for which we want to calculate the average value is crucial. If the scale level is nominal, only the mode is available. If the scale level is ordinal, we can calculate the mode and the median. With a metric scale level, we can determine both the mode and the median as well as the arithmetic mean. If we want to calculate growth rates, we need the geometric mean.

Freak-Knowledge
Strictly speaking, we need a ratio scale for the calculation of the geometric mean. The scale level of metric data can be further divided into interval scale and ratio scale. The interval scale requires clearly defined distances between two points. In the ratio scale, an absolute zero point must also be defined. This is not the case, for example, with the calendar year, where the zero point is the year zero, Christ's birth. However, we could also choose the year zero differently, e.g., the death of Ramses II. On the other hand, there is an absolute zero point for sales, no sales. Only when an absolute zero point exists two values of a variable can be meaningfully put into a ratio, and only then the geometric mean should be calculated.

It is also important that the arithmetic mean is very sensitive to outliers or extreme values, while the median and the mode are robust. We can illustrate this again with our cheeseburger example. Let's assume we observe the weight of seven cheeseburgers with the following result in grams:

$$x_1 = 251$$
$$x_2 = 245$$
$$x_3 = 252$$
$$x_4 = 248$$
$$x_5 = 255$$
$$x_6 = 249$$
$$x_7 = 47$$

We see that the seventh observed cheeseburger deviates a lot from the other cheeseburgers. To illustrate the influence of this extreme value on the arithmetic mean and the median, we want to calculate both once with and once without the extreme value:

$$\bar{x}_{\text{with outlier}} = \frac{\sum x_i}{n} = \frac{251 + 245 + 252 + 248 + 255 + 249 + 47}{7} = 221$$

$$\bar{x}_{\text{without outlier}} = \frac{\sum x_i}{n} = \frac{251 + 245 + 252 + 248 + 255 + 249}{6} = 250$$

Before we can calculate the median, we must first sort in ascending order

$$
\begin{aligned}
x_7 &= 47 \\
x_2 &= 245 \\
x_4 &= 248 \\
x_6 &= 249 \\
x_1 &= 251 \\
x_3 &= 252 \\
x_5 &= 255
\end{aligned}
$$

and then look for the value that divides the values into 50% smaller and 50% larger values. Including outliers, the median is 249, without outliers it is 250. The example clearly shows us that an extreme value has a large influence on the arithmetic mean, but only a very small influence on the median.

This fact can be used to manipulate opinions. Imagine a lobbyist of a business association is lobbying for higher subsidies. He knows that very few companies in his association report very high profits, while the vast majority make small or no profits. Will he choose the arithmetic mean or the median? He will choose the median, as the arithmetic mean is sensitive to extreme values and is higher than the median. This does not meet his goal of getting higher subsidies. If the lobbyist is smart, he will even conceal that he worked with the median, but will only speak of an average value. This does not make him formally wrong, as the mode the median and the arithmetic amean are referred to as average values. He just conceals from us which average value he used. A person working seriously, on the other hand, will present both averages and draw our attention to the outliers.

3.7 Calculation of Averages with the R Commander

To calculate the average values with R, R and then the R Commander must be started first. Then the data must be read in. Once the data is loaded, we can begin. R or the R Commander offers several ways to display the average values. If we go via the graphical interface, we use, for example, the Statistics menu, the Summaries submenu, and then the Active data set command.

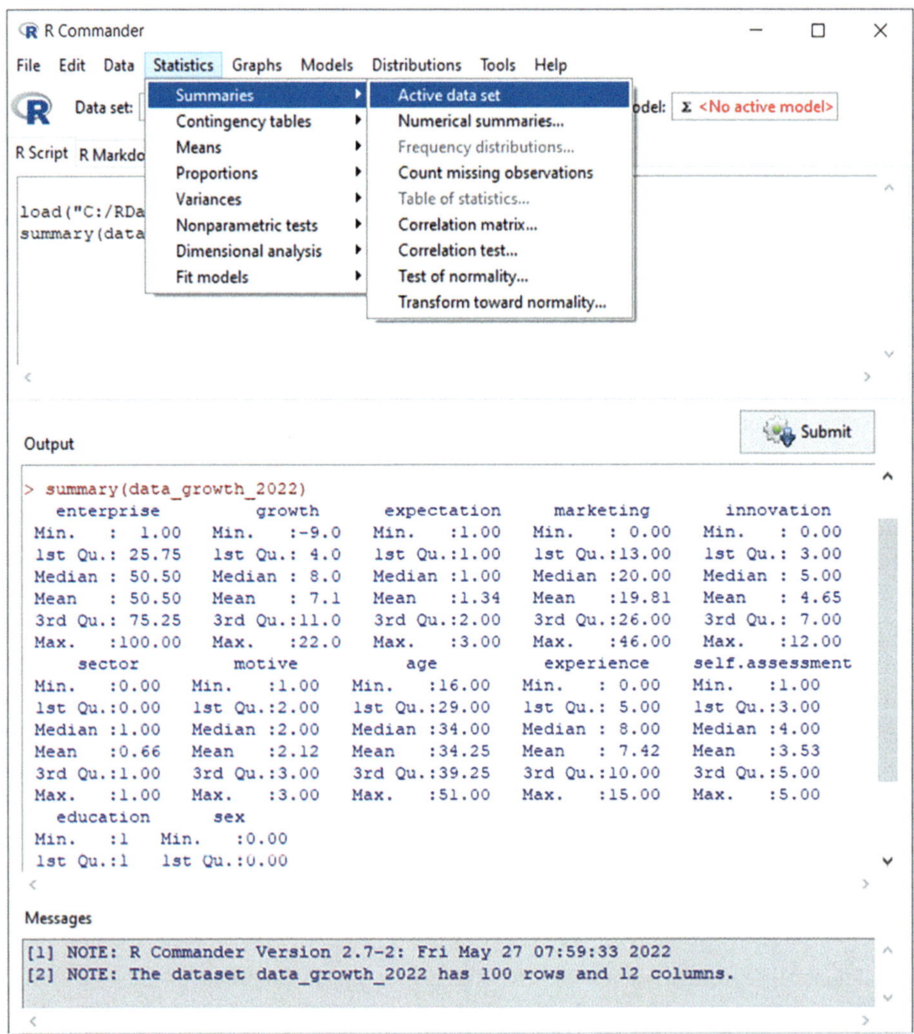

This command gives us a summary overview of our variables, which also includes the arithmetic mean and the median of each variable. It should be noted that the measures are calculated for each variable, even if a calculation is not correct due to the scale level.

Alternatively, we can enter the corresponding commands in the command line. For the arithmetic mean and the variable growth, we enter, for example, the command mean(data_growth_2022$growth). Then we place the cursor behind the command and confirm it either with the Submit button or by pressing the Ctrl and r keys simultaneously (see Chap. 2).

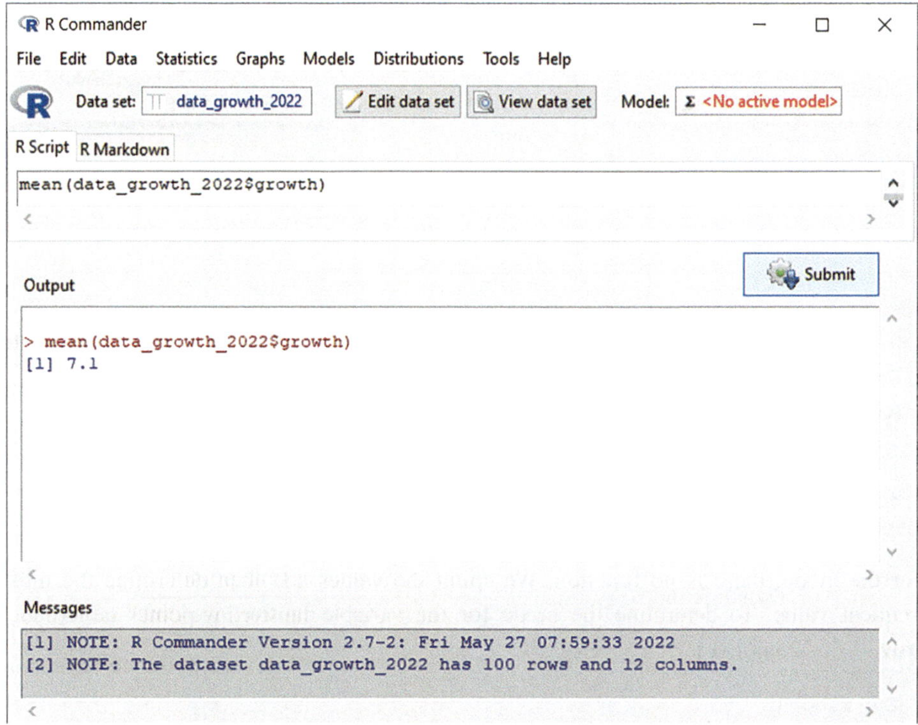

For the median and the variable self-assessment, we enter median(data_growth_2022$self.assessment).

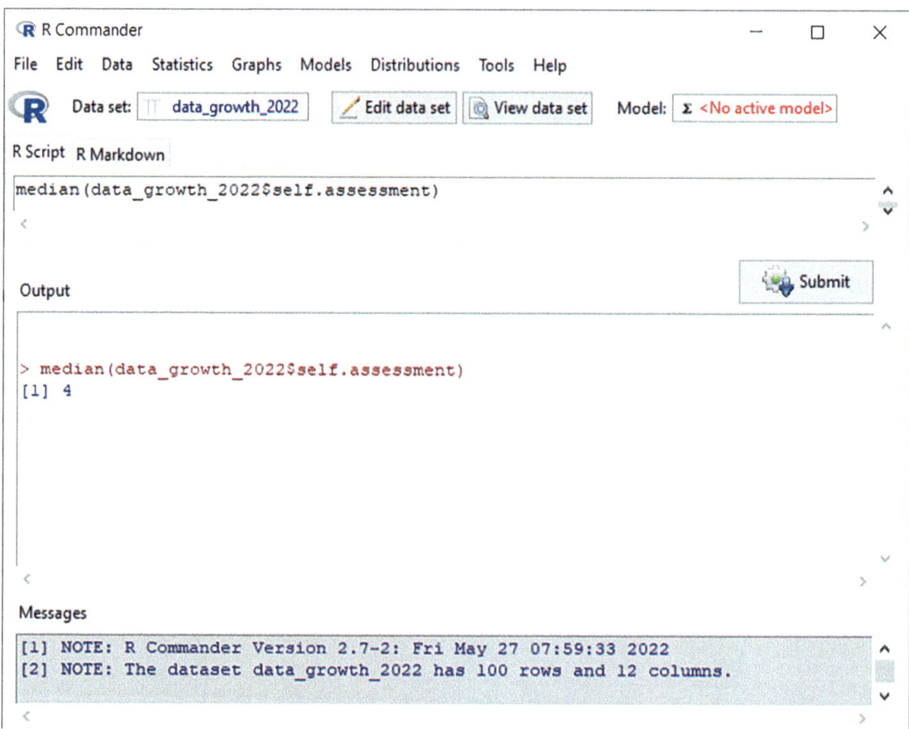

For the mode, there is no function. We count the values and then determine the most frequent value. To determine the mode for the variable motive, we enter table(data_growth_2022$motive).

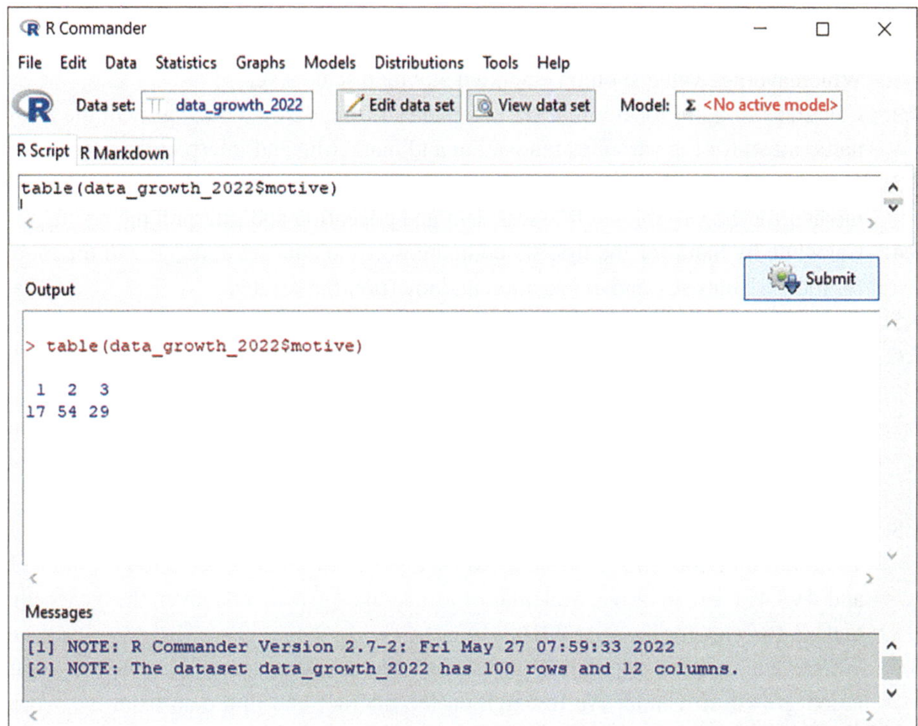

The most frequent value and thus the mode is 2.

Let's take a closer look at the commands. These consist of three parts, the function, the name of the dataset, and separated by a dollar sign, the variable name: Function(Name_Dataset$Name_Variable).

3.8 Checkpoints

- *The mode is used for nominal variables.*
- *For ordinal variables, we can calculate both the mode and the median.*
- *For metric variables, we have the mode, the median, and the arithmetic mean available.*
- *The arithmetic mean is sensitive to outliers, while the mode and the median are insensitive to extreme values.*
- *The geometric mean is used when we calculate average growth rates over time.*
- *People who work seriously indicate which average value they have used.*

3.9 Application

3.1. Which average value should be applied at which scale level?

3.2. Calculate by hand for the first six companies of the data set data_growth the arithmetic mean for the variables innovation and marketing and interpret the results.

3.3. Calculate by hand for the first six companies of the data set data_growth the median for the variables self-assessment and education and interpret the results.

3.4. Calculate by hand for the first six companies of the data set data_growth the mode for the variables sex and expectation and interpret the results.

3.5. Can the median be meaningfully calculated for the variable motive?

3.6. Calculate the corresponding averages for all variables in the data set data_growth using the R Commander.

3.7. Our third company in the data set grew by 16% in the first year, 11% in the second year, 28% in the third year, 13% in the fourth year, and 23% in the fifth year. What is the average growth rate?

3.8. The turnover of our fourth company is supposed to have been 120,000 Fr. in 2007, 136,800 Fr. in 2008, 151,848 Fr. in 2009, 176,144 Fr. in 2010, 202,566 Fr. in 2011, and 193,451 Fr. in 2012. Calculate the average growth rate over the years and make a forecast for the year 2016.

3.9. We are interested in population growth in India and want to compare it with population growth in China. We research on the internet and find data from the World Bank (https://data.worldbank.org/indicator). Calculate the average growth rate for both countries and make a forecast for the year 2025.

Absolute population at mid-year (estimates)

Year	India	China
2013	1,280,846,129	1,357,380,000
2014	1,295,604,184	1,364,270,000
2015	1,310,152,403	1,371,220,000
2016	1,324,509,589	1,378,665,000
2017	1,338,658,835	1,386,395,000
2018	1,352,617,328	1,392,730,000

Source: World Bank 2020, data downloaded from https://data.worldbank.org/ on 11.04.2020

3.10. Which average value is sensitive to outliers, which two are not, and why?

4 Variation: The Deviation from Average Behavior

4.1 Variation—The Flip Side of Average Behavior

Averages are only one side of the coin. They are only as good as they represent the individuals or objects being studied. The other side of the coin is the deviation of the individual observed values from their average value. If the individual values are closer to the average value, then the average value better represents the values, if they are further away from the average value, the average value is not as meaningful.

Let's assume we observe the salaries of employees from two small companies, each with eight employees:

Employee								
	1	2	3	4	5	6	7	8
Company 1	6000	6000	6000	7000	7000	8000	8000	8000
Company 2	4000	4000	4000	4000	4000	8000	8000	20000

Employee 1 of Company 1 earns CHF 6000 per month, Employee 2 of Company 1 also, etc. If we calculate the average salary of the employees for both companies using the arithmetic mean, it turns out that the average salary in both companies is 7000 francs:

$$\bar{x}_1 = \frac{\sum x_i}{n} = \frac{6000 + 6000 + 6000 + 7000 + 7000 + 8000 + 8000 + 8000}{8} = 7000$$

$$\bar{x}_2 = \frac{\sum x_i}{n} = \frac{4000 + 4000 + 4000 + 4000 + 4000 + 8000 + 8000 + 20000}{8} = 7000$$

Now if we compare the average salary with the individual salaries of the employees, we immediately see that the average value of 7000 francs reflects the salaries at the first

F. Kronthaler, *Statistics Applied with the R Commander*,
https://doi.org/10.1007/978-3-662-69107-6_4

company well, that is, the average value is meaningful here. At the second company, however, the salaries deviate significantly from the average value. The average value no longer has much significance about the salaries of the individual employees.

The variation, the deviation of the individual observations from the average value, is a central concept in statistics, on which almost all methods of statistics are based. It is worth dealing with the concept of variation. Once we have understood the variation, many things will be easier for us later. To analyze the variation, we have the range, the standard deviation, the variance, the coefficient of variation, and the interquartile range available.

4.2 The Range

The simplest measure of variation is the range. The range is the distance between the largest and smallest value of a variable:

$$\mathrm{ra} = x_{\max} - x_{\min}$$

It can be calculated for ordinal and metric data and gives us a quick overview of the range within which the observed values move. In our example of growth, the range of the growth rate for the first six companies is 13 percentage points (see Table 1.5).

$$\mathrm{ra} = x_{\max} - x_{\min} = 18 - 5 = 13$$

Whether a difference of 13 percentage points in the growth rate is a lot or not so much is a matter of interpretation, and we will put that aside for now. But we now know that the values move within this range. The range might be more impressive when looking at observed salaries. Let's say we observe the monthly salaries of 100 people. The observed minimum salary is CHF 3500, but by chance, Bill Gates is also included, his monthly salary is completely fictitiously 50 million francs. Then the range is an impressively high number.

$$\mathrm{ra} = x_{\max} - x_{\min} = 50{,}000{,}000 - 3500 = 49{,}996{,}500$$

The range provides a quick overview over the area within which the values move. It also contains hints about unusual numbers, possible outliers (Bill Gates). At the same time, it is very sensitive to outliers, as only the smallest and largest value are considered, all other values are ignored. A measure that includes all values in the calculation of the deviation from the average value is the standard deviation.

4.3 The Standard Deviation

The standard deviation can be interpreted as the average deviation of the individual observations from the average value. The calculation requires metric data. We first calculate the deviations of the individual values from the mean, square these deviations (so that negative and positive deviations do not add up to zero), divide by the number of observations, and then take the square root:

$$sd = \sqrt{\frac{\sum (x_i - \bar{x})^2}{n-1}}$$

For our example of company growth rates, we have the following growth rates:

$$\begin{aligned} x_1 &= 5 \\ x_2 &= 8 \\ x_3 &= 18 \\ x_4 &= 10 \\ x_5 &= 7 \\ x_6 &= 12 \end{aligned}$$

with the arithmetic mean:

$$\bar{x} = 10$$

The standard deviation can be calculated as follows:

$$\begin{aligned} sd &= \sqrt{\frac{\sum (x_i - \bar{x})^2}{n-1}} \\ &= \sqrt{\frac{(5-10)^2 + (8-10)^2 + (18-10)^2 + (10-10)^2 + (7-10)^2 + (12-10)^2}{6-1}} \\ &\sqrt{\frac{(-5)^2 + (-2)^2 + 8^2 + 0^2 + (-3)^2 + 2^2}{5}} \\ &= \sqrt{\frac{25+4+64+0+9+4}{5}} = \sqrt{\frac{106}{5}} = \sqrt{21.2} = 4.6 \end{aligned}$$

The average growth rate is 10%, with an average deviation of the observed values of 4.6 percentage points. The deviation from the height of the average value is therefore 46%. Is that a lot or a little? The interpretation depends on our understanding of the growth rates of young companies. A banker who deals with the financing of young companies, for example, might conclude that the growth rates do vary considerably, which would then have implications for his financing decisions. He might also wonder about the reasons for the fluctuations. More on that later.

Freak-Knowledge

As with the arithmetic average value $\bar{x}$ we have calculated the standard deviation for the sample and denoted it with *sd*. When we calculate the standard deviation for the entire population, i.e., all relevant objects or persons, we do not divide by $n-1$ but only by n. In addition, the standard deviation is then denoted by the Greek letter σ and the number of observations with N. One might ask why we divide by n−1 in the sample. Especially with a small sample, the standard deviation is underestimated. If we divide by n−1 instead of n, we correct for this circumstance.

If the data is already prepared in a classified form, as we have already discussed in the calculation of the arithmetic mean, then we need to slightly adjust the formula. For this, we look again at Table 3.1, which is shown again below.

Class	Growth rate from over … to … [in percent]	Number of enterprises
1	−10 to −5	3
2	−5 to 0	8
3	0 to 5	27
4	5 to 10	34
5	10 to 15	23
6	15 to 20	4
7	20 to 25	1

The formula is then:

$$sd \approx \sqrt{\frac{\sum (\bar{x}_j - \bar{x})^2 \times n_j}{n-1}}$$

$\bar{x}_j$ are the mean values of the respective classes,

n_j the number of companies in the respective class and

n the total number of observed companies

In class one with three companies, the mean of the growth rate is −7.5, in class two with eight companies −2.5, etc. We have already calculated the mean in Chap. 3 with $\bar{x} \approx 6.6\%$.

We can calculate the standard deviation as follows:

$$sd \approx \sqrt{\frac{\sum (\bar{x}_j - \bar{x})^2 \times n_j}{n-1}}$$

$$\approx \sqrt{\frac{(-14.1)^2 \times 3 + (-9.1)^2 \times 8 + (-4.1)^2 \times 27 + 0.9^2 \times 34 + 5.9^2 \times 23 + 10.9^2 \times 4 + 15.9^2 \times 1}{100-1}}$$

$$\approx \sqrt{\frac{596.43 + 662.48 + 453.87 + 27.54 + 800.63 + 475.24 + 252.81}{99}} \approx \sqrt{\frac{3269}{99}} \approx 5.8$$

The average growth rate of the companies over the last 5 years is therefore approximately 6.6% with a standard deviation of 5.8 percentage points.

4.4 The Variance

There is something else we need to know. If we do not take the root in the formula of the standard deviation, we get the variance instead of the standard deviation. Of course, we can also square the standard deviation:

$$\text{var} = \frac{\sum (x_i - \bar{x})^2}{n - 1}$$

The variance, the squared standard deviation, is more difficult to interpret than the standard deviation. We square not only the values but also the units. In the above example with unclassified values, the variance (4.6 percentage points)^2 is equal to 21.16 percentage points^2. However, percentage points^2 can be interpreted as little as, for example, CHF^2. Nevertheless, the variance is important. As already mentioned, variation is a central concept used in statistical procedures. When applying these procedures, we usually do not bother to take the root, but use the variance as a measure of variation.

Whenever you hear the term variance, you should think of variation. In one of the next chapters, the chapter on correlation, we will see the term covariance. Here too, variation is the term of the hour, with "co" indicating the joint deviation of two variables.

4.5 The Coefficient of Variation

The coefficient of variation is used when we want to compare the variation of two variables, i.e., when we ask ourselves which variable has the greater variation. It measures the deviation relatively, in percent from the average value:

$$\text{cv} = \frac{sd}{\bar{x}} \times 100$$

cv is the coefficient of variation,

sd is the standard deviation and

$\bar{x}$ is the mean.

Let's take the example of the growth rates of our six companies again. We had calculated the following mean and the following standard deviation:

$$\bar{x} = 10 \quad \text{und} \quad sd = 4.6$$

Similarly, we can calculate the mean and the standard deviation for the variable age. If we do this, we get:

$$\bar{x} = 38 \quad \text{und} \quad sd = 8.2$$

Do the values for the variable age vary more than those for the variable growth rate? At least the standard deviation is larger, but also the average value. If we calculate the coefficient of variation for both variables, the following values result:

$$\text{cv}_{\text{growth rate}} = \frac{sd}{\bar{x}} \times 100 = \frac{4.6}{10} \times 100 = 46.0$$

$$\text{cv}_{\text{age}} = \frac{sd}{\bar{x}} \times 100 = \frac{8.2}{38} \times 100 = 21.6$$

When comparing the values, it becomes apparent that the standard deviation for the variable age is 21.6% of the average value, while for the variable growth rate it is 46%. With the help of the coefficient of variation, one can see a larger deviation of the individual observations from their average value in the variable growth rate than in the variable age. The mean of the variable age thus better represents the individual persons.

4.6 The Interquartile Range

Another important measure of variation is the interquartile range. The interquartile range is used for both metric data and ordinal data. It tells us the range within which the middle 50% of the values move.

For simplicity's sake, we expand our case study from six to eight companies. The growth rates of the first eight companies in our dataset are as follows:

$$x_1 = 5$$
$$x_2 = 8$$
$$x_3 = 18$$
$$x_4 = 10$$
$$x_5 = 7$$
$$x_6 = 12$$
$$x_7 = 16$$
$$x_8 = 2$$

To determine the interquartile range, we need to sort the values in ascending order, just like with the median:

$$x_8 = 2$$
$$x_1 = 5$$
$$x_5 = 7$$
$$x_2 = 8$$
$$x_4 = 10$$
$$x_6 = 12$$
$$x_7 = 16$$
$$x_3 = 18$$

The interquartile range is the value that reflects the range within which the four middle companies move. In our case, these move within a range of 6% to 14%. These two values are also called the 1st quartile and the 3rd quartile. They are calculated from the arithmetic mean of the two values closest to the respective quartile. The interquartile range is therefore 8 percentage points. The median is also referred to as the 2nd quartile. All of this can also be well represented on a line (Fig. 4.1).

If we look at the quartiles again, we notice that between the quartiles there are always two out of a total of eight possible values or 25% of the values. Accordingly, there must be 50% of the values between the 1st quartile and the 3rd quartile.

We have learned about measures of variation for metric and ordinal variables, but not for nominal variables. We do not need measures of variation for nominal variables. We can illustrate this using the example of gender. Men and women like to be near the opposite sex, but they do not scatter around the opposite sex. Technically speaking, we have zeros (men) and ones (women), but no intermediate values.

4.7 The Boxplot

The boxplot combines the information about the median, the 1st and 3rd quartile, as well as the minimum and maximum value. The boxplot is therefore a popular tool for the graphical analysis of the variation of ordinal and metric variables. The following figure shows the boxplot for our first eight companies (Fig. 4.2).

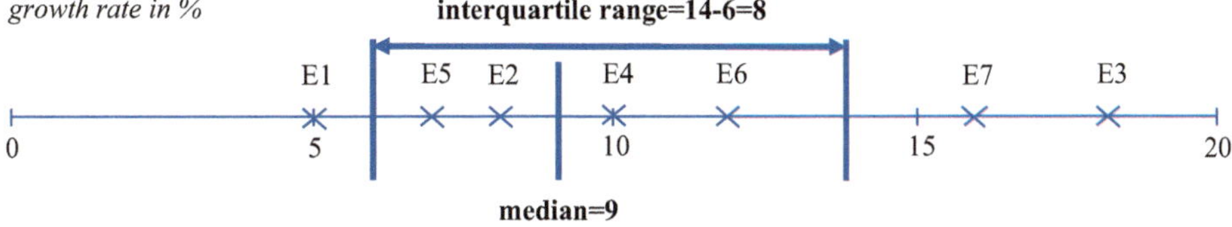

Fig. 4.1 The interquartile range for the company's growth rate

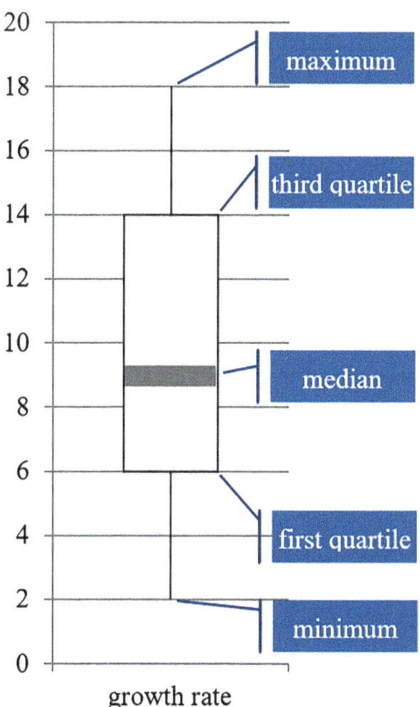

Fig. 4.2 Boxplot growth rate companies

In the companies considered, the lowest growth rate achieved by a company is 2%, the highest is 18%, the 1st quartile is 6%, the 3rd quartile is 14% and the median is 9%. This information is represented in the boxplot. The height of the box is defined as the difference between the 3rd quartile and the 1st quartile. Within this range are the growth rates of the middle 50% of the companies. The ends of the lines mark the minimum and maximum growth rate. Between the 1st quartile and the end of the line are thus the 25% of companies with the lowest growth rates. Between the 3rd quartile and the end of the line are the 25% of companies with the highest growth rates. The median divides our companies exactly into the 50% with the higher growth rates and the 50% with the lower growth rates.

Freak Knowledge

The position of the median also gives us information about whether the variable is symmetrical or skewed. If the median divides the box exactly in the middle, we speak of a symmetrical variable. If the median is in the upper half of the box, the variable is left-skewed. Accordingly, the variable is right-skewed if the median is found in the lower half of the box.

In addition, the boxplot often marks unusual values with a star or a dot. The idea is that all values outside a certain range are unusual observations and should not be integrated into the displayed range from minimum to maximum value. One option often used is to limit the lines from the quartiles to the minimum or maximum value (also called whiskers) to 1.5 times the length of the box.

4.8 Calculation of Measures of Variation with the R Commander

To calculate the measures of dispersion with the R Commander, we again use the graphical interface. We select Summaries from the Statistics menu and go to the command Numerical summaries. A window opens in which we can select the variable. We then select the Statistics tab and specify the required measures. We have the options of standard deviation, interquartile range, coefficient of variation, and quantiles available.

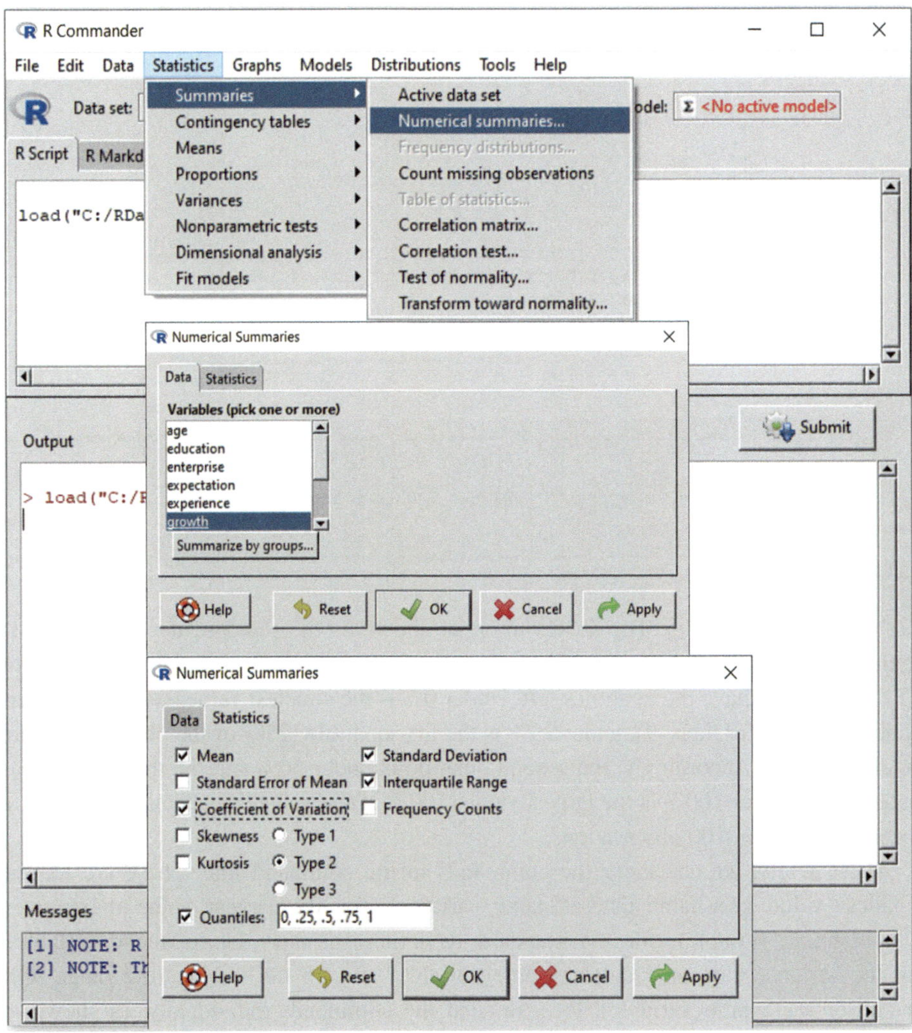

We confirm the command with OK and receive the results.

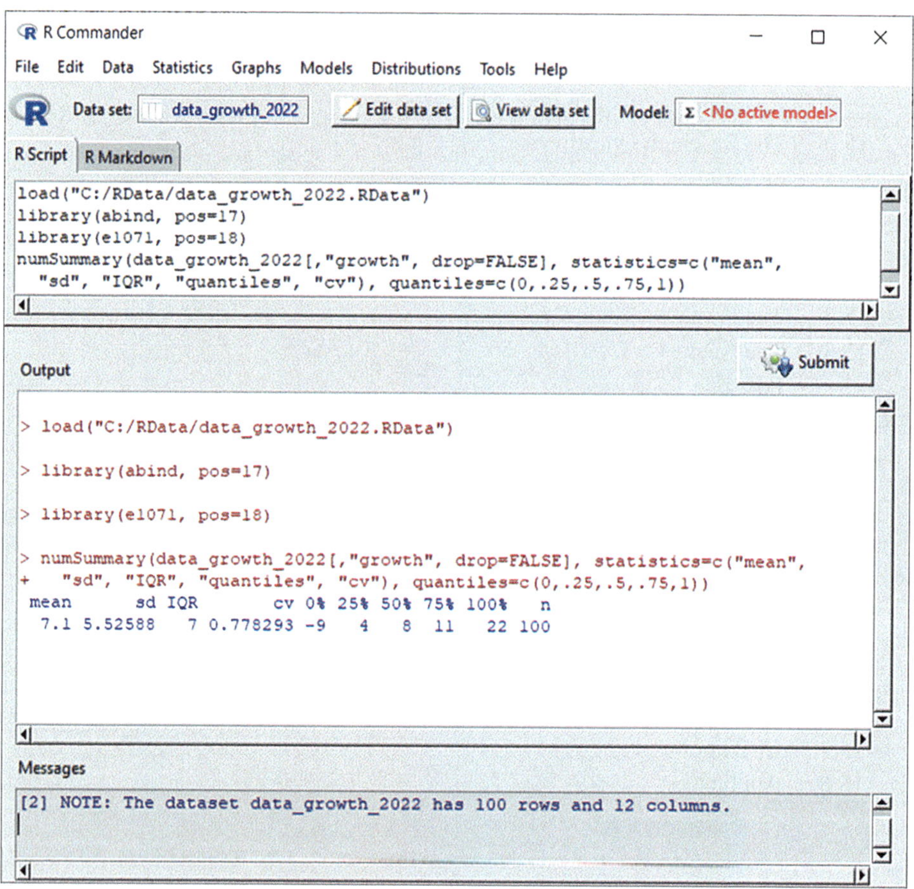

Let's look at the results. The abbreviation sd shows us the standard deviation. This is rounded 5.53, the interquartile range is 7 (IQR), the coefficient of variation is 0.778 or 77.8%. Now we have the quantiles left. Under 0% is the smallest value (0% of the observations are smaller than -9), under 25% is the first quantile (25% of the observations are smaller than 4). Accordingly, the second quantile is under 50% and the third quantile is under 75%. Under 100% is the largest value (100% of the values are smaller than 22). In total, we have n = 100 observations.

Alternatively, we can enter the commands in the command line. These are for the smallest value min(name_dataset$name_variable), for the largest value max(name_dataset$name_variable), for the standard deviation sd(name_dataset$name_variable), for the variance var(name_dataset$name_variable) and for the interquartile range IQR (name_dataset$name_variable). We can send the commands individually, as shown in Chap. 2, or select all commands and confirm them together.

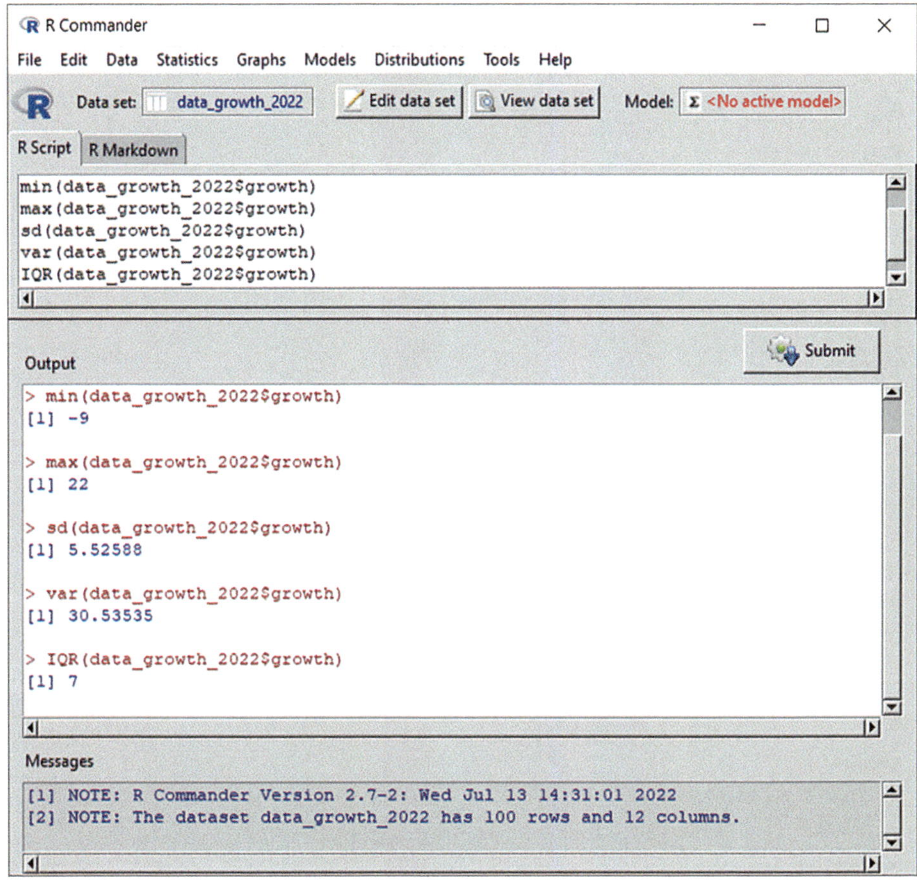

4.9 Creating the Boxplot with the R Commander

To create the boxplot with the R Commander, we use the Graphs menu. Here we select the Boxplot command. In the window that opens, we then select the variable.

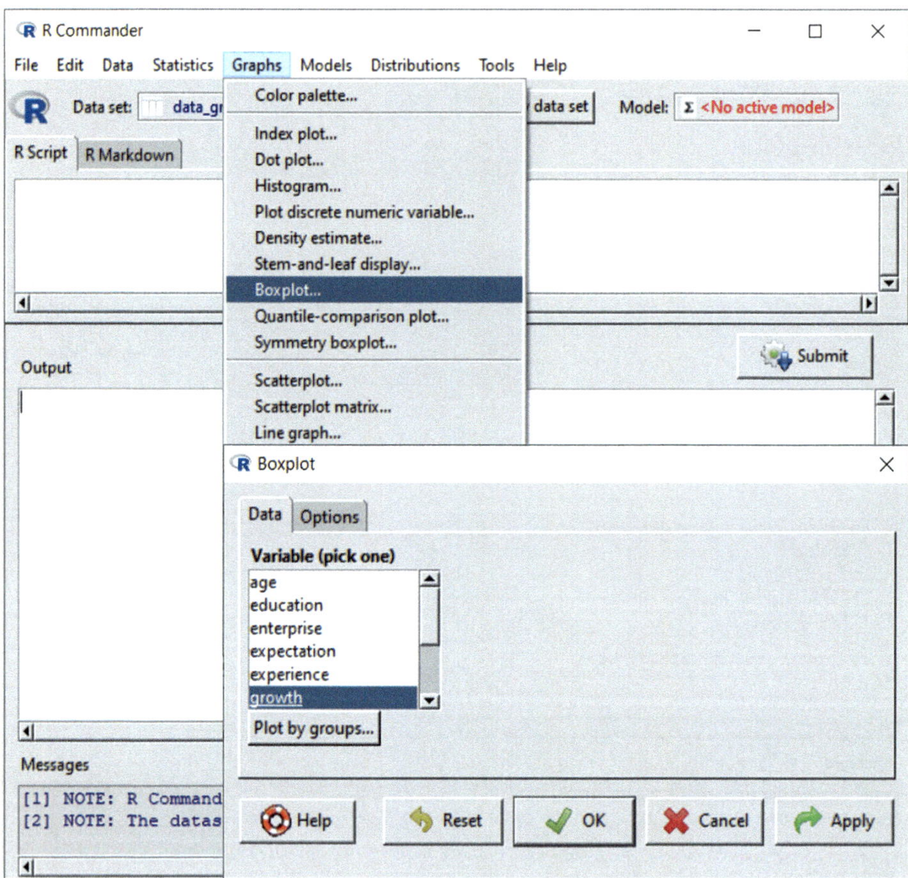

After we have confirmed the command with OK, our boxplot for the selected variable appears in a new window.

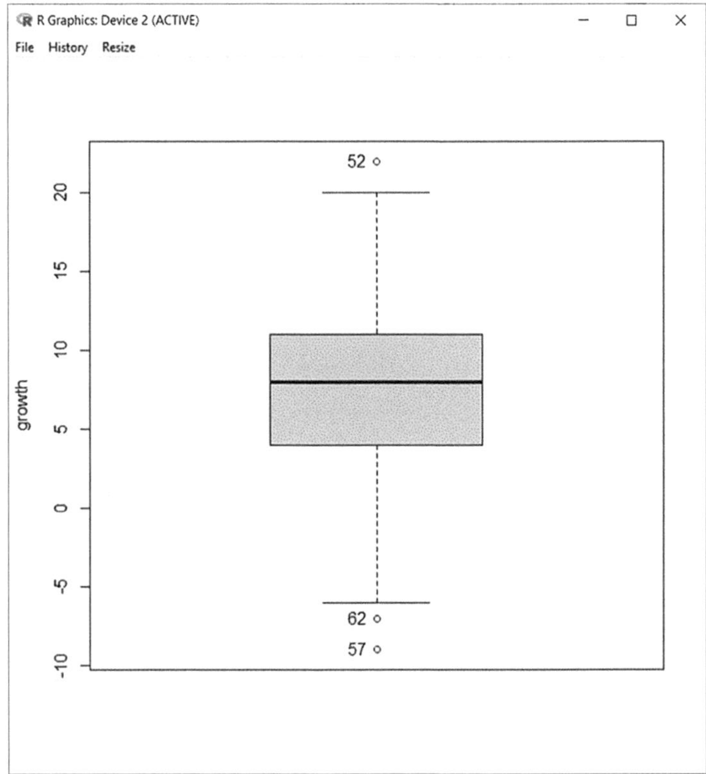

The boxplot contains the information described above. In addition, it shows us observations outside the ends. In our case, these are observations 52, 57, and 62. This is an indication of exceptionally large or small observations, i.e., outliers.

More interesting than drawing a boxplot for one variable is usually drawing the boxplot for the variable by groups, e.g., we might be interested in whether companies founded by men or women differ in terms of the variation in the growth rate. For this purpose, we first need to tell R that our gender variable is the grouping variable, in the language of R a factor. We accomplish this by selecting Manage variables in active data set in the Data tab and then clicking on Convert numeric variables to factors. We then select the variable, tell R that digits are used in our data set, and possibly rename the variable.

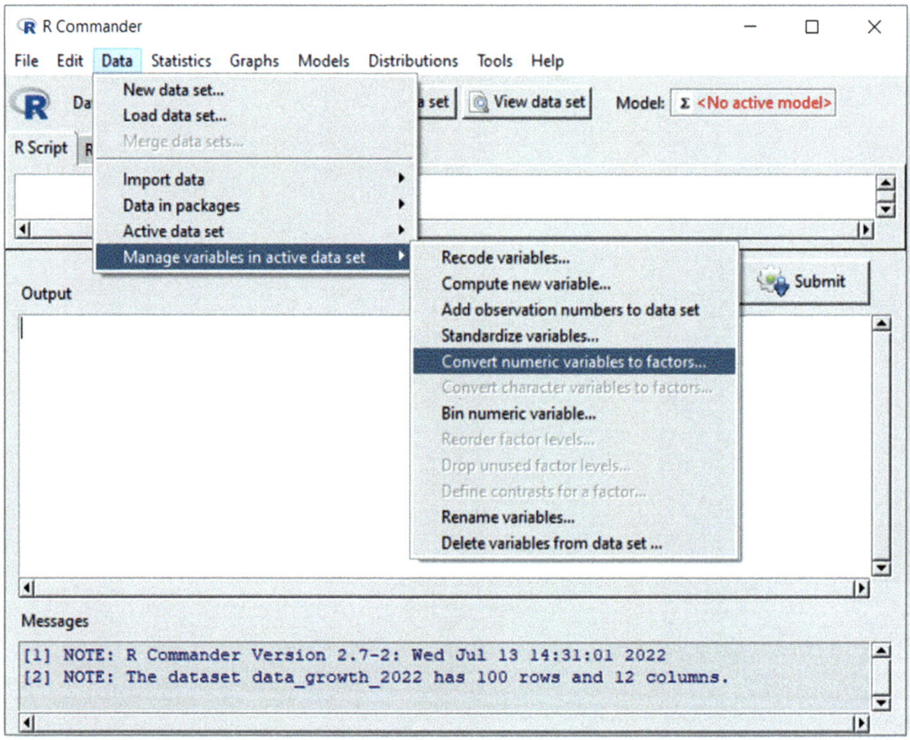

We click OK and have generated the factor. Now we can draw the boxplot for the growth rate by the sex of the company founders and get the following result.

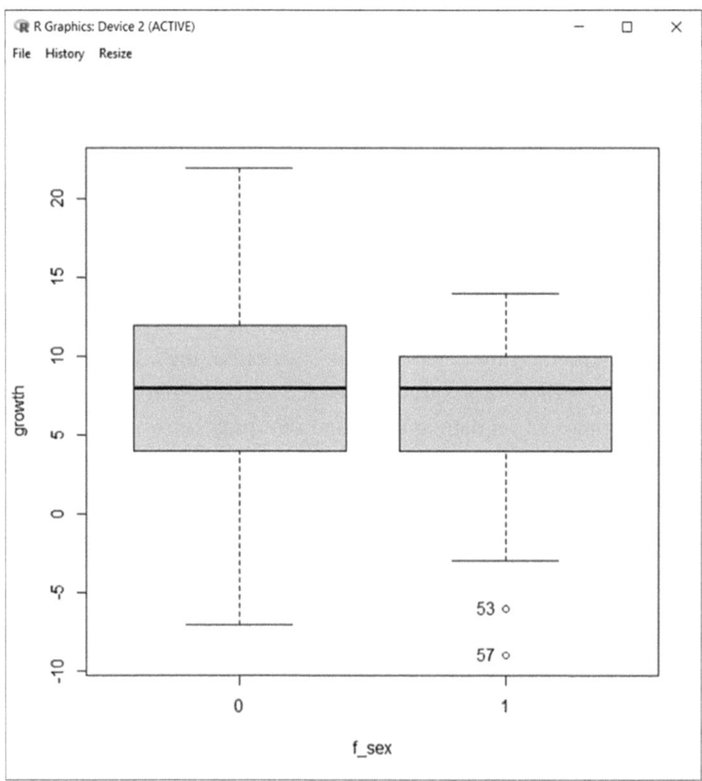

4.10 Checkpoints

- *Measures of variation complement our information about the average value, the smaller the variation, the more meaningful is the average value.*
- *The range is the difference between the maximum and minimum of a variable, it is very sensitive to outliers.*
- *The standard deviation can be interpreted as the average deviation of all values from the mean. The variance is the squared standard deviation. Both values can be meaningfully calculated for metric variables.*
- *The interquartile range gives us the range within which the middle 50% of the values move. It can be used with both metric and ordinal data.*
- *The coefficient of variation is used with metric data to analyze which variables have greater variation.*
- *The boxplot is a graphical tool for analyzing variation.*

4.11 Application

4.1. Which measure of variation is applicable at which scale level?

4.2. Can a measure of variation be calculated for nominal data, why or why not?

4.3. Calculate by hand the range, the standard deviation, the variance, the interquartile range, and the coefficient of variation for the variable age for the first eight companies of our data set data_growth and interpret the result.

4.4. Calculate by hand the range, the standard deviation, the variance, the interquartile range, and the coefficient of variation for the variable marketing for the first eight companies of our data set data_growth and interpret the result.

4.5. Calculate by hand the range, the standard deviation, the variance, the interquartile range, and the coefficient of variation for the variable innovation for the first eight companies of our data set data_growth and interpret the result.

4.6. Which variable has a greater variation, marketing or innovation (Application 4.4 & 4.5)? Which measure of variation is used to answer the question?

4.7. We are interested in the volatility of two stocks, i.e., how much the stock prices fluctuate over time. We search the internet for a financial platform that provides the data and look at the closing prices for a week (e.g., https://www.finanzen.ch). For the first stock (Novartis) we get the following values: 16.03. CHF 70.19, 17.03. CHF 73.49, 18.03. CHF 72.02, 19.03. CHF 75.70, 20.03. CHF 74.00. For the second stock (UBS) we get: 16.03. CHF 7.46, 17.03. CHF 7.76, 18.03. CHF 7.77, 19.03. CHF 8.73, 20.03. CHF 8.26. Calculate the range, the standard deviation, the variance, and the coefficient of variation and decide which stock had higher volatility during the given period.

4.8. We have the following values for the variable growth rate from our data set: minimum −9, maximum 22, 1st quartile 4, median 8, 3rd quartile 11. Draw the boxplot by hand. Within which range do the middle 50% of the observations move?

4.9. Draw the boxplot for the variable growth of the data set data_growth using the R Commander.

4.10. We have the following values for the variable growth, grouped by the motive for company foundation, from our data set: Unemployment: minimum −3, maximum 22, 1st quartile 2, median 7, 3rd quartile 10; Implementing idea: minimum −7, maximum 20, 1st quartile 4.25, median 8, 3rd quartile 11; Higher income: minimum −9, maximum 13, 1st quartile 3, median 7, 3rd quartile 11. Draw the boxplots and compare the variation.

4.11. Draw the boxplots for the variable growth by the motive for company foundation using the R Commander.

4.12. Calculate the range, the interquartile range, the standard deviation, and the variance for all variables in our data set data_growth using the R Commander (if possible). Also, draw the corresponding boxplots.

5 Charts: The Possibility to Display Data Visually

5.1 Charts: Why do We Need Them?

We have already learned some statistical tools, the arithmetic mean, the median, the mode, the geometric mean, the range, the standard deviation, the variance, the coefficient of variation, and the interquartile range. With these measures, we have calculated how individuals or objects behave on average with different variables and how the individual persons or objects deviate from it. We already know a lot about our individuals or objects. However, experience shows that numbers are sometimes difficult to convey. Listeners quickly get bored when only numbers are presented. This is an important reason why we need charts. Charts help us to convey numbers and create lasting impressions.

Charts also help us to recognize patterns in the data or give us clues about the behavior of our examined individuals or objects, e.g., how many people really like eating hamburgers or how many companies have a growth rate between 10 and 15%, how many of the company founders are very experienced in their industry, etc. To discuss such questions, we can, for example, use the frequency table.

5.2 The Frequency Table

The frequency table is the starting point of frequency analysis. The frequency table shows us how often values occur absolutely and relatively. We can distinguish frequency tables according to the scale level of the data. With nominal and ordinal data, only a few specific values usually occur and we can enter the values directly into the table. With metric data, we usually have many different values that we first have to classify. Let's start with the frequency table for nominal and ordinal data.

F. Kronthaler, *Statistics Applied with the R Commander*,
https://doi.org/10.1007/978-3-662-69107-6_5

In Table 5.1, the frequency table for the (ordinal) variable self-assessment for our 100 companies is shown. The observed values range from 1 (very experienced) to 5 (not experienced).

x_i	are in this case the values that the variable can take,
n_i	is the number of occurrences of the respective value,
$f_i = \frac{n_i}{n}$	is the number of the respective value to the total number of observations and
cf_i	are the added relative shares (Table 5.1).

If we multiply f_i or cf_i by 100, we get percentage values. We can easily interpret the table. $x_1 = 1$ occurs six times. Six percent of companies are led by a very experienced founder according to self-assessment. $x_2 = 2$ occurs 17 times, thus representing 17% of all companies. If we add up the values, we get the cumulative share of 0.23 or 23%, i.e., 23% of companies are led by a very experienced or experienced founder.

As already mentioned, we need to form classes with metric data first in order to create a frequency table. The reason for this is that metric values usually do not occur more than once and forming frequencies without classification makes no sense. To design classes meaningful, we first look at the range and then consider how many classes we should form and how wide they should be. Too many classes are not good, too few are not either. As a rule, we work with 5 to 15 classes. For our variable growth, the range is from −9 to 22, i.e., it is 31 percentage points. We quickly see that we can work well with class widths of 5 percentage steps, resulting in seven classes (Table 5.2).

We can interpret the table like Table 5.1. Three companies have a growth rate between −10 and −5%. That is 3% of the companies. Another eight companies, or 8% of the companies, have a growth rate between −5 and 0%, etc. If we add up both values, we see that a total of 11% of the companies have a growth rate less than or equal to 0%. The fact that we count from over one value to a certain value (see the first column of Table 5.2) is to avoid double counting.

Table 5.1 Frequency table for the variable self-assessment

Self-assessment	Frequency, absolute	Frequency, relative	Frequency, cumulative
x_i	n_i	f_i	cf_i
1	6	0.06	0.06
2	17	0.17	0.23
3	21	0.21	0.44
4	30	0.30	0.74
5	26	0.26	1
	$n = 100$		

Table 5.2 Frequency table for the variable growth

Growth	Frequency, absolute	Frequency, relative	Frequency, cumulative
from over … to …	n_i	f_i	cf_i
−10 to −5	3	0.03	0.03
−5 to 0	8	0.08	0.11
0 to 5	27	0.27	0.38
5 to 10	34	0.34	0.72
10 to 15	23	0.23	0.95
15 to 20	4	0.04	0.99
20 to 25	1	0.01	1
	$n = 100$		

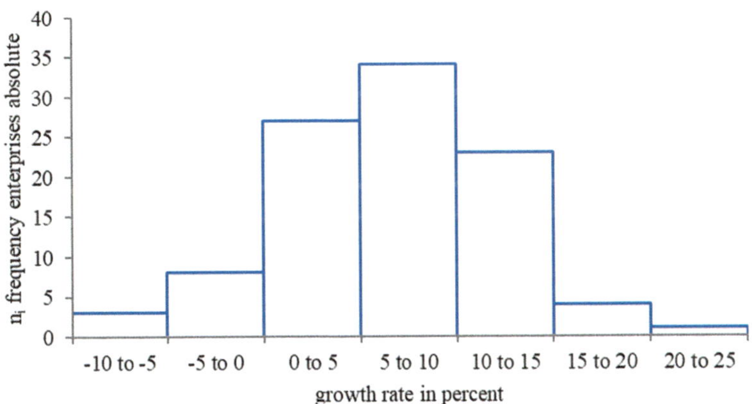

Fig. 5.1 Absolute frequency chart for the growth rate

5.3 The Frequency Charts

For the graphical display of a frequency table, we have three options. Either we draw an absolute frequency chart, a relative frequency chart, or a histogram.

In the absolute frequency chart (Fig. 5.1), we plot the absolute frequencies n_i on the y-axis and the classes for metric data or the possible values for ordinal or nominal variables on the x-axis. We see that 3 companies reflect the height of the first bar, 8 companies the height of the second bar, etc.

In the relative frequency chart (Fig. 5.2), we plot the relative frequencies f_i on the y-axis and the classes for metric data or the possible values for ordinal or nominal vari-

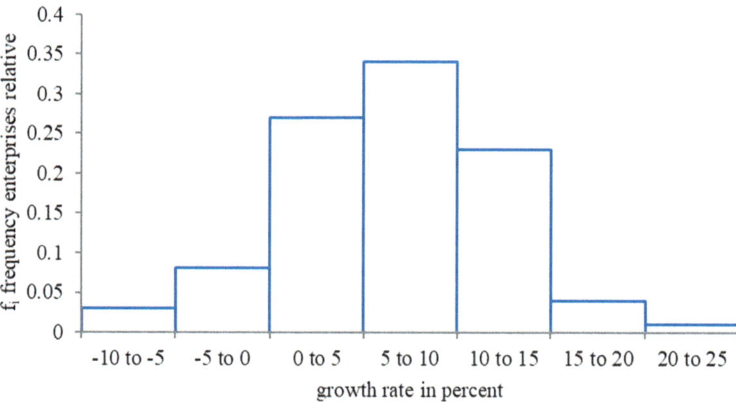

Fig. 5.2 Relative frequency chart for the growth rate

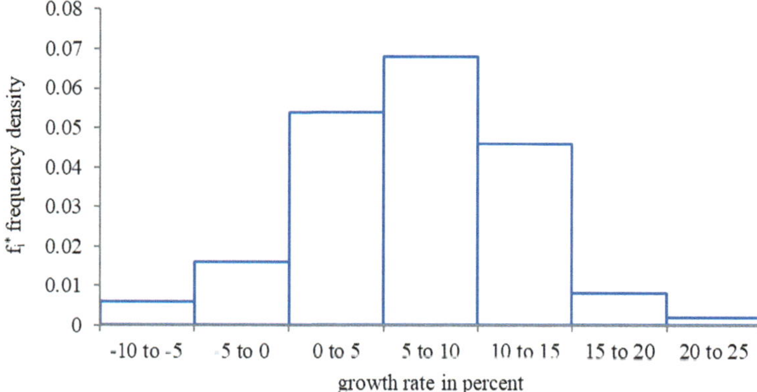

Fig. 5.3 Histogram for the growth rate

ables on the x-axis. We can read from the height of the bars that 0.03 or 3% are included in the first class, 0.08 or 8% of the companies in the second class, etc.

In the histogram (Fig. 5.3), we plot the frequency density

$$f_i^* = \frac{f_i}{w_i}$$

on the y-axis, where w_i is the class width of the respective class. On the x-axis, we plot the classes or the possible values.

The difference to the relative frequency chart is that it is no longer the height of the bar that indicates what percentage of the companies are in the respective class, but the size of the area. We can easily imagine this if we solve the formula of the frequency density for f_i. This clearly shows that the relative frequency is the frequency density times the respective class width $f_i = f_i^* \times w_i$.

5.4 Absolute Frequency Chart, Relative Frequency Chart, or Histogram?

Which frequency chart should we choose? When we look at the three charts, we notice that they look identical. This is always the case when all classes are of equal width. Here, visually, it makes no difference which chart we choose. However, the absolute frequency chart and the relative frequency chart are easier to interpret, as the height of the bars reflects the number or proportion of companies.

If we have different class widths, the histogram is preferable. The histogram takes into account the different class widths by dividing the relative frequency by the class width. If we didn't do this, the bar would only become higher or lower because the class is wider or narrower. So, if we apply the absolute or relative frequency chart, a distorted image is created with different class widths, while the histogram gives us the undistorted, correct visual image. We have different class widths especially when our range is influenced by extreme values. Including all values would require us to form many classes with equal widths. If we want to avoid this, we either have to exclude individual values or vary the class width.

Freak-Knowledge

The frequency charts, like the box plot, are tools that can be used to assess whether a variable is symmetrical. It is symmetrical if we cut out the frequency chart, fold it in the middle, and both sides come to rest on top of each other. Furthermore, we use the frequency charts to assess whether a variable is normally distributed. Normal distribution is present when the frequency chart resembles the Gaussian bell curve.

5.5 More Ways to Display Data

The absolute frequency chart, the relative frequency chart, and the histogram are not the only ways to display data. Before we delve into other methods, here are some general tips to help create meaningful statistical charts.

To create a good chart, the following rules should be applied:

1. Creating a good chart is not easy, but an art.
2. Before creating a chart, it should be clear what information you want to display.
3. As a rule, only one piece of information should be presented in a chart.
4. A good chart is as simple and understandable as possible.
5. A good chart is not misleading.
6. A good chart complements the text and facilitates interpretation.
7. If more than one chart are used in a text, they should be consistent in formatting, e.g., size, font size, font type.

Next, we will look at the pie chart, the bar chart, and the line chart. Of course, these are not the only ways to create charts. Anyone interested in graphically representing data beyond this book should engage with the work of Edward R. Tufte (Tufte E. R. (2001), The Visual Display of Quantitative Information, 2ed, Graphics Press, Ceshire, Connecticut). Edward Tufte was referred to by the New York Times as the "Leonardo da Vinci of data".

We can use the pie chart, for example, to show how many people or objects have a certain characteristic. Figure 5.4 shows us the motives behind the companies in our data set being founded. The majority of founders, 54%, wanted to implement an idea with their company foundation, 29% pursued the goal of earning a higher income with the company foundation, and unemployment was the decisive factor in 17% of company foundations.

What we see here too is the two-dimensional chart. In science, we avoid unnecessary formatting as it merely distracts from the facts. In consulting, however, charts are often drawn in three dimensions to make the information more appealing. But this tends to distract from the actual information.

We can use the bar chart well when we want to compare groups, e.g., when we look at men and women regarding their motive for founding a company. Figure 5.5

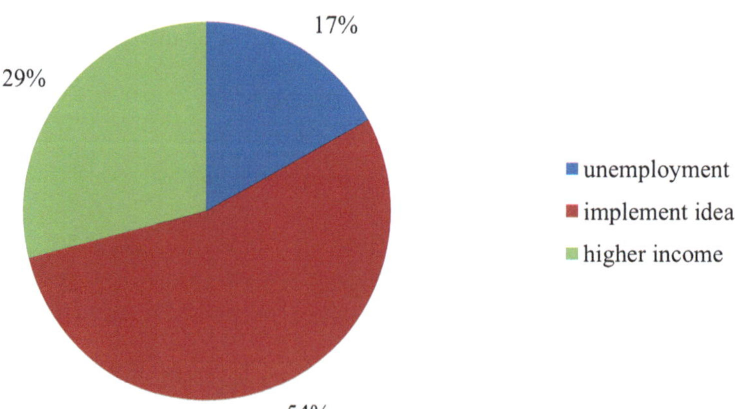

Fig. 5.4 Motive for founding a company

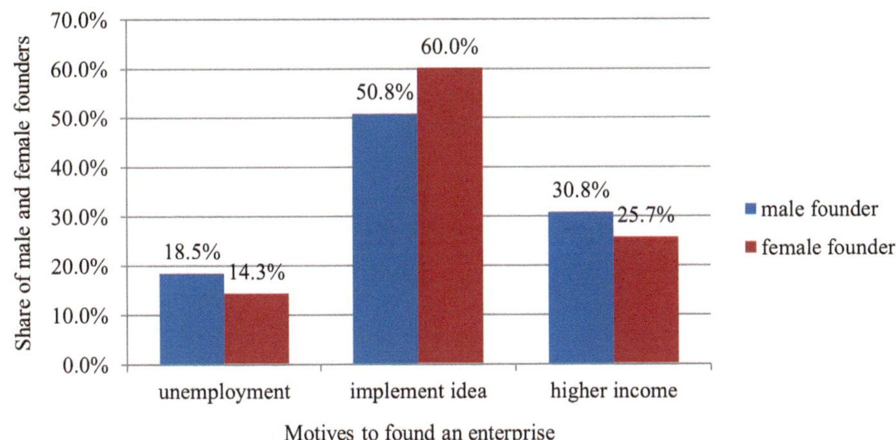

Fig. 5.5 Motive for founding a company by gender

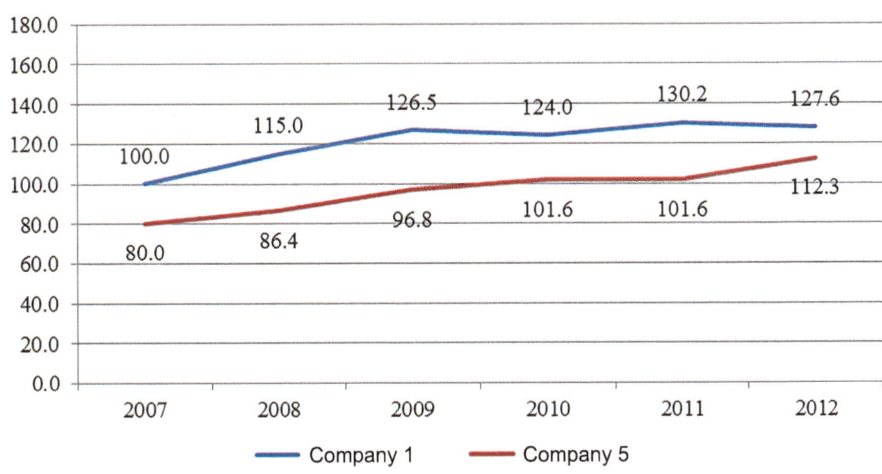

Fig. 5.6 Turnover development of companies from 2007 to 2012 (in CHF 1000)

shows that the founders differ slightly in their motivation to start a company. 18.5% of the male founders start out of unemployment, while it's only 14.3% among the female founders. On the other hand, 60% of the female founders wanted to implement an idea, while this is only the case for 50.8% of the male founders. The motive to earn a higher income is represented by more men (30.8%) than women (25.7%).

Finally, we want to look at the line chart. We use this especially when we want to show a trend. Figure 5.6 depicts the turnover development of two companies in our dataset from 2007 to 2012. We see how the turnover of company 1 have developed compared to company 5. Both show a generally positive development. However, it seems that the turnover of company 1 have been stagnating since 2009, while the turnover of company 5 have been increasing over the entire period, with one small exception.

5.6 Creating the Frequency Charts with the R Commander

Before we discuss the creation of frequency charts with the R Commander, here's a general note. Creating publication-ready charts with R requires some formatting work and knowledge of the corresponding commands. These contents go far beyond the scope of this book and will not be discussed here. For those who want to use R to create publication-ready charts, I recommend at this point to engage more intensively with R. For others, I recommend creating charts, e.g., for publications, simply with Excel and only using R to generate charts that are helpful for data analysis. For the latter, it will often be necessary to create frequency charts. Drawing frequency charts with the R Commander is incredibly simple. We use the Graphs menu and select the Histogram command. In the window that appears, we select the variable and with the Options tab we can specify whether we want to create the absolute frequency chart, the relative frequency chart, or the histogram.

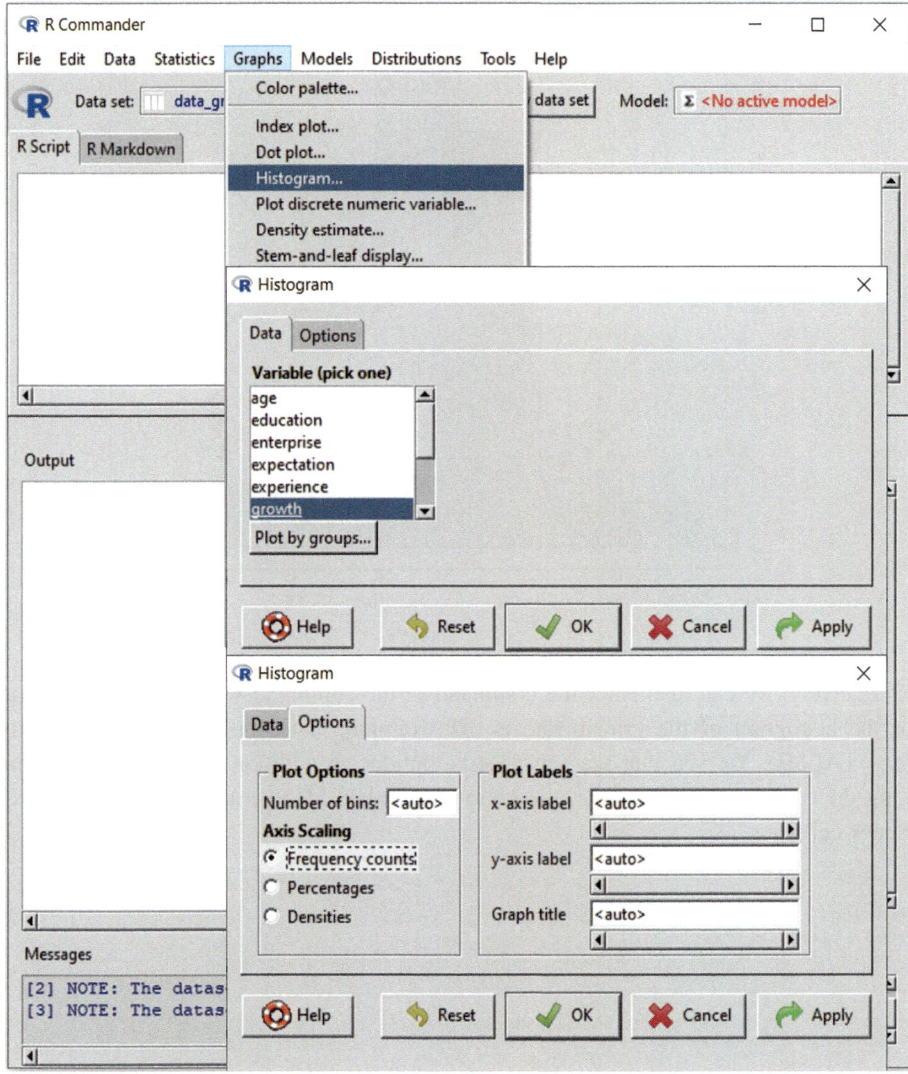

After we have specified which frequency chart we want to create, we confirm with OK. The frequency chart appears in a new window. As we already know, with the same class width, the graphical display of the three frequency charts is identical. In our case, the absolute frequency chart for the growth rate is shown.

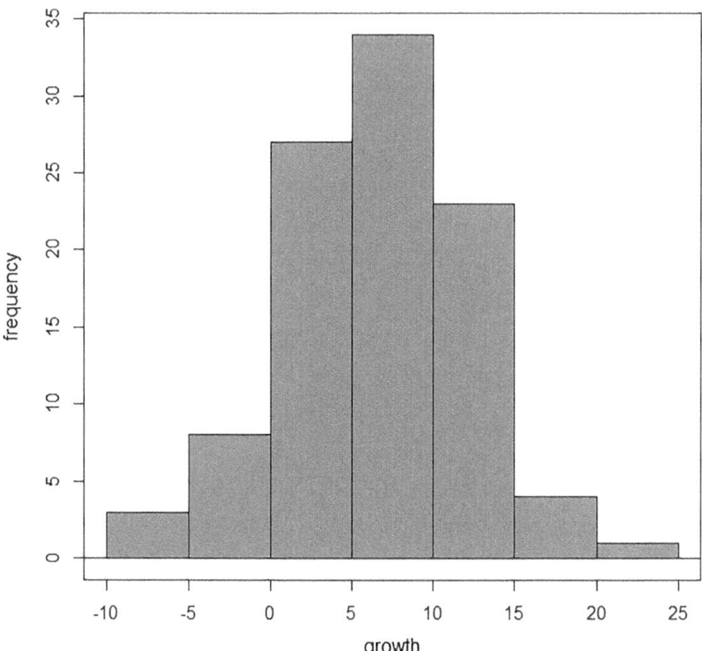

Alternatively, we can also enter the command in the command window. The command for the histogram of the growth rate is, for example, hist(data_growth_2022$growth, freq = FALSE). We see that we can specify options in addition to the variable name using the command. In this case, the option freq = FALSE instructs R to draw the frequency densities.

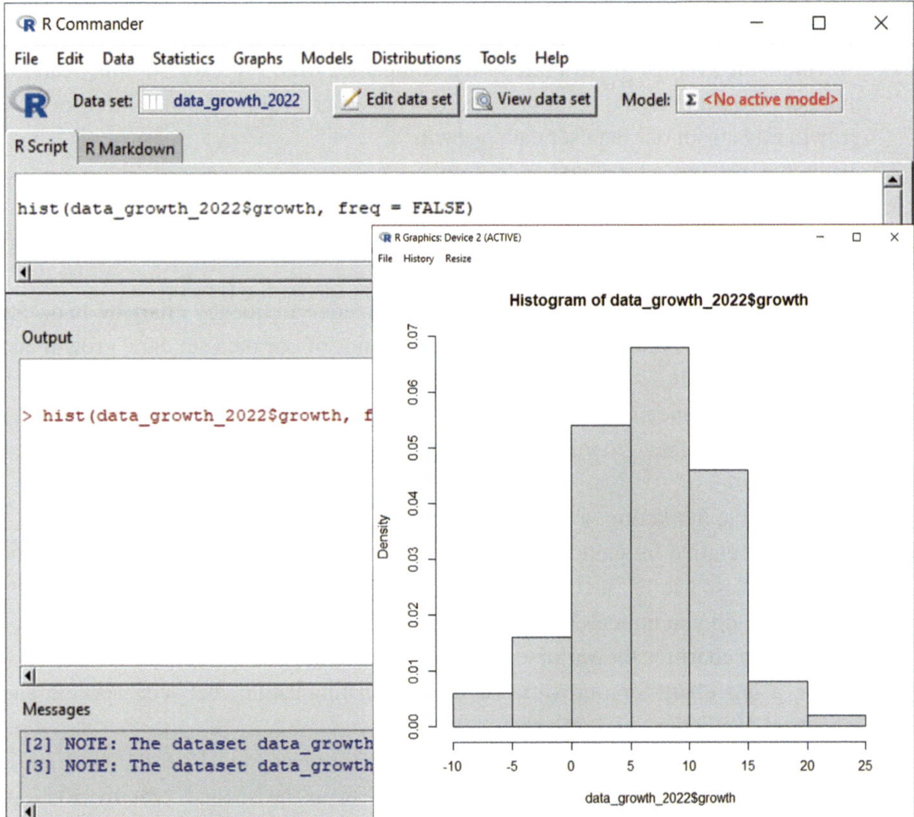

A note at this point. The aim of this book is not to learn the command language of R. However, I try to give hints on the command language at one point or another.

5.7 Checkpoints

- *The graphical presentation of data helps us to explain facts.*
- *The graphical presentation of data often gives us clues about the behavior of individuals or objects as well as patterns in the data.*
- *We use frequency charts to find out how often certain values occur.*
- *The frequency table tells us how often certain values occur absolutely and relatively.*
- *Graphically, we can present the frequency distribution absolutely, relatively, or with the histogram.*
- *If we have equal class widths, the absolute or relative frequency charts are preferable, with different class widths the histogram.*
- *A good chart is as simple as possible and clearly presents the information.*

5.8 Application

5.1. Calculate the average growth rate of the companies from Fig. 5.6. For this, read the numbers from the figure. Compare the calculated growth rates with the respective growth rates from our data set data_growth.
5.2. Name two reasons why charts are helpful.
5.3. What purpose do we need frequency charts for?
5.4. Is the absolute frequency chart, the relative frequency chart, or the histogram preferable for different class widths? Why?
5.5. Create the frequency table, the absolute and relative frequency chart by hand for the variable expectation for the first six companies of our data set data_growth and interpret the result.
5.6. Create the frequency table and the relative frequency chart with the R Commander for our data set data_growth for the variables marketing and innovation and interpret the result.
5.7. Is the variable marketing or the variable innovation more symmetric? Why?
5.8. Create the relative frequency chart and the histogram for the growth rate with the following class widths: −10 to −5, −5 to 0, 0 to 5 and 5 to 25. Draw the charts by hand. What do you notice? Which chart is preferable and why?
5.9. Create a pie chart for the variable education and interpret the result.
5.10. Create a bar chart for our variable motive, distinguishing between service and industrial companies, and interpret the result.
5.11. The second company in our data set achieved a turnover of CHF 120,000 in 2007. In the first year of its existence, it grew by 9%, in the second year by 15%, in the third year by 11%, in the fourth year by 9% and in the fifth year by −3%. Based on this information, create a line chart that represents the company's turnover development.
5.12. Search the internet for the number of overnight stays for the Swiss holiday destinations Graubünden, Lucerne/Lake Lucerne, Bernese Oberland and Valais for the years 2005 to 2019 at the Swiss Federal Statistical Office. Display the development of overnight stays using a line chart. We go a little deeper into the development, standardize the numbers to the year 2008 (value 2008 equals 100) and redraw the line chart (this is advanced). Briefly describe in a paragraph what is striking.

Correlation: From Relationships 6

6.1 Correlation—The Joint Movement of Two Variables

Correlation or the calculation of relationships is, in addition to averages, variation, and frequencies, another central concept of statistics. Here we ask for a relationship between two variables, e.g., the number of cheeseburgers that people eat per week and the amount of bacon they carry around their waist. There could be a relationship, think of the self-experiment by Morgan Spurlock. He ate only fast food for a while. Or think of the former mayor of New York Michael Bloomberg, who wanted to ban the sale of XXL soft drinks. In these cases, a relationship is suspected. The more cheeseburgers people eat, the more bacon they carry around their waist. But it could also be the other way around, the more bacon people carry around their waist, the more cheeseburgers they eat. With the last statement, we have received an important information. The correlation does not give us any information about the direction of the relationship, about causality. When we have calculated a correlation, we do not know whether people who eat many cheeseburgers have more bacon or whether people who have more bacon eat more cheeseburgers. We only know that the variables move in the same direction.

Correlations can be positive or negative. This is not a judgment, but only a statement about the direction of the relationship. Positive means that two variables move in the same direction. If the value of one variable increases, then the value of the other variable also increases and vice versa. Negative means that two variables move in the opposite direction. If the values of one variable increase, the values of the other variable decrease and vice versa (Table 6.1).

Much of the knowledge we possess is based on the correlation between variables. So it is once again worth dealing with it. If we want to understand the relationships our knowledge is based on, we need to understand the correlation coefficient. Depending on the scale level, different coefficients are used. For metric data, it is the correlation

F. Kronthaler, *Statistics Applied with the R Commander*,
https://doi.org/10.1007/978-3-662-69107-6_6

Table 6.1 Correlation between two variables

Variable 1	Variable 2	Relationship	Example
Value of X increases	Value of Y increases	Positive correlation	Number of cheeseburgers per week increases, amount of bacon around the waist increases
Value of X increases	Value of Y decreases	Negative correlation	Number of carrots per week increases, amount of bacon around the waist decreases
Value of Y increases	Value of X increases	Positive correlation	Amount of bacon around the waist increases, number of cheeseburgers per week increases
Value of Y increases	Value of X decreases	Negative correlation	Amount of bacon around the waist increases, number of carrots per week decreases

coefficient of Bravais-Pearson, for ordinal data the correlation coefficient of Spearman, and for nominal data the phi coefficient or the contingency coefficient.

6.2 The Correlation Coefficient of Bravais-Pearson for Metric Variables

The correlation coefficient of Bravais-Pearson is used to determine the strength of the linear relationship between two metric variables. It is calculated as follows:

$$r = \frac{\sum (x_i - \bar{x})(y_i - \bar{y})}{\sqrt{\sum (x_i - \bar{x})^2 \sum (y_i - \bar{y})^2}}$$

We already know all the components of the formula.

$\bar{x}$ and $\bar{y}$ are the arithmetic means and
x_i and y_i are the observed values,

i.e., we calculate the Bravais-Pearson correlation coefficient from the deviations of the observed values from their mean (concept of variation, compare Chap. 4).

Freak-Knowledge

The upper part of the formula, the numerator, gives us the joint deviation of the related x_i and y_i values from their respective mean, we also call it covariance σ_{xy}. If the deviations each point in the same direction, always positive or always negative, then the amount of the covariance is large and we have a relationship. If the deviations go in different directions, the number becomes small and there is no relationship. The lower part of the formula, the denominator, is used to normalize the Bravais-Pearson correlation coefficient to the range between −1 and 1.

We want to calculate the correlation coefficient between our variables marketing and innovation. We are interested, for example, in whether companies that spend a lot on marketing invest little in innovation and vice versa. Let's try this for the first six companies in our dataset. It is best to create a table for this. The variable X is our marketing variable, the variable Y is the innovation variable.

Company	x_i	y_i	$x_i - \bar{x}$	$(x_i - \bar{x})^2$	$y_i - \bar{y}$	$(y_i - \bar{y})^2$	$(x_i - \bar{x})(y_i - \bar{y})$
1	29	3	9	81	−3	9	−27
2	30	5	10	100	−1	1	−10
3	16	6	−4	16	0	0	0
4	22	5	2	4	−1	1	−2
5	9	9	−11	121	3	9	−33
6	14	8	−6	36	2	4	−12
$\bar{x}$	20	6					
Σ				358		24	−84

If we insert the three sums into our formula, we get our correlation coefficient of −0.91:

$$r = \frac{\sum (x_i - \bar{x})(y_i - \bar{y})}{\sqrt{\sum (x_i - \bar{x})^2 \sum (y_i - \bar{y})^2}} = \frac{-84}{\sqrt{358 \times 24}} = -0.91$$

How should this value be interpreted? First, we need to know that the Bravais-Pearson correlation coefficient is defined in the range of $-1 \leq r \leq 1$. Values outside −1 and 1 are not possible. −1 means that we have a perfect negative correlation, 1 means a perfect positive correlation. To interpret values in between, the following rule of thumb is helpful (Table 6.2).

For us, this means that we have a strong negative correlation between the two variables. Companies that spend a lot on marketing invest little in innovation, or companies that invest a lot in innovation spend little money on marketing.

Table 6.2 Rule of thumb to interpret the correlation coefficient

Value of r	Relationship between two variables
$r = 1$	Perfect positive correlation
$1 > r \geq 0.6$	Strong positive correlation
$0.6 > r \geq 0.3$	Weak positive correlation
$0.3 > r > -0.3$	No correlation
$-0.3 \geq r > -0.6$	Weak negative correlation
$-0.6 \geq r > -1$	Strong negative correlation
$r = -1$	Perfect negative correlation

6.3 The Scatterplot

We can graphically display the correlation in a scatterplot. A scatterplot is a point diagram in which we plot our individuals or objects according to pairs of values. For our companies, we have the following pairs of values for the variables marketing and innovation:

Company	Marketing x_i	Product Improvement y_i	Pair of Values $(x_i; y_i)$	Statement
1	29	3	29;3	Company 1 spends 29% of revenue on marketing and 3% on product improvement
2	30	5	30;5	Company 2 spends 30% and 5%
3	16	6	16;6	Company 3 spends 16% and 6%
4	22	5	22;5	Company 4 spends 22% and 5%
5	9	9	9;9	Company 5 spends 9% and 9%
6	14	8	14;8	Company 6 spends 14% and 8%

If we plot our companies on the point diagram, we get our scatterplot (Fig. 6.1).

We see that enterprises that spend a lot on innovation spend little on marketing, and vice versa. If we also draw a straight line into the scatterplot, we recognize the strength of the linear relationship or how strongly the variables move together (Fig. 6.2).

If the points, as in our case, are close to the straight line, there is a strong relationship. If the points scatter widely around the straight line, as constructed in the following case, it is a weak relationship (Fig. 6.3).

In the scatterplot, we can also read whether we are dealing with a positive or a negative correlation. If the straight line rises, there is a positive correlation, if the straight line falls, a negative correlation (Fig. 6.4).

In addition, we can see in the scatterplot whether there is a correlation between two variables at all. We have no correlation between two variables when the variables behave independently of each other. Graphically, this looks like Fig. 6.5, for example.

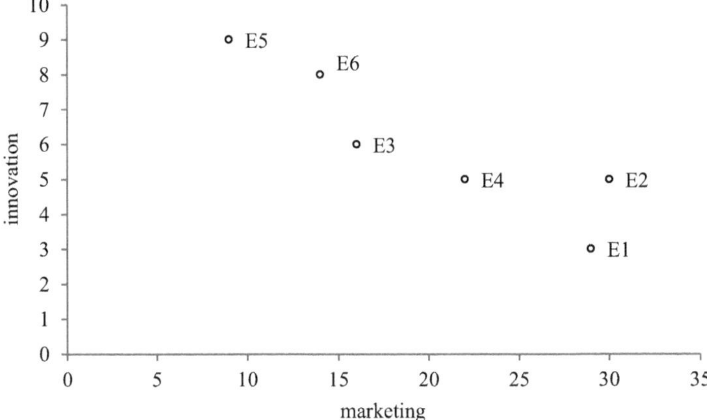

Fig. 6.1 Scatterplot for the variables marketing and innovation

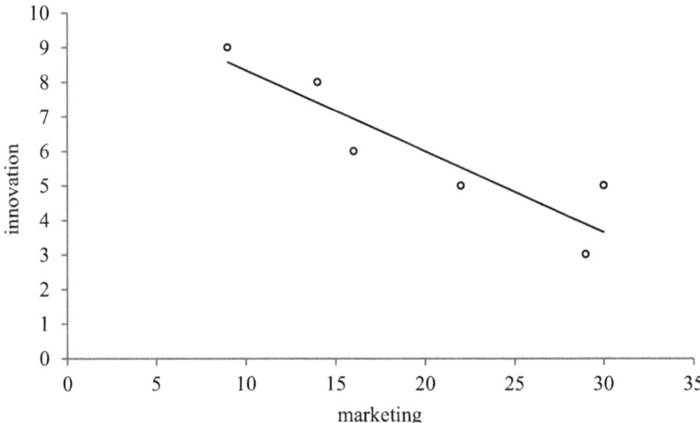

Fig. 6.2 Scatterplot for the variables marketing and innovation with a strong relationship

In the left image, the x_i values change, the y_i values remain constant. In the middle image, the y_i values vary, the x_i values do not change. In the right image, both the x_i values and the y_i values vary. However, this change is unsystematic, i.e., for large x_i values we have both small and large y_i values, for small x_i values we have both small and large y_i values.

The scatterplot is also important for another reason, namely for the calculation of the Bravais-Pearson correlation coefficient. It shows us whether there is a linear relationship between two variables. Such a relationship is a prerequisite for the Bravais-Pearson correlation coefficient.

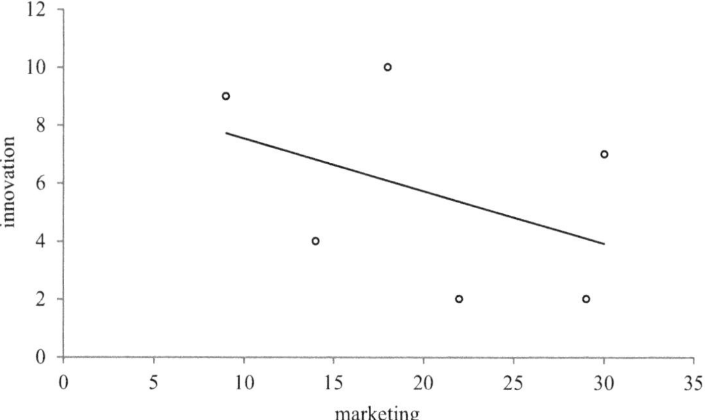

Fig. 6.3 Scatterplot for the variables marketing and innovation with a weak relationship

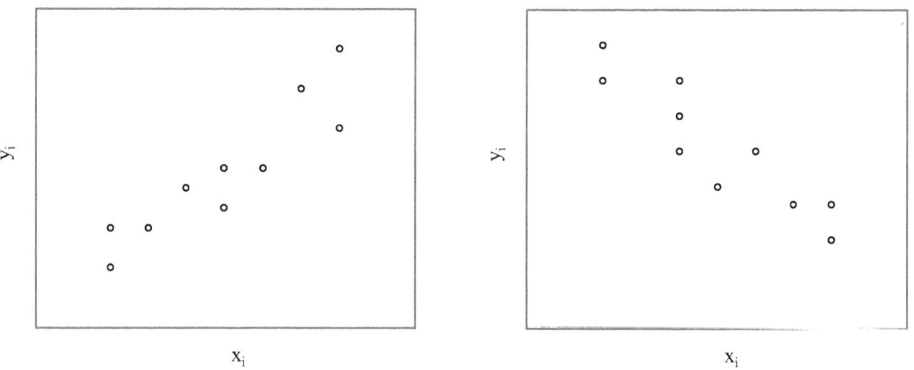

Fig. 6.4 Positive and negative correlation between two variables

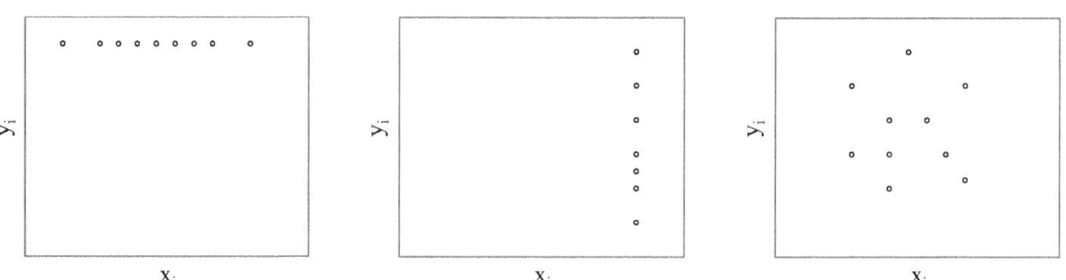

Fig. 6.5 No correlation between two variables

Furthermore, it should be noted that the correlation coefficient of Bravais-Pearson can be strongly influenced by unusual observations or outliers. We also recognize these in the scatterplot. If we have discovered such an observation in the scatterplot, then we can treat it accordingly. We can calculate the correlation coefficient with and without outliers and thus determine the influence of the unusual observation. We can correct the unusual observation if it is a data entry error, we can remove it if it is irrelevant for our analysis, or we can keep it if we consider it important.

As an example, we look at the scatterplot from Fig. 6.2 again in the following image. However, we have added a constructed outlier (Fig. 6.6). If we now calculate the correlation coefficient without or with outliers, we get completely different values.

For all the reasons mentioned, we should always first draw the scatterplot before calculating the Bravais-Pearson correlation coefficient. We recognize whether a linear relationship exists, we understand whether the relationship is positive or negative, we recognize how strong the relationship is and we see whether our relationship is influenced by outliers.

Freak Knowledge
It often happens that the relationship between the variables of interest to us is of a non-linear nature, e.g., exponential. In such a case, we can often establish a linear relationship by mathematically transforming one of the two variables. Subsequently, the correlation coefficient of Bravais-Pearson can be used.

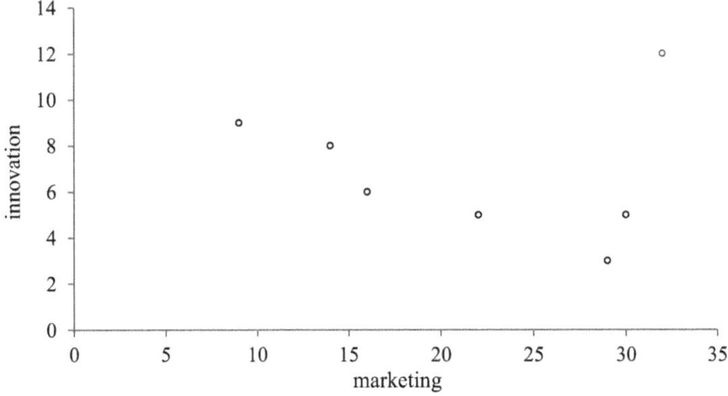

Fig. 6.6 The scatterplot from Fig. 6.2 with constructed outlier

6.4 The Correlation Coefficient of Spearman for Ordinal Variables

If we have ordinal data instead of metric data, we use the rank correlation coefficient of Spearman. As with Bravais and Pearson, Spearman is the one who developed this coefficient. Behind our measures are real people and considerations. It is calculated as follows:

$$r_{\text{Sp}} = \frac{\sum \left(R_{x_i} - \bar{R}_x\right)\left(R_{y_i} - \bar{R}_y\right)}{\sqrt{\sum \left(R_{x_i} - \bar{R}_x\right)^2 \sum \left(R_{y_i} - \bar{R}_y\right)^2}}$$

with

R_{xi} and R_{yi} are equal to the ranks of the observed x values and y values,
$\bar{R}_x$ and $\bar{R}_y$ are equal to the means of the respective rank numbers.

As we can see, Spearman's rank correlation coefficient is very similar to Bravais-Pearson's for metric data. The difference is that we now have to assign a rank to each observed value of the respective variable before we can calculate it. Usually, we assign the rank one to the highest value and the highest rank to the lowest value. If values occur more often, we assign the average rank for the values and then continue with the rank at which we would be without multiple nominations. After we have assigned ranks to all values of the variables in this way, we can calculate the required values and insert them into the formula. We best explain this using the example of our first six companies. We are interested in whether there is a relationship between the two ordinal variables expectation X and self-assessment Y.

Company	x_i	y_i	R_{x_i}	R_{y_i}	$R_{x_i} - \bar{R}_x$	$\left(R_{x_i} - \bar{R}_x\right)^2$	$R_{y_i} - \bar{R}_y$	$\left(R_y - \bar{R}_y\right)^2$	$\left(R_{x_i} - R_x\right)\left(R_{y_i} - R_y\right)$
1	2	5	3	1.5	−0.5	0.25	−2	4	1
2	1	4	5.5	3.5	2	4	0	0	0
3	3	5	1	1.5	−2.5	6.25	−2	4	5
4	2	3	3	5.5	−0.5	0.25	2	4	−1
5	1	3	5.5	5.5	2	4	2	4	4
6	2	4	3	3.5	−0.5	0.25	0	0	0
$\bar{R}$			3.5	3.5					
Σ						15		16	9

For the variable X, we are looking for the highest value and assign it rank 1. The highest value is held by company 3, which is therefore assigned the rank 1. Next, we look for the second highest value. This is the number 2, which appears three times, i.e., for companies 1, 4, and 6. These companies would therefore receive rank 2 if we were to assign it multiple times. However, we only ever assign a rank once. Since companies 1,

4, and 6 are equivalent, we resort to using the mean rank. This is calculated from the sum of the ranks we need to assign, divided by the number of ranks we assign. We have three ranks to assign, rank 2, 3, and 4 for the three companies, so the rank to be assigned is then 9 divided by 3, which equals rank 3. Now we have assigned the first four ranks, we continue with the remaining companies 2 and 5, giving them the mean rank from ranks 5 and 6, which is rank 5.5. We proceed in the same way with the variable Y. We then calculate the values needed for the formula and insert them into the formula. Now we can calculate the rank correlation coefficient. This is:

$$r_{\mathrm{Sp}} = \frac{\sum \left(R_{x_i} - \bar{R}_x\right)\left(R_{y_i} - \bar{R}_y\right)}{\sqrt{\sum \left(R_{x_i} - \bar{R}_x\right)^2 \sum \left(R_{y_i} - \bar{R}_y\right)^2}} = \frac{9}{\sqrt{15 \times 16}} = 0.58$$

We interpret the correlation coefficient of Spearman in the same way as the correlation coefficient of Bravais-Pearson. Thus, we have found a positive relationship between the two variables. If the value for the variable X increases, the value for the variable Y also increases, and vice versa. In our case, this means that less experienced individuals tend to slightly underestimate future developments and vice versa. The interpretation is a bit tricky. To carry out the interpretation, one must look very carefully in the legend to see how the data are defined.

Freak Knowledge

If no values occour more often within our respective variables, i.e., if in the variable X as well as in the variable Y each value only occurs once, then we can simplify the formula used above. The formula for Spearman's rank correlation coefficient is then

$$r_{Sp} = 1 - \frac{6 \times \sum d_i^2}{n^3 - n}$$

with n for the number of observations and d_i for the rank distances.

6.5 The Phi Coefficient for Nominal Variables with Two Characteristics

If you have nominal variables with only two characteristics, e.g., the variable gender with the characteristics male or female or 0 and 1, the correlation can be calculated using the phi coefficient:

$$r_\phi = \frac{a \times d - b \times c}{\sqrt{S_1 \times S_2 \times S_3 \times S_4}}$$

with

r_ϕ	is the phi coefficient,
a, b, c, d	are the fields of the 2×2 matrix,
S_1, S_2, S_3,S_4	are the row or column sums.

		Variable Y		
		0	1	
Variable X	0	a	b	S_1
	1	c	d	S_2
		S_3	S_4	

Let's look at this using the example of our data set data_growth for the first ten companies. Nominal variables with only two characteristics are sex and sector. Sex differentiates between woman and man, sector differentiates between industrial and service companies.

We now want to know if there is a relationship between the sex of the founders and the sector in which they founded. Our data shows that the first company is an industrial company founded by a man, the second company is a service company founded by a woman, the third again an industrial company, founded by a man, etc. (compare our data set for the first ten companies). In summary, we can present the information in the following *2 × 2 matrix*:

		Gender		
		0 = Man	1 = Woman	
Industry	0 = Industry	2	0	2
	1 = Service	3	5	8
		5	5	

If we insert the numbers accordingly into our formula, we obtain the phi coefficient:

$$r_\phi = \frac{a \times d - b \times c}{\sqrt{S_1 \times S_2 \times S_3 \times S_4}} = \frac{2 \times 5 - 0 \times 3}{\sqrt{2 \times 8 \times 5 \times 5}} = 0.50$$

We can interpret the phi coefficient almost like the correlation coefficients of Bravais-Pearson and Spearman. The phi coefficient is also defined between −1 and 1. The only difference is that the sign is meaningless. We could easily check this. If we, for example, swap the columns man and woman, the sign becomes negative. This shows that we have a medium relationship between the two variables. If we now want to know what relationship exists, we need to look at our *2 × 2 matrix* again. We see that relatively many service companies were founded by women and that women found relatively few industrial companies. This allows us to determine the trend. Women tend to found service companies more often.

6.6 The Contingency Coefficient for Nominal Variables

If we are confronted with nominal data with more than two characteristics, it is sufficient if one of the nominal variables has more than two characteristics, then the correlation can be calculated using the contingency coefficient as follows:

$$C = \sqrt{\frac{U}{U+n}}$$

with

C is the contingency coefficient,
n is the number of observations and
U the sum of deviations between the observed and theoretically expected values:

$$U = \sum\sum \frac{\left(f_{jk} - e_{jk}\right)^2}{e_{jk}}$$

with

f_{jk} equal to the observed row and column values,
e_{jk} equal to the theoretically expected row and column values,
j the rows and
k the columns.

The contingency coefficient is not defined between -1 and 1, but in the value range of $0 \leq C \leq C_{max}$. C_{max} is calculated from the number of columns and the number of rows, it is always less than 1:

$$C_{max} = \frac{1}{2}\left(\sqrt{\frac{z-1}{z}} + \sqrt{\frac{s-1}{s}}\right)$$

with

z is equal to the number of rows and
s is equal to the number of columns.

At this point, it is useful to start directly with an example to illustrate the said. Let's take our data set dat_growth again and analyze the relationship between the two variables motive and sector. This time we are referring to our 100 companies, as none of the following cells should be occupied less than five times. According to the phi coefficient, we can build a matrix, but in this case with two rows and three columns, as the variable sector has two characteristics and the variable motive has three characteristics. Theoretically, a matrix with two and three characteristics has the following structure:

Rows	Columns		
	1	2	3
1	f_{11}	f_{12}	f_{13}
2	f_{21}	f_{22}	f_{23}

In our example, it looks like this, with the counted observed frequencies.

Industry	Motive		
	Unemployment	Implementing an idea	Higher income
Manufacturing	8	15	11
Service	9	39	18

The table shows that eight industry foundations originated from unemployment, 15 industry founders wanted to implement an idea, etc. We have two rows and three columns, so we can calculate our maximum achievable contingency coefficient:

$$C_{\max} = \frac{1}{2}\left(\sqrt{\frac{z-1}{z}} + \sqrt{\frac{s-1}{s}}\right) = \frac{1}{2}\left(\sqrt{\frac{2-1}{2}} + \sqrt{\frac{3-1}{3}}\right) = 0.76$$

We can achieve a maximum contingency coefficient of 0.76. If we discover such a one, we have a perfect relationship. The minimum we can achieve is 0, in which case we would have no relationship between sector and founding motive.

The next step is to calculate the theoretically expected values. To accomplish this, we first need to calculate the row and column sums.

Industry	Motive			
	Unemployment	Implement idea	Higher income	Row sum
Industry	8	15	11	34
Service	9	39	18	66
Column sum	17	54	29	

To arrive at the theoretically expected values, we use the following formula:

$$e_{jk} = \frac{f_j \times f_k}{n}$$

with

f_j equal to row sums and
f_k equal to column sums.

This results in the following picture:

Industry	Motive			
	Unemployment	Implement idea	Higher income	Row sum
Industry	$e_{11} = \frac{34 \times 17}{100} = 5.78$	$e_{12} = \frac{34 \times 54}{100} = 18.36$	$e_{13} = \frac{34 \times 29}{100} = 9.86$	34
Service	$e_{21} = \frac{66 \times 17}{100} = 11.22$	$e_{22} = \frac{66 \times 54}{100} = 35.64$	$e_{23} = \frac{66 \times 29}{100} = 19.14$	66
Column sum	17	54	29	

We notice that the row and column sums have remained the same. However, the cells now contain the values we would theoretically expect if the respective rows ratios adjust to the ratios of the columns.

With this, we can calculate the sum of deviations U by subtracting the observed values from the theoretically expected values:

$$U = \sum\sum \frac{(f_{jk} - e_{jk})^2}{e_{jk}}$$

$$= \frac{(f_{11} - e_{11})^2}{e_{11}} + \frac{(f_{12} - e_{12})^2}{e_{12}} + \frac{(f_{13} - e_{13})^2}{e_{13}} + \frac{(f_{21} - e_{21})^2}{e_{21}} + \frac{(f_{22} - e_{22})^2}{e_{22}} + \frac{(f_{23} - e_{23})^2}{e_{23}}$$

$$= \frac{(8 - 5.78)^2}{5.78} + \frac{(15 - 18.36)^2}{18.36} + \frac{(11 - 9.86)^2}{9.86} + \frac{(9 - 11.22)^2}{11.22} + \frac{(39 - 35.64)^2}{35.64} + \frac{(18 - 19.14)^2}{19.14}$$

$$= 2.42$$

If we insert the sum of deviations into the formula for the contingency coefficient, we get coefficient:

$$C = \sqrt{\frac{U}{U + n}} = \sqrt{\frac{2.42}{2.42 + 100}} = 0.15$$

With this value we are closer to zero than to the maximum possible contingency coefficient, so we have no correlation between the motive for founding and the sector in which it was founded.

6.7 Correlation, Causality, Third Variables, and Other Correlation Coefficients

If we have calculated a correlation, we still know nothing about causality, we do not know whether

X ⟶ Y Variable X influences the variable Y,Y ⟶ X Variable Y influences the variable X,

X ⟷ Y Variable X and variable Y mutually influence each other.

We should look at this again using an example. Between the variables marketing and innovation, we had calculated a very high negative correlation of −0.91. This implies that companies that spend a lot of their revenue on marketing spend little of their revenue on innovation. However, we do not know in which direction the relationship runs, whether, for example, companies that spend a lot on marketing have little money left for innovation, or vice versa, companies that spend a lot on innovation have little money for marketing. Perhaps there is also a mutual relationship. If more money is spent on marketing, there is less for innovation. If less money is then spent on innovation, more is spent on marketing again, and so on.

Causality is a question of theory or theoretical considerations. We have knowledge about causality when we can clearly deduce from theory that a variable *X* influences a variable *Y*. Only then do we know something about the direction. An example of this is a stone falling into water. We hold a stone, let it go, it falls into the water, and waves are created in the water as a result. The waves on the water surface are a direct consequence of the stone. Once again, correlation does not imply causality.

In addition, one more thing is important. Many correlations are based on the influence of third variables, variables that underlie the correlation between variables *X* and *Y*.

In Fig. 6.7, such a situation is depicted. There is a correlation between the variables *X* and *Y*, but this is not based on a correlation between *X* and *Y*, but variable *X* is correlated with variable *Z* and *Z* with variable *Y* as well. The correlation between *X* and *Y* comes about through the correlation between *X* with *Z* and *Y* with *Z*. We speak of a spurious correlation. A classic example of a spurious correlation is the positive correlation observed during industrialization between the number of storks and the birth rate. The number of storks decreased and the birth rate also decreased. But this does not mean that there is a relationship between storks and births. The variable behind this is industrialization. Industrialization led to people becoming richer and having fewer children. At the same time, the accompanying increase in environmental pollution reduced the number of storks. Another example is a found positive correlation between the amount of ice cream consumed each summer and the number of people who drown in a summer. The more ice cream is consumed, the more people drown and vice versa. This does not mean that eating ice cream increases the risk of drowning or that "drowned people consume more ice cream". The underlying factor is the quality of the summer. If it is hotter, people eat more ice cream, but they also go swimming more and accordingly there are more deaths by drowning. So when we discover a correlation between two variables, we need to check very carefully whether we have not found a spurious correlation. This brings us back to theory. Only this can tell us whether it is a real correlation.

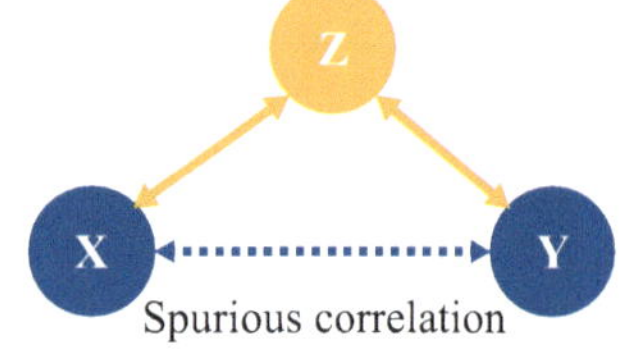

Fig. 6.7 The influence of third variables on the correlation between two variables

Table 6.3 Correlation coefficients and scale level of the variables

Variable X	Variable Y	Correlation coefficient	Example from the data set Data_ Growth
Metric	Metric	Bravais-Pearson correlation coefficient	Correlation between growth rate and marketing
Metric	Ordinal	Spearman rank correlation coefficient	Correlation between growth rate and self-assessment
Metric	Nominal	Point-biserial correlation	Correlation between growth rate and gender
Ordinal	Ordinal	Spearman rank correlation coefficient	Correlation between self-assessment and education
Ordinal	Nominal	Biserial rank correlation	Correlation between self-assessment and gender
Nominal	Nominal	Contingency coefficient	Correlation between motive and gender
Nominal (0/1)	Nominal (0/1)	Phi coefficient	Correlation between industry and gender

When we calculate a correlation, we use the Bravais-Pearson correlation coefficient for metric data, the Spearman rank correlation coefficient for ordinal data, and either the phi coefficient or the contingency coefficient for nominal data. But what do we do when we have mixed data, for example, when one variable is metric and the other is ordinal? If one variable is metric and the other is ordinal, we can use the Spearman rank correlation coefficient. If one variable is metric and the other is nominal, then the point-biserial correlation is to be calculated. The only situation left is when one variable is ordinal and the other is nominal. In such a case, we use the biserial rank correlation (Table 6.3).

6.8 Calculation of Correlation Coefficients with the R Commander

The calculation of the Bravais-Pearson correlation coefficient and the Spearman correlation coefficient is done with the R Commander via the Statistics menu with the Summaries submenu and the Correlation matrix command. In the window that opens, the variables for which the correlation is to be calculated are selected. To select multiple variables, we use the control key together with the mouse. We can also select more than two variables. In addition, we specify the correlation coefficient. We then confirm with OK and a matrix with the correlation coefficients appears as a result.

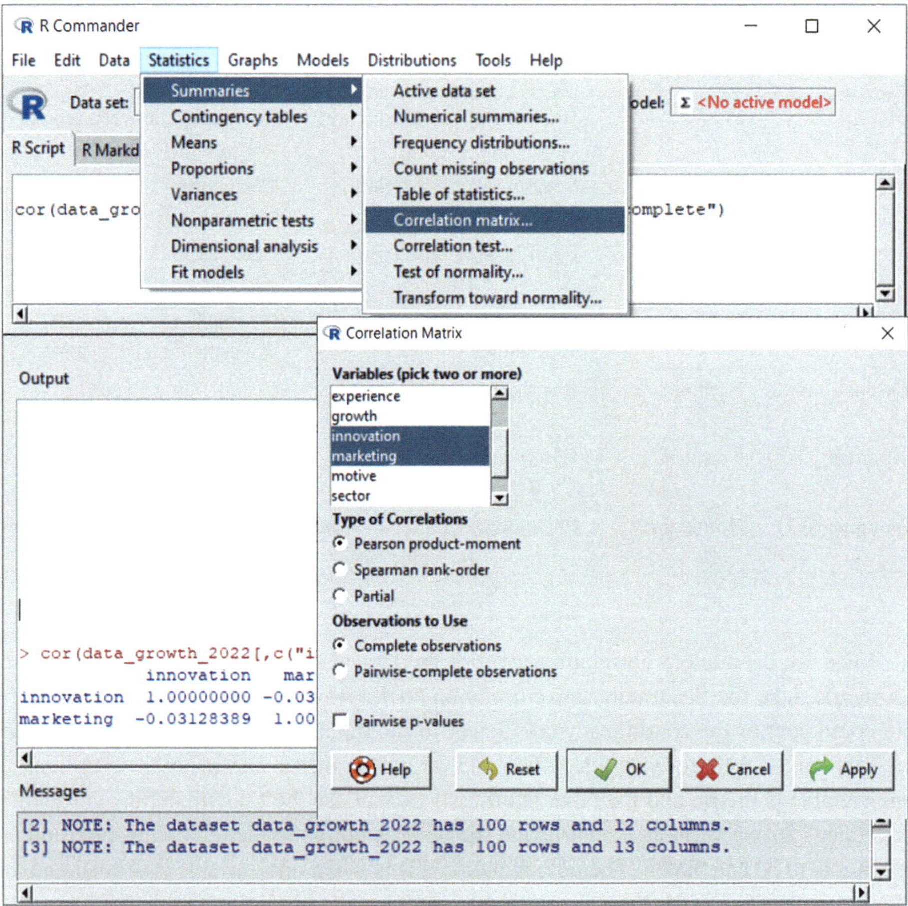

The corresponding command commands are cor(name_dataset$name_variable1, name_dataset$name_variable2, method = "pearson") for the Bravais-Pearson correlation coefficient and cor(name_dataset$name_variable1, name_dataset$name_variable2, method = "spearman") for the Spearman rank correlation coefficient.

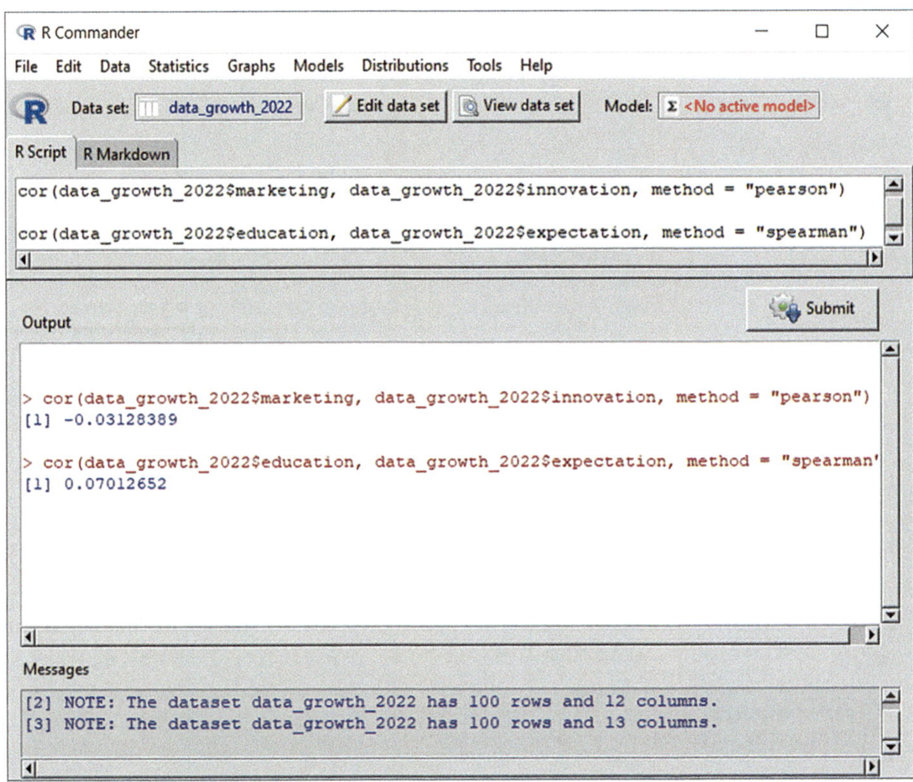

Before we calculate the Bravais-Pearson correlation coefficient, we should always first create the scatterplot and check for a linear relationship. We do this with the Graphs menu and the Scatterplot command. In the window that appears, we select the variables and in the Options tab we remove all check marks. The visual impression should not be disturbed by additional information and lines.

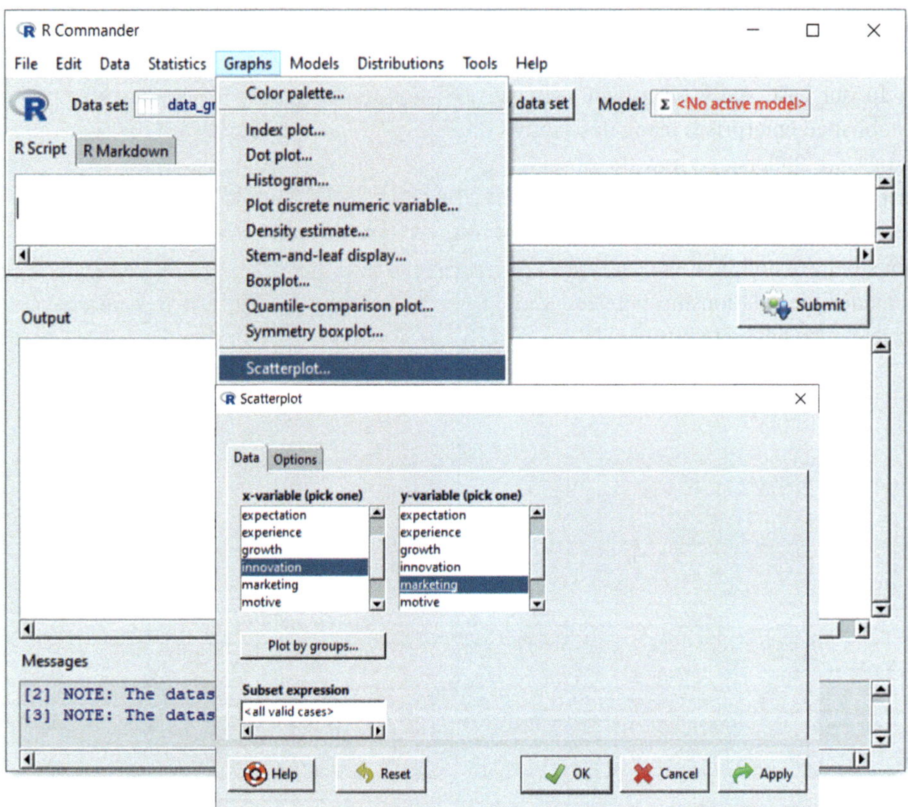

We then confirm the command with OK. We get the scatterplot. In this case, there seems to be neither a linear nor a non-linear relationship.

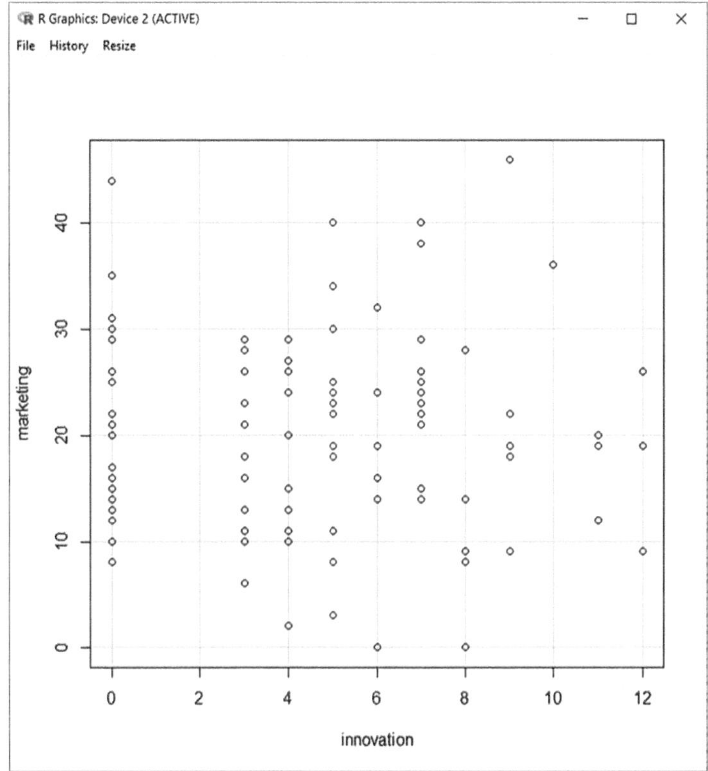

For the calculation of the correlation coefficients for nominal data, we first have to manipulate our data set. The reason is as follows: R initially always treats the imported data as numerical values. If a variable is scaled nominally, we have to inform R first. We do this with the help of the Data menu. In this, we select the Manage variables in active data set submenu and then Convert numeric variables to factors. A window opens in which we specify the variable, select Use numbers for Factor Levels, and possibly rename the variable. We confirm with OK and thus create the factor variable.

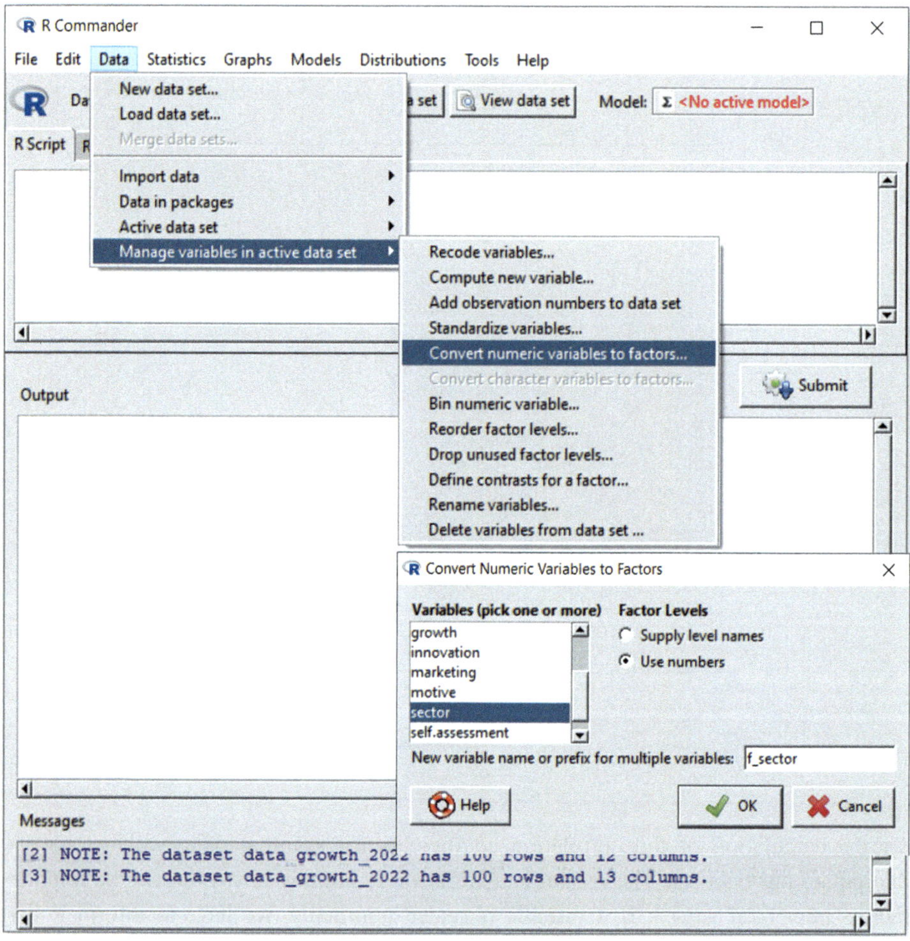

After we have informed R which are the nominal variables, we can start calculating the contingency coefficient. We use the Statistics menu again for this, with the Contingency tables submenu and select the Two-way table command. In the window that opens, we mark the row and column variable and in the Statistics tab we check the boxes for Chi-square test of independence and Print expected frequencies. We then confirm with OK.

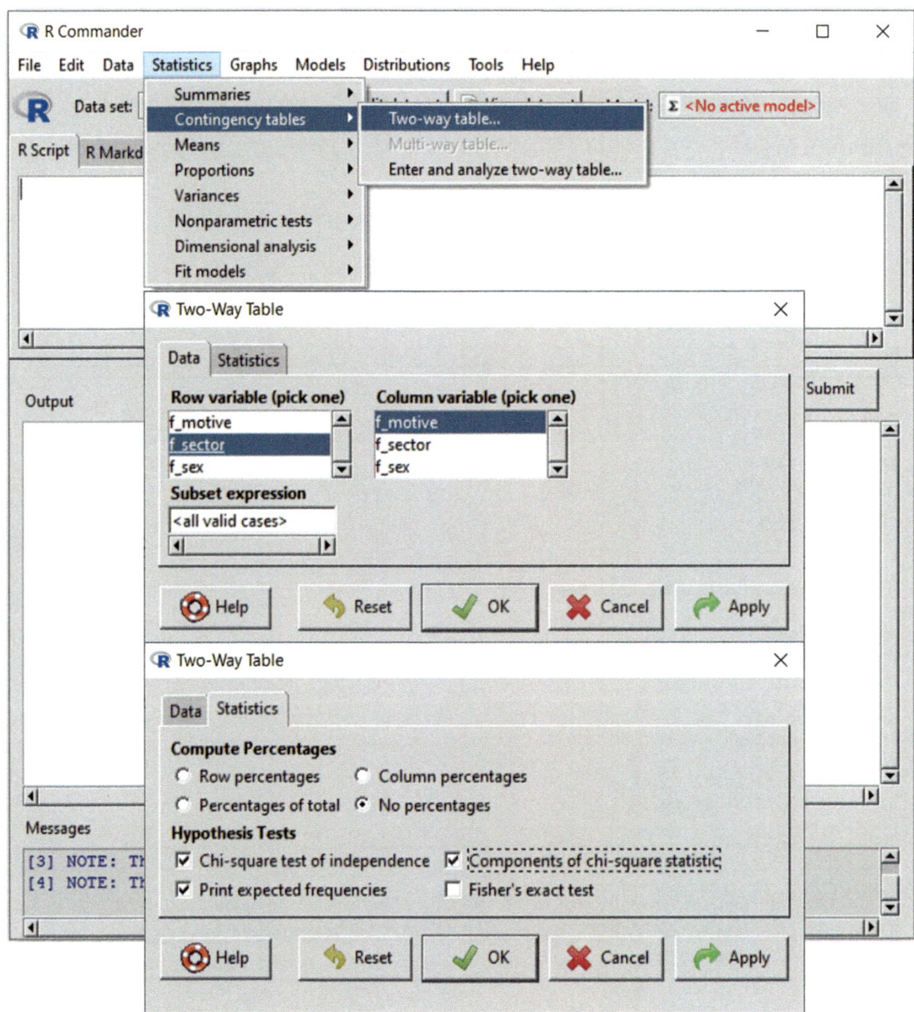

We get the following results:

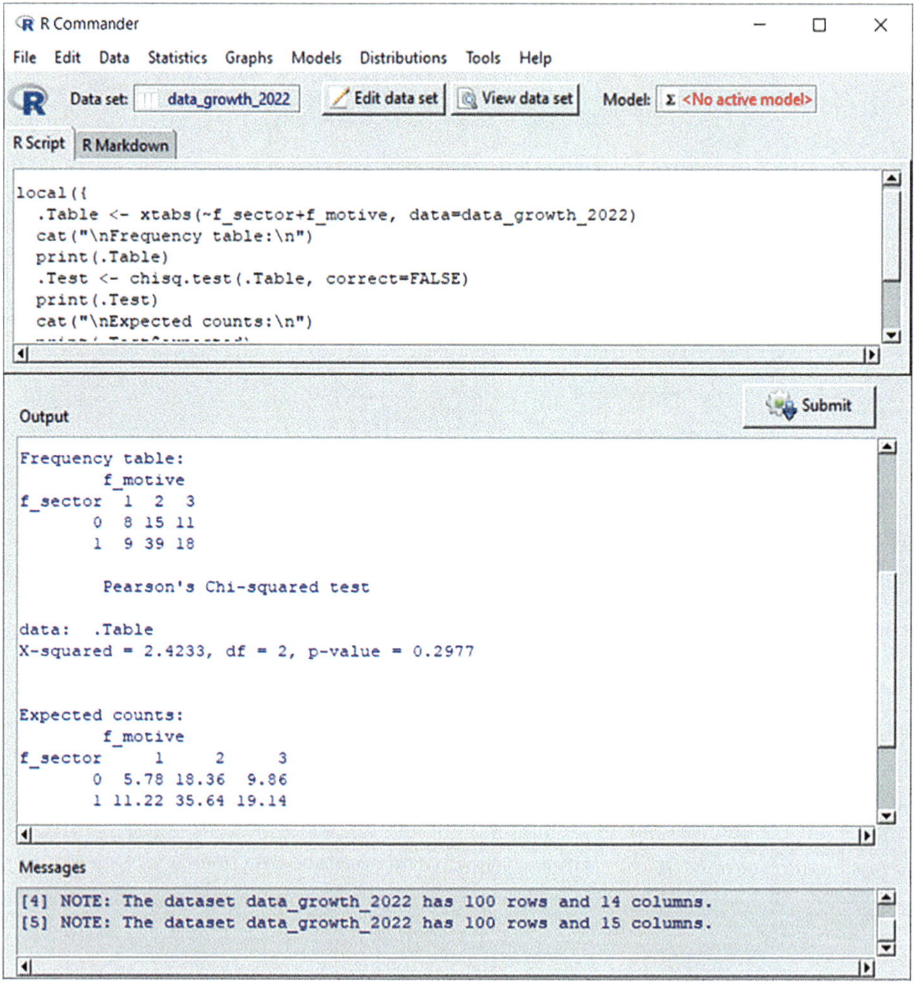

First, we can determine that we do not directly receive the contingency coefficient. We see the observed frequencies (Frequency table) and the expected frequencies (Expected counts). The numbers should be familiar to us, we have already calculated these by hand above. Furthermore, we see a result, the Pearson's Chi-squared test, which we cannot yet interpret (we will get to this later in the course of the book). The value of 2.4233 should be familiar to us. It is our sum of deviations U. All that remains is to calculate the contingency coefficient using this value. We can also do this directly in the R Commander. The instruction for this is sqrt(2.4233/(2.4233 + 100)), sqrt() stands for the square root function. As we have already calculated by hand above, we get a value of 0.15.

6.9 Checkpoints

- *Correlations provide information about whether there is a relationship between two variables, they do not say anything about causality.*
- *It is often the case that correlations are due to the influence of a third variable. Correlations must always be theoretically justified.*
- *The correlation coefficient of Bravais-Pearson is suitable for calculating the linear relationship between two metric variables.*
- *Before calculating the correlation coefficient of Bravais-Pearson, the scatterplot should be drawn to check whether a linear relationship exists.*
- *The correlation coefficient of Bravais-Pearson is sensitive to unusual observations or outliers.*
- *The correlation coefficient of Spearman is used to calculate the relationship between two ordinal variables.*
- *If the correlation between two nominal variables is to be calculated, the phi coefficient or the contingency coefficient should be used.*
- *For the calculation of correlations between variables with different scale levels, further specialized correlation coefficients are available.*

6.10 Application

6.1. Which correlation coefficient is to be used at which scale levels?

6.2. Why are theoretical considerations necessary before a correlation between two variables is calculated?

6.3. Why does a correlation coefficient say nothing about causality?

6.4. Consider the following three scatter plots (after Anscombe 1973). In which of the cases can the correlation coefficient of Bravais-Pearson be calculated without problems? Why, why not?

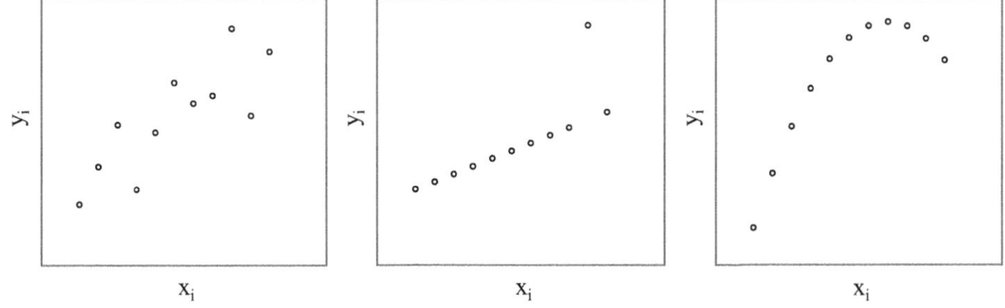

6.5. Draw and calculate by hand the scatterplot and the correlation coefficient for the variable growth and marketing for the first eight companies of our data set data_growth and interpret the result.
6.6. Draw and calculate using the R Commander for our data set data_growth for the variable pairs growth and marketing, growth and innovation, growth and experience as well as growth and age the scatterplot and the correlation coefficient of Bravais and Pearson and interpret the result.
6.7. Calculate by hand the correlation coefficient of Spearman for the variables growth and self-assessment for the first ten companies of our data set data_growth and interpret the result.
6.8. Calculate using the R Commander the correlation coefficient of Spearman for the variables growth and self-assessment of our data set data_growth and interpret the result.
6.9. Calculate for our data set data_growth for the variables gender and industry the phi coefficient and interpret the result.
6.10. Calculate for our data set data_growth for the variables gender and motive the contingency coefficient and interpret the result.
6.11. We are interested in the relationship between the per capita CO_2 emissions and the gross domestic product per capita in the European Union. In the Eurostat database (https://ec.europa.eu/eurostat/de/home) we find the following numbers:

Country	GDP per capita 2017 [in 1000 Euro/capita]	CO_2 per capita 2017 [in t/capita]
Belgium	35.3	9.0
Bulgaria	6.3	6.8
Czech Republic	17.2	10.2
Denmark	47.4	6.6
Germany	35.4	10.0
Estonia	14.4	14.3
Ireland	54.2	8.7
...		

Calculate and interpret the correlation coefficient.
6.12. We are interested in whether there is a correlation between three stock values. Once again, we go on the internet, visit a financial platform (this time perhaps to https://www.boerse-online.de/) and get historical values for VW, Daimler and SAP. We obtain the following values:

Date	VW	Daimler	SAP
01.04.2020	99.27	25.74	97.40
31.03.2020	105.48	27.14	101.48
30.03.2020	105.50	27.21	101.70
27.03.2020	106.50	27.32	100.78
26.03.2020	112.90	29.21	103.88
25.03.2020	115.50	30.25	100.00

Subsequently, we calculate the possible correlations between the stocks. Why might be this of interest to us?

7 Ratios and Indices: The Opportunity to Generate New Knowledge from Old Ones

So far, we have analyzed variables that are available. But what if the information is not yet available, but is hidden in our data. We can then generate the information from our existing variables by relating or weighting variables to each other. Let´s look at the profit of an enterprise. It can make a difference whether we measure the profit in absolute terms or in relative terms. Let´s assume that we have a small enterprise with ten employees and an enterprise with 100 employees. Both enterprises should make a profit of CHF 500,000 per year. In absolute numbers, the profit is the same. But in relative terms, per employee, it is CHF 50,000 per employee for the smaller company, and only CHF 5,000 per employee for the larger company. We have formed a ratio that allows us to compare the companies:

$$\text{Profit per capita} = \frac{\text{Profit in CHF}}{\text{Employee}}$$

Ratio s are quotients of two numbers. With them, we can compare people and objects of study in terms of content, space, or time. In statistics, we distinguish between reference numbers, quotas, and dynamic numbers. In the following, we will get to know these and discuss their application. We do not use our data set, but show various situations where ratio numbers are used. Furthermore, we will discuss index numbers, which we need when price or quantity developments of several products are to be analyzed simultaneously.

F. Kronthaler, *Statistics Applied with the R Commander*,
https://doi.org/10.1007/978-3-662-69107-6_7

7.1 The Reference Number—The Quotient of Two Different Sizes

A reference number is formed when we relate two different sizes to each other:

$$\text{Reference number} = \frac{\text{Parameter A}}{\text{Parameter B}}$$

If it makes sense, we multiply the number by 100 and thus obtain a percentage value.

One of the most well-known ratio numbers is probably the gross domestic product per capita. This is formed by dividing a country's gross domestic product by the population of the country. This makes the average wealth of the population in different countries comparable. The following table shows the gross domestic product, the population, and the resulting reference number gross domestic product per capita for selected countries (Table 7.1).

In absolute terms, the USA appears to be the richest nation in the table, followed by China, Germany, Switzerland, and South Africa. Relative to the population, a different picture emerges, the Swiss population is the richest, followed with a large difference by the USA, Germany, China, and South Africa. Depending on whether we operate with absolute or relative numbers, a different picture emerges.

Further examples of reference numbers are

- the vehicle density, number of registered vehicles to population,
- the birth rate, number of live births to population,
- the doctor density, number of doctors to population or to the area of a country,
- the return on equity, profit to equity,
- productivity, work result to hours worked, or
- beer consumption per capita (for Bavarian or interested readers, look in the internet for "Munich beer heart").

The possibilities to form a reference number are diverse and it depends on the question at hand which two numbers are related to each other. If we want to make a statement about

Table 7.1 Gross domestic product per capita of selected countries—comparison using a reference number (World Bank 2020, data downloaded from https://data.worldbank.org/ on 10.04.2020)

Country	GDP (in $, 2018)	Population (2018)	GDP/capita (in $, 2018)
China	13,608,151,864,638	1,392,730,000	9771
Germany	3,947,620,162,503	82,927,922	47,603
South Africa	368,288,939,768	57,779,622	6374
Switzerland	705,140,354,166	8,516,543	82,797
USA	20,544,343,456,937	327,167,434	62,795

the doctor density of a country, we have to consider whether it makes sense to calculate the number of doctors per area or per population. Depending on which approach we choose, we will probably come to different conclusions.

7.2 The Quota—The Quotient of a Partial Number to a Total Number

We form a quota by relating a partial size to its total size:

$$\text{Quota} = \frac{\text{Share parameter}}{\text{Total parameter}}$$

As a rule, we multiply the number by 100 and we get the percentage share of the part to its total.

In the next table, we see the export quota of selected countries as an example (Table 7.2).

The export quota puts the value of exported goods in relation to the gross domestic product. The number illustrates which country exports more of the goods produced in the country. We see that the export quota in Switzerland is the highest, followed by Germany. The export quota of the USA is the smallest. One of the reasons for this is that the USA has a large domestic market and therefore companies can sell relatively many goods domestically. They are thus less dependent on export markets.

Further examples of quotas are

- Odds of having of boys, number of newborn boys to total number of newborns,
- Company's export ratio, export turnover to total turnover,
- National saving rate, national savings to gross domestic product,
- Equity ratio of a company, equity to its capital,
- Marketing share, expenditure for marketing to total expenditure.

Here too, we see that the possibilities for forming quotas are manifold and depend on the question.

Table 7.2 Export quota of selected countries—comparison using a quota (World Bank 2020, data downloaded from https://data.worldbank.org/ on 09.04.2020)

Country	Exports (in $, 2018)	GDP (in $, 2018)	Exports/GDP (in % 2018)
China	2,655,609,104,875	13,608,151,864,638	19.5
Germany	1,871,807,732,127	3,947,620,162,503	47.4
South Africa	110,144,477,441	368,288,939,768	29.9
Switzerland	466,305,408,689	705,140,354,166	66.1
USA	2,510,250,000,000	20,544,343,456,937	12.2

7.3 The Dynamic Index

The last ratio, we want to discuss is the dynamic index. We use this to look at developments over time and compare them with each other:

$$\text{Dynamic index}_t = \frac{\text{Value of the parameter at time point t}}{\text{Value of the parameter at the starting point}}$$

We usually multiply this number by 100 and get the change to a base point we have chosen, in percent.

In the upcoming table, we see the population development of China, Germany, and Switzerland measured absolutely and with the help of the dynamic index to the base point, the year 2005 (Table 7.3).

With the help of the dynamic index, it is easy to compare the development of the population of the three countries. We see that the population in China has increased by 6.8% since 2005 compared to the base year, the population in Germany has remained almost the same with a growth of 0.6%, and the population in Switzerland has increased by 14.5%. Switzerland has thus recorded the highest population growth in this period. In the same way, other time series can be compared with the help of the dynamic index, e.g. the economic growth of countries, the company growth of different companies, the development of stocks, and much more.

Table 7.3 Population growth of selected countries—comparison using a dynamic index (World Bank 2020, data downloaded from https://data.worldbank.org/ on 09.04.2020)

Year	China		Germany		Switzerland	
	Population	I_{2005}	Population	I_{2005}	Population	I_{2005}
2005	1,303,720,000	100.0	82,469,422	100.0	7,437,115	100.0
2006	1,311,020,000	100.6	82,376,451	99.9	7,483,934	100.6
2007	1,317,885,000	101.1	82,266,372	99.8	7,551,117	101.5
2008	1,324,655,000	101.6	82,110,097	99.6	7,647,675	102.8
2009	1,331,260,000	102.1	81,902,307	99.3	7,743,831	104.1
2010	1,337,705,000	102.6	81,776,930	99.2	7,824,909	105.2
2011	1,344,130,000	103.1	80,274,983	97.3	7,912,398	106.4
2012	1,350,695,000	103.6	80,425,823	97.5	7,996,861	107.5
2013	1,357,380,000	104.1	80,645,605	97.8	8,089,346	108.8
2014	1,364,270,000	104.6	80,982,500	98.2	8,188,649	110.1
2015	1,371,220,000	105.2	81,686,611	99.1	8,282,396	111.4
2016	1,378,665,000	105.7	82,348,669	99.9	8,373,338	112.6
2017	1,386,395,000	106.3	82,657,002	100.2	8,451,840	113.6
2018	1,392,730,000	106.8	82,927,922	100.6	8,516,543	114.5

7.4 The Price and Quantity Index of Laspeyres and Paasche

If we want to analyze the development of one quantity, then the dynamic index is a good possibility. But let's imagine we want to look at the development of several quantities, e.g. of a shopping basket or several stocks as in a stock index. The dynamic index is then no longer suitable. We need a number that analyzes the development of several quantities over time. Such numbers are available to us with the indices of Laspeyres and Paasche.

Let's briefly look at the example of the development of the cost of living. To calculate the development of the cost of living, we could take the prices of the different goods, calculate the average prices for several points in time, e.g. years, and then show the price development with it. However, this would not take into account the imprtance of the individual goods. Think of our own basket of goods that we buy weekly, monthly or annually. Some goods, like bread, we buy almost daily, other goods like a television or a car we buy only every few years. To determine the importance of the price development of the goods for our budget, we therefore have to weight the goods according to their importance. It makes a difference whether gasoline becomes 10% more expensive or a television.

This was noticed earlier and so a man named Laspeyres developed a price index in the nineteenth century, which takes into account the importance of the goods in the price development. This index is still used today, as already indicated e.g. in the calculation of the development of consumer prices or in the calculation of stock indices.

The price index according to Laspeyres is calculated with the following formula:

$$P_{Lt} = \frac{\sum p_{ti} \times q_{0i}}{\sum p_{0i} \times q_{0i}} \times 100$$

where

P_{Lt} denotes the price index (P) according to Laspeyres (L) at time t,
p_{ti} is the price of good i at time t,
p_{0i} is the price of good i at the base time 0,
q_{0i} is the quantity of good i at the base time 0.

Let's illustrate the calculation with a small example. Let's assume a café owner wants to adjust the prices of the breakfast menu after five years. To make the adjustment, the owner looks at the purchased quantities and the prices today and five years ago. For simplicity's sake, let's assume that breakfast consists only of coffee, bread, cheese, and ham. The numbers could look like this:

Goods	Price 2014	Quantity 2014	Price 2019	Quantity 2019
Coffee	5 CHF/kg	150 kg	7 CHF/kg	150 kg
Bread	2 CHF/piece	5000 pieces	3 CHF/piece	6000 pieces
Cheese	15 CHF/kg	500 kg	17 CHF/kg	750 kg
Ham	25 CHF/kg	250 kg	35 CHF/kg	150 kg

We insert these into the formula for the price index according to Laspeyres and thus obtain the price development since 2014.

$$P_{L2019} = \frac{\sum p_{2019i} \times q_{2014i}}{\sum p_{2014i} \times q_{2014i}} \times 100$$
$$= \frac{7\frac{\text{CHF}}{\text{kg}} \times 150\text{ kg} + 3\frac{\text{CHF}}{\text{piece}} \times 5000\text{ piece} + 17\frac{\text{CHF}}{\text{kg}} \times 500\text{ kg} + 35\frac{\text{CHF}}{\text{kg}} \times 250\text{ kg}}{5\frac{\text{CHF}}{\text{kg}} \times 150\text{ kg} + 2\frac{\text{CHF}}{\text{piece}} \times 5000\text{ piece} + 15\frac{\text{CHF}}{\text{kg}} \times 500\text{ kg} + 25\frac{\text{CHF}}{\text{kg}} \times 250\text{ kg}} \times 100$$
$$= 135.9$$

From the base year 2014, prices have increased by 35.9% by 2019 (135.9–100). So, we have to pay 35.9% more for the goods, i.e., perhaps we should adjust our breakfast prices accordingly upwards.

Firstly, it doesn't matter in which units we measure the goods, the units cancel out of the formula. Secondly, it is noticeable that we have used the quantities of the base year in both the denominator and the numerator. The quantities of the current year are not taken into account. This implies that the Laspeyres price index shows how much more we would have to pay today for the basket of goods from back then. This works well if the basket of goods does not change significantly, but if the basket is different today than in the base year, then the Laspeyres price index is inaccurate.

Mr. Paasche also noticed this. He suggested not calculating the price development to the base year, but to the observation year t. The formula adjusts accordingly. In the denominator, the quantities of the base year are no longer represented, but the quantities of the observation year. The formula is as follows:

$$P_{Pt} = \frac{\sum p_{ti} \times q_{ti}}{\sum p_{0i} \times q_{ti}} \times 100$$

where

P_{Pt} denotes the price index (P) according to Paasche (P) at time t,
p_{ti} is the price of good i at time t,
p_{0i} is the price of good i at the base time 0,
q_{ti} is the quantity of good i at time t.

If we now insert the values, we get a different value for the price development.

$$P_{P2019} = \frac{\sum p_{2019i} \times q_{2019i}}{\sum p_{2014i} \times q_{2019i}} \times 100$$
$$= \frac{7\frac{\text{CHF}}{\text{kg}} \times 150\text{ kg} + 3\frac{\text{CHF}}{\text{piece}} \times 6000\text{ piece} + 17\frac{\text{CHF}}{\text{kg}} \times 750\text{ kg} + 35\frac{\text{CHF}}{\text{kg}} \times 150\text{ kg}}{5\frac{\text{CHF}}{\text{kg}} \times 150\text{ kg} + 2\frac{\text{CHF}}{\text{piece}} \times 6000\text{ piece} + 15\frac{\text{CHF}}{\text{kg}} \times 750\text{ kg} + 25\frac{\text{CHF}}{\text{kg}} \times 150\text{ kg}} \times 100$$
$$= 133.5$$

According to the Paasche index, the basket of goods purchased today costs 33.5% (133.5–100) more than at the base time. The advantage of the Paasche index is that it takes into account the current situation. The disadvantage is that the values cannot be compared over time for several points in time, as the quantities are adjusted each time.

The advantage of the Laspeyres price index lies in its comparability over time. We observe the price development of a basket of goods over time, with the quantities always remaining the same. The disadvantage is that the weighting scheme (the quantities) slowly becomes outdated and the Laspeyres price index leads to results that become increasingly irrelevant the more the weighting scheme becomes outdated. The advantage of the Paasche price index, on the other hand, is that it always works with the current weighting scheme. However, it is not comparable over time, as the weighting scheme changes again and again.

If comparability over time is important, then it is better to work with the Laspeyres price index. However, the weighting scheme should be adjusted from time to time, especially when it has changed significantly.

Just like the price development, the quantity development can also be determined with the Laspeyres index and the Paasche index. The only difference is that the importance of the goods is expressed by their prices, i.e., the weighting scheme is the price. Otherwise, the formulas are constructed identically, except for the difference that now the quantities and not the prices change.

The formula for the Laspeyres quantity index is:

$$Q_{Lt} = \frac{\sum q_{ti} \times p_{0i}}{\sum q_{0i} \times p_{0i}} \times 100$$

The formula for the Paasche quantity index is:

$$Q_{Pt} = \frac{\sum q_{ti} \times p_{ti}}{\sum q_{0i} \times p_{ti}} \times 100$$

where

Q_{Lt} the quantity index (Q) according to Laspeyres (L) at time t and
Q_{Pt} the quantity index (Q) according to Paasche (P) at time t is.

The procedure for calculation and interpretation is identical and we refrain from presenting it here. However, it can generally be said that quantity indices are used when the quantitative development is of interest, e.g. the development of the quantities of all products produced at the company level or the development of production by industrial companies at the macroeconomic level.

7.5 Generating New Variables with the R Commander

At this point, we want to show how new variables can be generated from an existing data set using the R Commander. The possibilities are diverse, we can relate two variables to each other or apply other mathematical operations, such as squaring a variable, taking the root, logarithmizing or others. At this point, we relate two variables to each other, the variable marketing and the variable innovation, and thus obtain a reference number that reflects the ratio of marketing expenses to expenses for innovation for the companies.

In the R Commander, we select the Data menu and then in the submenu Manage variables in active data set the command Compute new variable. In the window that opens, we enter the new variable name and the instruction for the calculation. When we confirm with OK, we have added the new variable to our data set. We see in the message window that the number of variables in the data set has increased from 12 to 13 variables.

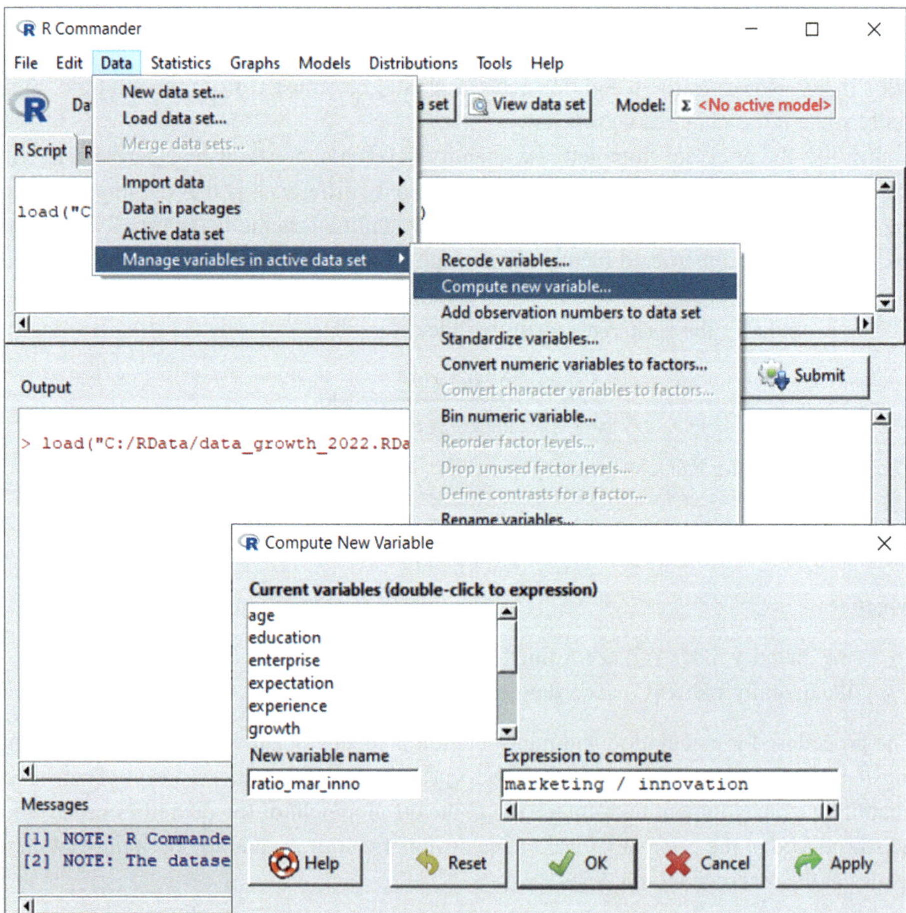

7.6 Checkpoints

- *Ratio numbers relate two quantities to each other. We can use them to generate new knowledge from existing data.*
- *Reference numbers relate different quantities to each other.*
- *Quotas relate sub-quantities to a total quantity.*
- *We can use dynamic indices to analyze the development of a quantity over time.*
- *Index numbers are used when the development of several quantities over time is to be analyzed.*

7.7 Application

7.1. Find three facts that can be analyzed with the help of a reference number.
7.2. Find three facts that can be analyzed with the help of a quota.
7.3. Find three facts that can be analyzed with the help of a dynamic index.
7.4. Analyze the development of the gross domestic product per capita for the following countries. Use the year 2007 as the base year.

Gross domestic product per capita (in $, current prices)			
Year	China	Germany	Switzerland
2007	2694	41,587	63,555
2008	3468	45,427	72,488
2009	3832	41,486	69,927
2010	4550	41,532	74,606
2011	5618	46,645	88,416
2012	6317	43,858	83,538
2013	7051	46,286	85,112
2014	7651	47,960	86,606
2015	8033	41,140	82,082
2016	8079	42,099	80,172
2017	8759	44,240	80,450
2018	9771	47,603	82,797

Source: World Bank 2020, data downloaded from https://data.worldbank.org/ on 11.04.2020.

7.5. A company needs energy from oil, gas, and electricity for the production of its goods. The company wants to analyze the price and quantity development of the energy required for production. In particular, it wants to know the effect of efforts to produce more energy-efficient. Calculate and analyze the price and quantity indices of Laspeyres and Paasche.

Energy source	Price 2015	Quantity 2015	Price 2019	Quantity 2019
Oil	0.12 [CHF/liter]	70,000 [liter]	0.30 [CHF/liter]	64,000 [liter]
Gas	0.28 [CHF/m3]	10,000 [m3]	0.42 [CHF/m3]	14,000 [m3]
Electricity	0.08 [CHF/kWh]	280,000 [kWh]	0.06 [CHF/kWh]	220,000 [kWh]

7.6. As a student, you are fortunate in possessing a small fortune and have invested it in stocks. Naturally, you are interested in the performance of your stock portfolio and you want to depict the development, similar to the Swiss Market Index SMI or the German Stock Index DAX, using an index value. Develop an index for the stock portfolio using the Laspeyres price index, which represents the price development.

	Stock	Daily closing prices in euro (June 2020)										
	Number	10.6	11.6	12.6	15.6	16.6	17.6	18.6	19.6	22.6	23.6	24.6
German Bank	50	8.71	8.10	8.22	8.29	8.44	8.42	8.22	8.26	8.34	8.51	8.12
BMW	12	58.81	55.00	56.17	56.32	57.11	57.34	57.08	56.97	57.68	58.49	56.40
Tesla	10	886.10	874.60	818.20	856.30	881.80	885.90	895.30	894.20	885.00	887.10	853.20
Allianz	30	185.40	175.68	177.66	177.80	183.02	181.90	181.00	180.40	179.64	182.98	178.36
Tui	100	5.50	4.88	5.08	5.12	5.29	5.25	5.06	4.95	4.72	4.69	4.30

Source: Finanzen.net, data downloaded from https://www.finanzen.net/ on 26.06.2020.

Part III
From Few to All

From Few to All or from the Sample to the Population

We already know a lot about our observed companies. We understand their average behavior, we know about their variation, we have calculated correlations between variables and thus are informed about the relationship between different variables. But does this knowledge about the observed companies apply to all newly established companies? So far, we have only analyzed a few companies. It could be that the results only pertain to the observed companies. The central question is therefore whether our results can be applied to all newly established companies. Technically speaking, the question is whether the sample results also apply to the population. In the next four chapters, we will address this question precisely. How can we, with the help of knowledge about a few, the sample, make statements about all, the population? Providing the answer to this is the objective of the next four chapters. We will once again engage with our data, clarify this little, crazy thing called hypothesis-many people, including lecturers often have a incorrect understanding of it-, examine how chance is distributed and think outside the box with the hypothesis test. Things are about to get really exciting now.

8 About Data and Truth

8.1 How do we Get Our Data: Or Primary or Secondary Data?

It is now time to talk about our data again. Why do we have it and how do we get it? We have data to solve a problem or answer a question. Therefore, at the beginning there is always a problem or a question. We want to know something, for example:

- What is the average growth rate of newly founded companies?
- What is the average age of company founders?
- Do more women or more men start a company?
- Is there a relationship between marketing activities and the growth of newly founded companies?

or

- How many grams of meat does a McBurger cheeseburger contain?

First, we need to define the problem and the question. Then we can think about what data we need to solve the problem and answer the question. If the problem and the question are not clearly defined, we usually fall into two traps. Either we collect data that we do not need, have extra work and create data graveyards, or we collect too little data and find out in the end that the data we need has not been collected.

Once we have clearly defined the problem and the question and we know what data we need, we need to get this data. There are two possibilities for this. Either someone else has already collected the data and we can use it, or we have to collect the data

F. Kronthaler, *Statistics Applied with the R Commander*,
https://doi.org/10.1007/978-3-662-69107-6_8

ourselves. In the first case, we speak of secondary data, in the second case of primary data or also of a primary data collection, we collect the data specifically for our question.

Data collection is expensive and time-consuming, i.e. whenever data is already available, we should use it. There are a number of reputable data providers, not least state and supranational institutions. At the state level, these are primarily the statistical offices in the EU the Statistical Office of the European Union, and at the supranational level the OECD, the World Bank, and the International Monetary Fund (Table 8.1).

Let's take a brief look at an example with the World Bank. This shows us the wealth of information that is available. The World Bank provides online via the internet address https://data.worldbank.org/ over 600 variables in time series from 1960 for almost all countries in the world. The range extends from poverty, climate change, agriculture and rural development to economic and social development (Fig. 8.1).

For example, we are interested (to stay close to our data set) in the founding conditions of companies in different countries worldwide. We want to know how the founding conditions in Switzerland compare to other countries in the world. In the World Bank's database, we find the variable "Time required to start a business (days)". We download the data as an Excel file and now "just" have to analyze it (Table 8.2).

In addition to state and supranational institutions, there are a number of semi-state and semi-private institutions that offer data. Well-known examples include Transparency International, Freedom House, or the World Economic Forum. For example, Transparency International collects the Corruption Perception Index, which gives us indications of the extent of the corruption problem in the individual countries of the world.

Before we collect data ourselves, it is always worth researching data. If there is no data that we can use, we have to collect the data ourselves. We have to carry out a primary data collection.

Table 8.1 Sources of secondary data

Institution	Internet address	Data
Federal Statistical Office Switzerland BfS	https://www.bfs.admin.ch/bfs/de/home.html	Data on Switzerland and the regions of Switzerland
Statistical Office of the European Union	https://ec.europa.eu/eurostat/de/home	Data on the European Union and its member states
OECD	https://data.oecd.org/	Mainly data on the member states of the Organisation for Development and Cooperation
World Bank	https://data.worldbank.org/	Data on almost all countries in the world
International Monetary Fund IMF	https://www.imf.org/en/Data	Mainly financial and economic data on almost all countries in the world

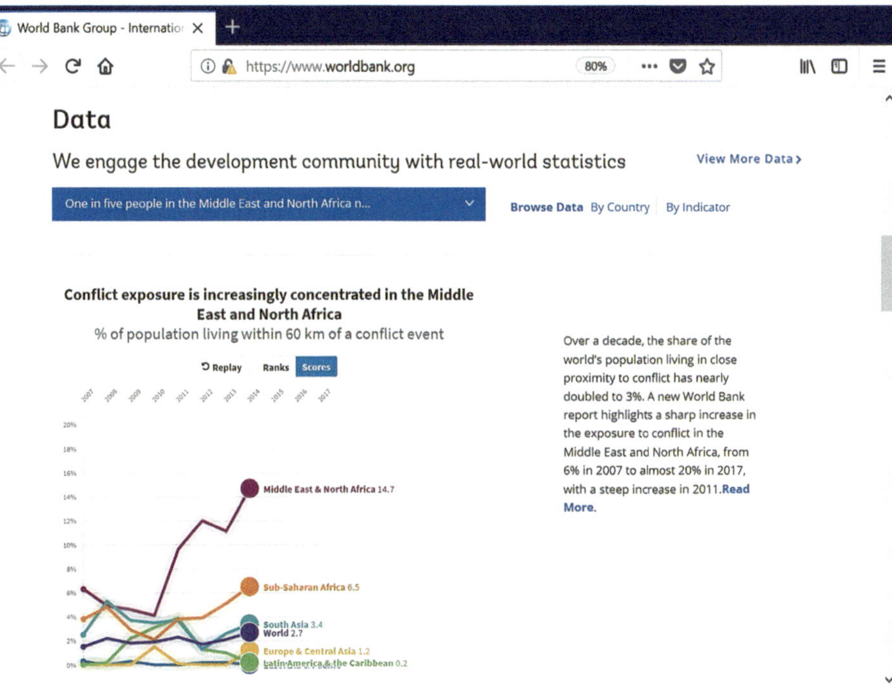

Fig. 8.1 The data of the World Bank (14.04.2020)

8.2 The Random Sample—The Best Estimator for Our Population

When we are faced with the task of collecting data ourselves, the following aspects need to be clarified first: 1) who is the population or about whom do we want to make a statement, 2) do we carry out a full survey or a sample survey, 3) if we carry out a sample survey, how do we ensure that the sample is representative for the population, and 4) how large should the sample be.

First, it needs to be clarified who or what the population is. The population contains all persons or objects about which we want to make a statement. This could be, for example, all cheeseburgers that are produced or only the cheeseburgers that McBurger produces. In our case, the population contains all newly founded companies.

Once we know the population, we have two options for data collection. First, we can collect data on all persons or objects of the population, we carry out a full survey. Second, we can limit ourselves to a subset of all persons or objects, we carry out a sample survey. As a rule, typically it will not be possible to carry out a full survey for reasons of time and cost. In Switzerland alone, we have about 40.000 company foundations per year. If we collect data on all 40.000 foundations, this is firstly very expensive and secondly very

Table 8.2 WDI data on the founding conditions in the world from 2010. *Source* World Bank 2020, data downloaded from https://data.worldbank.org/indicator/IC.REG.DURS on 14.04.2020

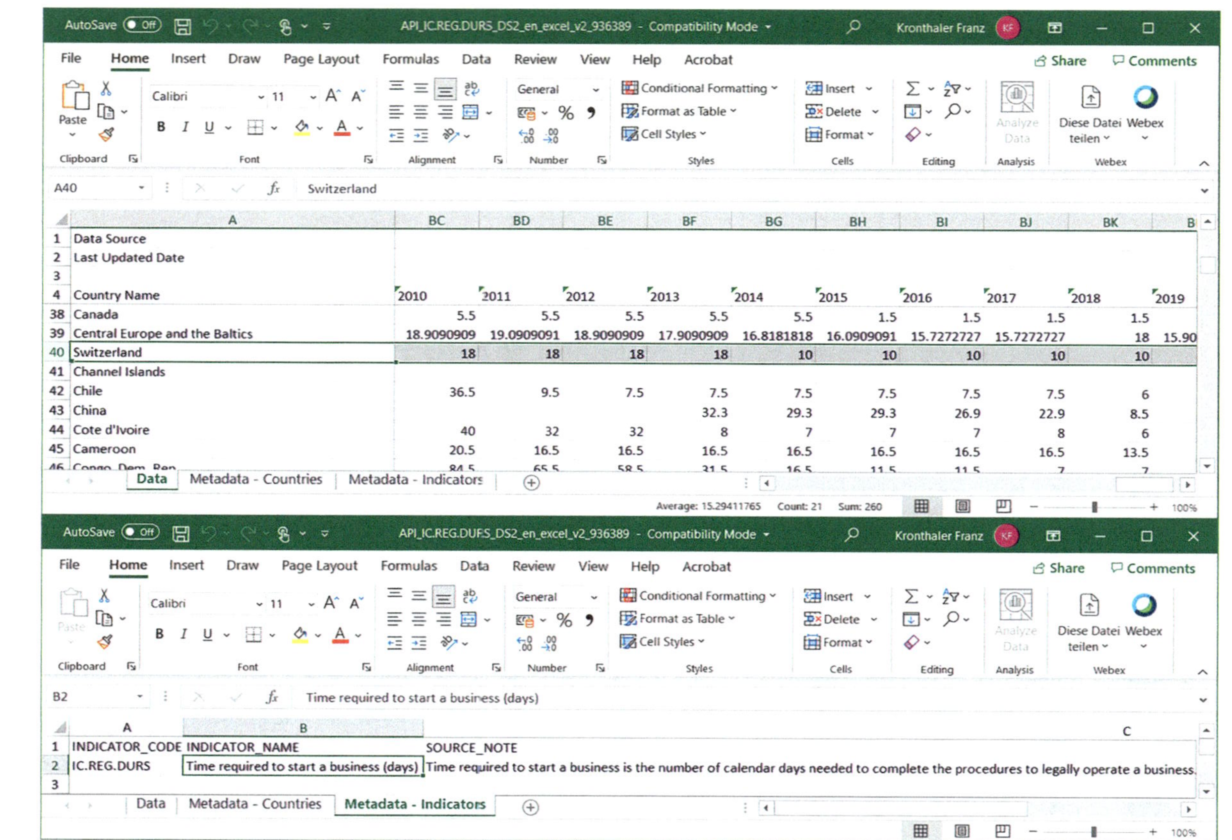

	A	BC	BD	BE	BF	BG	BH	BI	BJ	BK	B
1	Data Source										
2	Last Updated Date										
3											
4	Country Name	2010	2011	2012	2013	2014	2015	2016	2017	2018	2019
38	Canada	5.5	5.5	5.5	5.5	5.5	1.5	1.5	1.5	1.5	
39	Central Europe and the Baltics	18.9090909	19.0909091	18.9090909	17.9090909	16.8181818	16.0909091	15.7272727	15.7272727	18	15.90
40	Switzerland	18	18	18	18	10	10	10	10	10	
41	Channel Islands										
42	Chile	36.5	9.5	7.5	7.5	7.5	7.5	7.5	7.5	6	
43	China				32.3	29.3	29.3	26.9	22.9	8.5	
44	Cote d'Ivoire	40	32	32	8	7	7	7	8	6	
45	Cameroon	20.5	16.5	16.5	16.5	16.5	16.5	16.5	16.5	13.5	
46	Congo, Dem. Rep.	84.5	65.5	58.5	31.5	16.5	11.5	11.5	7	7	

	A	B	C
1	INDICATOR_CODE	INDICATOR_NAME	SOURCE_NOTE
2	IC.REG.DURS	Time required to start a business (days)	Time required to start a business is the number of calendar days needed to complete the procedures to legally operate a business.
3			

time-consuming, it takes at least one year. Moreover, even then we have not collected all foundations, but only for a section of one year. For all these reasons, we usually resort to a sample survey, i.e., we take a subset from the population and collect the required data for this. In our case, we have drawn a sample of 100 companies from our population.

However, as mentioned at the beginning, we do not want to describe the sample, but to make a statement about the population. For this to be possible, the sample must behave in the same way as the population with regard to the object of interest, i.e., it must represent the population or be representative for the population. The question that arises, therefore, is how do we achieve representativeness of the sample? The central criterion for this is that we randomly draw the sample out of the population. Technically speaking, every element of the population must have the same probability of being included in the sample. Let's imagine 20 people from whom we want to select 5 people. If we choose these five randomly, i.e., if each person of the population of 20 people has the same probability of being included in the sample, then we have fulfilled this condition. If we make a mistake in the random selection, e.g., if women have a higher probability of being selected, then the sample is biased and we can no longer make a statement about the population.

Freak-Knowledge

We have various options for selecting a representative sample from a population. The simplest is the drawing of a simple random sample. Each element of the total population has the same probability of being chosen for the sample. In addition, there is the stratified random sample and the random drawing according to the cluster method. In the former, the population is first divided into subgroups, e.g., by age or gender, and then a random sample is drawn from each subgroup. In the latter, the total population is first divided into subsets and then random subsets are selected. We can also combine and then come to multi-stage procedures.

If each element of the population has the same probability of being included in the sample, then the prerequisite for a representative sample is fulfilled and the values obtained from the sample, e.g., the sample mean values, are the best estimators we have for the population. For us, this means, for example, that the average age of the company founders of 34.25 years, which we have calculated for our 100 companies, is the best estimated value for the population. Technically speaking, the expected value of the sample is equal to the average value of the population:

$$E(\bar{x}) = \mu$$

where

$E(\bar{x})$ is the value that we would expect when drawing a sample and
μ is the mean of the population.

Let's think about this again briefly. The value we get from our sample is the best value we can get. However, it is not the only value, but it varies with the sample. To understand this, we need to go back to the example of our 20 people. Let's imagine we randomly select 5 people from our population of 20 people. In total, we have 15.504 possibilities for this. The following figure illustrates four possibilities. We could randomly get the people marked in red, the blue ones, the yellow ones, or the green ones, or we could mix the colors. If we think about it briefly, we quickly realize that there are a wide variety of combination possibilities, as mentioned, there are 15.504 (Fig. 8.2).

Now we want to think further and calculate the average body height for the different samples, e.g. the average body height of the red ones, the yellow ones, etc. It is relatively clear that we get a different average body height depending on the sample. Since 15.504 different samples are possible, there are also 15.504 possible sample means. However, we do not draw all possible samples, but only one. From this one, we say that it is the best estimator for the population. If we were to get a different sample, we would accordingly also get a different estimator. This presents us with a problem. We are not sure what the true value of the population is. The solution is to integrate the uncertainty by not making a point estimate, but by specifying a range or an interval in which the true value of the population is located with a certain degree of certainty. To calculate the interval, we first need to clarify something. If we draw all 15.504 samples and accordingly calculate all possible 15.504 sample means, we can calculate a "super" mean from the 15.504 sample means. All means of the samples vary around this "super" mean. If all sample means are knwon, we can calculate the standard deviation of the sample means. However, we do not need all sample means, but only the values of one sample. It could be shown (we refrain from doing so here), that the standard deviation of the sample means can be estimated as follows (the small hat over the σ stands for estimate):

$$\hat{\sigma}_{\bar{x}} = \frac{sd}{\sqrt{n}}$$

Fig. 8.2 Thought experiment on the sampling of 5 out of 20

with

$\hat{\sigma}_{\bar{x}}$ is the estimated standard deviation of the sample means, we also refer to this as the standard error of the mean,

$sd = \sqrt{\frac{\sum (x_i - \bar{x})^2}{n-1}}$ is equal to the standard deviation of the sample, and

n is equal to the number of observations.

From the the sample mean and the standard error of the mean, the interval can be determined in which the true value of the population is estimated to lie. We speak of the confidence interval. As a rule, we apply three intervals, the 90, 95 and 99% confidence interval:

$$CI_{90\%} = \left[\bar{x} - 1.64 \times \hat{\sigma}_{\bar{x}};\ \bar{x} + 1.64 \times \hat{\sigma}_{\bar{x}}\right]$$

$$CI_{95\%} = \left[\bar{x} - 1.96 \times \hat{\sigma}_{\bar{x}};\ \bar{x} + 1.96 \times \hat{\sigma}_{\bar{x}}\right]$$

$$CI_{99\%} = \left[\bar{x} - 2.58 \times \hat{\sigma}_{\bar{x}};\ \bar{x} + 2.58 \times \hat{\sigma}_{\bar{x}}\right]$$

What is meant here is that the confidence interval covers the true value of the population with a 90, 95, 99% probability. Or to put it better, in 90, 95 or 99 out of 100 samples, the value lies within this interval. The confidence interval can be illustrated graphically as shown in Fig. 8.3.

Let's calculate the 90, 95 and 99% confidence interval for our example. We already know that the average age of the company founders, i.e. the sample mean, is 34.25 years. We have also already calculated the standard deviation as 7.62 years and the number of observations is 100. This allows us to specify the confidence intervals as follows:

$$CI_{90\%} = \left[34.25 - 1.64 \times \frac{7.62}{\sqrt{100}};\ \bar{x} + 1.64 \times \frac{7.62}{\sqrt{100}}\right] = [33.00: 35.50]$$

33.00 34.25 35.50

$$CI_{95\%} = \left[34.25 - 1.96 \times \frac{7.62}{\sqrt{100}};\ \bar{x} + 1.96 \times \frac{7.62}{\sqrt{100}}\right] = [32.76: 35.74]$$

32.76 34.25 35.74

$$CI_{99\%} = \left[34.25 - 2.58 \times \frac{7.62}{\sqrt{100}};\ \bar{x} + 2.58 \times \frac{7.62}{\sqrt{100}}\right] = [32.28: 36.22]$$

32.28 34.25 36.22

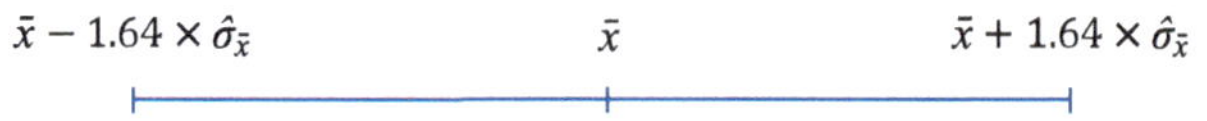

$\bar{x} - 1.64 \times \hat{\sigma}_{\bar{x}}$ $\bar{x}$ $\bar{x} + 1.64 \times \hat{\sigma}_{\bar{x}}$

Fig. 8.3 Graphical illustration of the 90% confidence interval for means

We can therefore make the following statement. The calculated confidence interval from 33.00 to 35.50 covers the true value of the population with a 90% probability. If we want to increase the certainty, the confidence interval becomes wider. If we want to be 99% sure that the confidence interval covers the true value, then the confidence interval ranges from 32.28 to 36.22.

We can also use this information to say something about the required sample size, i.e. about how many people or objects we need to observe. For the confidence intervals described above, the width of the confidence interval can be determined. This is twice the distance from the mean to the upper or lower interval limit. For the 95% confidence interval, for example, the width of the confidence interval is

$$CIW_{95\%} = 2 \times 1.96 \times \hat{\sigma}_{\bar{x}} = 2 \times 1.96 \times \frac{sd}{\sqrt{n}}$$

If we solve for n, we get the formula with which we can calculate the required sample size:

$$n = \frac{2^2 \times 1.96^2 \times sd^2}{CIW_{95\%}^2}$$

where

n is the sample size,
sd is the standard deviation of the sample, and
$CIW_{95\%}$ is the width of the 95% confidence interval.

With this, we observe the following. First, our sample size depends on the width of the confidence interval that we allow. Second, it depends on the standard deviation of the sample or the standard deviation in the population. Originally, we used the standard deviation of the sample as an estimate for the standard deviation in the population. The latter must be determined from experience or from preliminary studies. Third, the sample size depends on how large we choose our confidence level. At 90%, we take 1.64 instead of 1.96, at 99% we take 2.58. Fourth, and this is astonishing, the sample size does not depend on the size of the population. The size of the population can be neglected, as long as the population is not very small. If the population is very small, we will usually conduct a full survey, and the problem of sampling does not arise.

Let's apply the formula to our example. We want to estimate the average age of company founders with 95% certainty and the width of the confidence interval should not be more than 3 years. We know from preliminary studies that the standard deviation in the population is about 8 years. If we insert the values, we get the required sample size:

$$n = \frac{2^2 \times 1.96^2 \times s^2}{CIW_{95\%}^2} = \frac{2^2 \times 1.96^2 \times 8^2}{3^2} = 109.3$$

This means that to maintain the given framework, we need to draw a random sample of approximately 109 observations. Then we get the true value of the population with 95% certainty in an interval of 3 years.

The considerations just made about the mean can easily be transferred to proportion values. There are only a few small changes. In a representative sample, the expected value for the proportion value of the sample is equal to the proportion value of the population:

$$E(p) = \pi$$

with

$E(p)$ is the proportion value that we expect when drawing the sample, and
π is the proportion value of the population.

The standard deviation for the sample proportion values or the standard error of the proportion value is estimated as follows:

$$\hat{\sigma}_p = \sqrt{\frac{p(1-p)}{n}}$$

with

$\hat{\sigma}_p$ is equal to the standard error of the proportion value and
p is the proportion value of the sample.

The 90% confidence interval is calculated from the calculated sample proportion value p and the standard error of the proportion value (Fig. 8.4). If we want to calculate the 95% confidence interval instead of the 90% confidence interval, we use 1.96 instead of 1.64, or 2.58 for the 99% confidence interval.

Let's explain this with an example. From our data set, we can easily see that 35% of the companies were founded by a woman. In decimal notation, this means

$$p = 0.35.$$

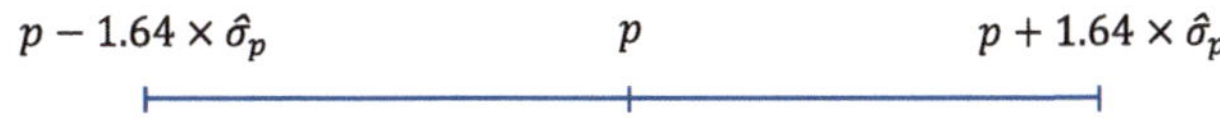

Fig. 8.4 Graphical illustration of the 90% confidence interval for proportion values

If we insert the proportion value into our formula for the standard error of the proportion value, we get a standard error of approximately 4.8 percentage points:

$$\hat{\sigma}_p = \sqrt{\frac{p(1-p)}{n}} = \sqrt{\frac{0.35(1-0.35)}{100}} = 0.048$$

The 90% confidence interval thus has interval limits of 27.1% and 42.9%.

$$CI_{90\%} = [0.35 - 1.64 \times 0.048; 0.35 + 1.64 \times 0.048] = [0.271; 0.429]$$

Therefore, the true proportion of female founders in start.ups is between 27.1% and 42.9% with a certainty of 90%. If we increase the certainty to 95% or 99%, the confidence interval widens. For the 95% confidence interval, the interval limits are 25.6% and 44.4%, and for the 99% confidence interval, they are 22.6% and 47.4%.

Looking at the confidence intervals, one might get the impression from their width that we are relatively imprecise regarding the proportion value in the population. If we want a more precise statement, we need to increase the sample size.

How large the sample size needs to be depends again on the width of the confidence interval that we want to allow. So we use the formula mentioned above again, with the small difference that we insert the standard error of the proportion value. The following is the formula for the sample size of a 90% confidence intervall. If we want to achieve a certainty of 95% or 99% instead, we replace the number 1.64 with 1.96 or 2.58.

$$CIW_{90\%} = 2 \times 1.64 \times \hat{\sigma}_p = 2 \times 1.64 \times \sqrt{\frac{p(1 \quad p)}{n}}$$

If we solve the formula for n, we get the formula for the sample size for a confidence intervall of 90%.

$$n = \frac{2^2 \times 1.64^2 \times p(1-p)}{CIW_{90\%}^2}$$

Let's clarify this again with an example. Let's assume that the width of the 90% confidence interval should not exceed 6 percentage points. We still lack the proportion value p to be able to completely fill out the formula. Before we have drawn the sample, we do not yet know the proportion value. We have to determine this with the help of considerations and previous studies. For our example, we assume that the proportion of female founders is approximately 30%. If we now insert the values, we get a sample size of

$$n = \frac{2^2 \times 1.64^2 \times p(1-p)}{CIW_{90\%}^2} = \frac{2^2 \times 1.64^2 \times 0.3(1-0.3)}{0.06^2} = 627.6.$$

In order to be able to say with 90% certainty that the true proportion value of the total population lies within a 6 percentage point wide confidence interval, we therefore need to draw a sample of $n = 627.6$ company foundations.

We have seen that the size of the sample essentially depends on how precise we want to be in our statement. We have to determine this in advance, before we draw the sample.

8.3 On Truth—Validity and Reliability

Drawing a good sample is only one side of the coin when you want to learn something about the population. Equally important is that we actually collect what we want to know and that we do it reliably. Technically speaking, we are talking about the validity and reliability of the data.

In simple terms, validity means that the data actually measures what is the subject of the investigation. Sometimes this is easy, sometimes less so. For example, if we are interested in the profit development of young companies, we can collect the company's profit in monetary units and the matter is relatively clear. If we want to know something about the wealth difference between countries, things are not so simple. Often, the concept of gross domestic product is used to measure a country's wealth. The gross domestic product, simply put, measures the goods and services produced in a country and available for consumption. Therefore, it is often taken as a proxy for a country's wealth. Proxy is the technical term for a variable that gives us approximate information about the subject we are interested in. However, many people doubt that gross domestic product is a good variable for saying something about a country's wealth. Internationally, for example, the Human Development Index is also used. This indicator is a composite measure, which includes, among other things, the gross domestic product, life expectancy in a country, medical care, and education. The government of Bhutan uses the "Gross Happiness Product". Even if it may seem strange, but happiness can perhaps also be a proxy for wealth. It is relatively easy to see that depending on the indicator, different results are obtained when we want to say something about wealth in different countries. Imagine we have to measure happiness. The task would then be to define what happiness is. Then we have to find a proxy that actually measures happiness. That doesn't look easy, does it?

How can we ensure that we find a suitable measure that actually tells us what we want to know? The easiest way is to go to an expert and ask which measure can be used. So if we want to know something about the innovation activity of companies and we are not innovation experts ourselves, then we go to a person who has been dealing with corporate innovations for a long time. If no one is available, then there is no choice but to read up on the literature yourself to find out how innovation can be measured approximately in a corporate context. There will certainly not be just one possibility. Reading up on the literature yourself is always recommended, as it reduces dependence on others and allows you to better understand and assess their statements.

Once we have ensured that we have a valid variable or a valid construct to answer our question, we must ensure that we measure reliably, we must ensure the reliability of the data. Reliability, simply put, means that we measure the true result and that we come to the same result again and again, even if we measure more often. Let's take a simple example that we want to measure the intelligence quotients of company founders. We must first realize that every investigation is fraught with errors. So let's assume we pick out a company founder and measure his intelligence quotient for him. We now also have a construct, a written test, that allows us to measure. Let's further assume that the true intelligence quotient of this founder is 115 points. We don't know this value, we are trying to determine it. Our test result indicates that the intelligence quotient is 108. We then made an error of 7 points. This error can occur because the founder was tired, because he was distracted, because the room temperature was too high, etc. This means that we must create conditions that ensure that the error is as small as possible. The smaller the error, the more reliable the measurement. Ensuring reliability requires that we ensure that we collect the true value and that we do so with as small an error as possible.

Validity and reliability are closely related. If we do not consider one of the two, our investigation is worthless. We can then not answer our question. Imagine we take care of validity, but do not consider reliability. Then we measure the right thing, but we do not measure it correctly. If we consider the reliability, but not the validity of the data, we measure reliably and correctly, but the wrong thing. The data is then worthless with regard to our question. We could then at most answer another question that fits the data. With regard to our investigation, this means that we must first ensure validity and then reliability.

8.4 Checkpoints

- *Before we collect and analyze data the research problem and the research question have to be clearly defined.*
- *Before primary data collection is carried out, we should check whether the required data or useful data already exist.*
- *The population contains all persons or objects about which we want to make a statement.*
- *We achieve representativeness of the sample by randomly drawing the elements or objects from the population.*
- *The sample values of a representative sample are the best estimators for the population.*
- *The confidence interval makes the point estimate more precise by including uncertainty.*
- *The sample size is calculated from the width of the confidence interval, the standard deviation in the population, and the defined confidence level.*

- *Validity of a variable means that the variable is a good proxy for the object of interest.*
- *Reliability means that we measured the object of interest with an measurement error as small as possible.*

8.5 Application

8.1 What steps and what order of steps should we consider in data collection?

8.2 Search the current data on the number of inhabitants in the countries of the world on the Internet. How many inhabitants do China, India, Switzerland, Germany, and the USA currently have?

8.3 Explain what the population is, what a sample is, and what property must be fulfilled so that we can infer from the sample to the population.

8.4 What is the average growth rate of the companies in our data set data_growth and what is the 90%, 95%, and 99% confidence interval?

8.5 Calculate the required sample size if we want to estimate the average growth rate of the companies with 95% certainty and the confidence interval width should not be more than 2 percentage points. From preliminary studies, we know that the standard deviation in the total population is approximately 5 percentage points.

8.6 What percentage of company formations in our data set data_growth are industrial companies and what is the 90%, 95%, and 99% confidence interval?

8.7 How large would the sample have to be if we want to estimate the proportion of industrial foundations in the population with 90% certainty in a confidence interval width of 8 percentage points. From preliminary considerations, we know that the proportion of industrial foundations in company foundations is approximately 25%.

8.8 What does validity and reliability mean?

8.9 Which of the two statements is correct: 1) Reliability is possible without achieving validity or 2) Validity is possible without achieving reliability. Explain briefly.

8.10 Search for the article by Costin, F., Greenough, W.T. & R.J. Menges (1971): Student Ratings of College Teaching: Reliability, Validity, and Usefulness, Review of Educational Research 41(5), pp. 511–535 and discuss how the concepts of validity and reliability are discussed in this article.

Hypotheses: Just a Refinement of the Question 9

9.1 The Small, Big Thing of the (Research) Hypothesis

Hypothesis is a term that is used again and again in the research process, but it is unclear to many people. Even lecturers often have a false understanding of what a hypothesis is. A hypothesis is fundamentally something very simple. Loosely speaking, it is just a translation of the question into a testable statement. At the same time, a hypothesis is also something quite magnificent. With a hypothesis, we generate statistically verified knowledge.

Let's consider this with an example. We strain the burger example again. Our question is as follows: How many grams of meat does a burger of McBurger contain? With the previous chapter, we can answer this question with a point and interval estimate. If we have prior knowledge, the question can be specified and formulated as a statement. Let's assume that McBurger claims that a burger contains an average of 100 g of meat. With this information, we can translate our question into a statement. The statement or hypothesis that can be tested is then as follows: A burger from McBurger contains an average of 100 g of meat. A hypothesis is created by specifying a question.

> *We go from the question:*
> *How many grams of meat does a McBurger burger contain?*
> *through prior knowledge to the ...*
> *Hypothesis: A burger from McBurger contains an average of 100 g of meat.*

The advantage of the hypothesis over the question is that it is more precise and can therefore be tested. What does this mean in specifically? In the previous chapter, we made a point estimate and an interval estimate to answer the question. After data collection, we might have found that the average value of the sample is 98 g and the 95% confidence interval ranges from 96 g to 100 g.

F. Kronthaler, *Statistics Applied with the R Commander*,
https://doi.org/10.1007/978-3-662-69107-6_9

Would we have rejected McBurger's statement with this? We don't really know. We did not check the statement, but generated knowledge about how much meat a McBurger burger contains. With the help of a hypothesis, we can check the statement. We use the knowledge generated from the sample (as we will see in the following chapters). If we find that the statement is not true, we can confront McBurger with it and demand more or less meat depending on the result.

How do we get from the question to the hypothesis? Hypotheses do not arise from the gut. We form them based on existing knowledge in the form of studies and theories. We should look at this using the example of our company foundations. The question that interests us is as follows. How old are our company founders on average? We research the literature on company foundations and come across studies that have already been carried out in the past. We also come across theories on company foundations, which state that company founders have gathered sufficient professional experience before the respective foundation and are not at the beginning, but in the middle of their professional career. Based on this existing knowledge, we formulate our hypothesis. This could look like this: Company founders are on average 40 years old.

We go from the question:
How old are company founders on average?
through existing studies and theories to the ...
Hypothesis: Company founders are on average 40 years old.

Let's briefly summarize again: A hypothesis is a well-founded testable statement formulated with the help of theory and prior knowledge. In the research process, we can also refer to the hypothesis as a research hypothesis.

9.2 The Null Hypothesis H_0 and the Alternative Hypothesis H_A

We now know what a hypothesis is. In the research process, two different types of hypotheses are used, the null hypothesis and the alternative hypothesis. The null hypothesis is the testable statement, the alternative hypothesis is the opposite of the null hypothesis. Let's look at our burger example again for clarification:

Research hypothesis: A burger from McBurger contains on average 100 g of meat.

If we divide this statement into the null hypothesis and the alternative hypothesis, the following two statements result:

$H_{0:}$ *A burger from McBurger contains on average 100 g of meat.*
$H_{A:}$ *A burger from McBurger contains on average less or more than 100 g of meat.*

For the null hypothesis, we write briefly H_0 and for the alternative hypothesis H_A. Sometimes the alternative hypothesis is also designated with H_1.

For our company founders, the research hypothesis is as follows:

Research hypothesis: Company founders are on average 40 years old.

If we divide this statement again into the null hypothesis and the alternative hypothesis, we get the following two statements:

$H_{0:}$ *Company founders are on average 40 years old.*
$H_{A:}$ *Company founders are on average younger or older than 40 years.*

We need to distinguish between the null hypothesis and the alternative hypothesis for two reasons, firstly for reasons of precision. We need to know what happens if we have tested the hypothesis and reject or do not reject it. If we test for the second example H_0 and assume that we reject H_0. We can then make the statement that company founders are younger or older than 40 years. If we do not reject H_0 we stick to the statement that company founders are on average 40 years old.

Secondly, it is usually not possible to test our research hypothesis formed from prior knowledge and theories. We then have to take the detour via the null hypothesis. To illustrate this, we will expand the example of the average age of company founders. We are interested in the following question. Are male and female founders of the same age when they start a company? We research the literature, find studies and theories, and form our research hypothesis from them. This then reads as follows:

> *Research hypothesis: Male and female founders are not of the same age when they start a company.*

Can we test this hypothesis? What does not the same age mean? Is it enough if we find a difference of 0.5 years in our sample or does the difference have to be 5 years? You see, the statement is imprecise and therefore cannot be tested, not the same age is not precisely defined. However, the statement that male and female founders are of the same age when they start a company is testable. The same age is clearly defined, the difference is zero years. Our research hypothesis, male and female founders are not of the same age when they start a company, is not testable, but the opposite is testable. In this case, we formulate the opposite of the research hypothesis as the null hypothesis. If we reject the null hypothesis based on our sample, we proceed with the alternative hypothesis, which is equal to our research hypothesis.

$H_{0:}$ *Male and female founders are of the same age when they start a company.*
$H_{A:}$ *Male and female founders are not of the same age when they start a company.*

Next, we want to use our company founder example to explain the principle using an assumed correlation between two variables. For this, we use the following question. Is there a relationship between the proportion of sales that young companies spend on marketing and the proportion of sales that young companies spend on innovation? We research the literature and formulate the following hypothesis based on the results:

Research hypothesis: There is a relationship between the proportion of sales that young companies spend on marketing and the proportion of sales that young companies spend on innovation.

This hypothesis implies a correlation between the two variables marketing and innovation. But what exactly does correlation mean? Does the correlation coefficient found in the sample have to be 0.2, 0.5 or 0.7? The statement is imprecise. Precise, however, is the statement that there is no relationship between the proportion of sales that young companies spend on marketing and the proportion of sales that young companies spend on innovation. No relationship means that the correlation coefficient is exactly zero. The null hypothesis is thus again the opposite of the research hypothesis:

$H_{0:}$ *There is no relationship between the proportion of sales that young companies spend on marketing and the proportion of sales that young companies spend on innovation.*

$H_{A:}$ *There is a relationship between the proportion of sales that young companies spend on marketing and the proportion of sales that young companies spend on innovation.*

9.3 Hypotheses, Non-Directional or Directional?

The null hypothesis and the alternative hypothesis can be formulated as non-directional or directional. Whether we formulate a null hypothesis and an alternative hypothesis as directional or non-directional depends on how much we know about the subject of research. If we know a lot, we can formulate the null hypothesis and the alternative hypothesis directionally, if we know less, we are content with a non-directional null and alternative hypothesis. We best illustrate this again with our two examples just mentioned. So far, we have formulated the hypotheses non-directionally in both cases. In the first case we named a difference or no difference in age between male and female entrepreneurs. In the second case, we formulated no or a relationship between the effort for marketing and the effort for innovation. In both cases, we did not postulate a direction, i.e., we did not note whether men or women are older or whether the correlation is positive or negative. If we add this information, we arrive at a directional null hypothesis and alternative hypothesis. To formulate directional hypotheses, we therefore need more information or more knowledge. Let's imagine that our literature research has led to a higher degree of information in both cases, e.g., we reasonably assume that women are older at the time of founding than men or that there is a negative correlation between the two variables marketing and innovation, then we can formulate our research hypothesis, null hypothesis and alternative hypothesis as follows:

Research hypothesis: Female founders are older at the time of founding than male founders.

This results in the following null and alternative hypotheses, with the research hypothesis becoming the alternative hypothesis:

H_0: *Female founders are younger or the same age at the time of founding than male founders.*

H_A: *Female founders are older at the time of founding than male founders.*

In the second example, the research hypothesis is as follows:

Research hypothesis: There is a negative relationship between the proportion of sales that young companies spend on marketing and the proportion of sales that young companies spend on innovation.

Here too, the research hypothesis becomes the alternative hypothesis:

H_0: *There is no or a positive relationship between the proportion of sales that young companies spend on marketing and the proportion of sales that young companies spend on innovation.*

H_A: *There is a negative relationship between the proportion of sales that young companies spend on marketing and the proportion of sales that young companies spend on innovation.*

In summary, it can be said that with non-directional hypotheses the direction is not determined, while with directional hypotheses the direction is also specified. The null hypothesis includes not only the opposite of the null hypothesis, but also the equality in the relationship.

9.4 What Makes a Good Hypothesis?

Now we need to briefly clarify what makes a good hypothesis. Basically, a good hypothesis should meet the following criteria. First, it should be a direct extension of the research question. Hypotheses must always be related to the problem or question. With their help, we want to solve the problem or find answers to our question. Second, it must be formulated as a statement, not as a question. Questions are not verifiable, clear, precise statements are. If we formulate the statement that a burger contains 100 g of meat, this statement can be tested. Third, a good hypothesis reflects the existing literature and theory. Almost always, we are not the first to conduct a study or make considerations on a particular question. Many others have dealt with similar questions before us. The knowledge gained in the process helps us answer our question and formulate our hypothesis as precisely as possible. A good hypothesis always has a strong reference to the existing literature. Fourth, a hypothesis must be short and meaningful. If the hypothesis contains no statement, then nothing can be tested. The longer and usually more cumbersome a hypothesis is, the more difficult it is to formulate the hypothesis precisely and to formulate the statement unambiguously. Fifth and last, the hypothesis must be testable. Let's imagine we have formulated the following research hypothesis. Women are better entrepreneurs than men. With this hypothesis, it is not clear what is meant by better. Do women have a better business model? Do women master the process of founding a company better? Do companies founded by women grow faster? etc. We do not know exactly what we should test. In contrast, the following formulation of the research hypothesis is much more precise and therefore testable. Companies founded by women achieve higher sales growth in the first five years than companies founded by men. Here it is clear that we need to compare sales growth.

9.5 Checkpoints

- *A hypothesis is a statement about an issue based on theory or experience.*
- *In the research process, we distinguish between null hypothesis H_0 and alternative hypothesis H_A. The null hypothesis is the testable statement, the alternative hypothesis is the opposite of the null hypothesis.*
- *We can formulate our hypotheses non-directional and directional. The formulation of a directetional hypothesis requires a higher level of knowledge than the formulation of a non-directional hypothesis.*
- *A good hypothesis meets the following criteria: It is a direct extension of the question, is formulated as a statement, reflects the existing literature and it is short and precise as well as testable.*

9.6 Application

9.1 Formulate the non-directional null and alternative hypotheses for the following research questions.
  - Is there a relationship between the turnover growth of newly founded companies and the proportion of turnover spent on innovation?
  - Is there a difference in the industry experience between male and female founders that they bring with them at the time of founding?
  - Is there a relationship between smoking and life expectancy?

9.2 Formulate a directional null and alternative hypothesis for the above research questions.

9.3 Go to the library and find three scholarly articles (which include data) from your field. Answer the following questions for each article: What is the correct citation of the article? What is the research question of the article? What is the research hypothesis? Is it explicitly formulated? What is the null hypothesis? Is it explicitly formulated? Also formulate the research hypothesis for those articles where the hypothesis is not explicitly formulated.

9.4 Evaluate the hypotheses you found in task 3 with regard to the five criteria that a good hypothesis should meet.

9.5 Why does the null hypothesis usually assume no relationship between the variables or no difference between groups?

9.6 If interested, read the story "The Balek Scales" by Heinrich Böll and formulate a hypothesis about the story with regard to the scale. Think about how the boy tests his assumption.

Normal Distribution and Other Test Distributions

10

We now know what a hypothesis is and that we use it to answer research questions or test statements. However, we do not yet know how this testing works. To understand this, we first need to realize that chance is not entirely random. We will see that chance can be normally distributed, but also distributed differently. This means very specifically, we can make a statement about how randomly chance occurs. Interesting, isn't it? In the following, we will deal with probability distributions. We will learn about the normal distribution and about other test distributions. It remains exciting.

10.1 The Normal Distribution

The normal distribution is interesting for two reasons. First, many variables in reality are at least approximately normally distributed. Second, the normal distribution gives us an intuitive access to probability distributions. Once we have understood the normal distribution, dealing with other test distributions is no longer a problem.

Let's first describe the normal distribution. The normal distribution is bell-shaped, with a maximum. It is symmetrical around the maximum. Its tails run asymptotically towards the x-axis (Fig. 10.1).

The maximum of the normal distribution is in the middle of the curve. Symmetrical means that we can fold the normal distribution in the middle and then both sides lie exactly on top of each other. For both reasons, the arithmetic mean, the median, and the mode are equal. The mode is in the middle because it is the most common value. The median is in the middle because the distribution is symmetrical and it divides the distribution exactly in two halfs. Finally, the average value is in the middle because we have exactly the same amount and exactly the same values on the left and right. Asymptotic

F. Kronthaler, *Statistics Applied with the R Commander*,
https://doi.org/10.1007/978-3-662-69107-6_10

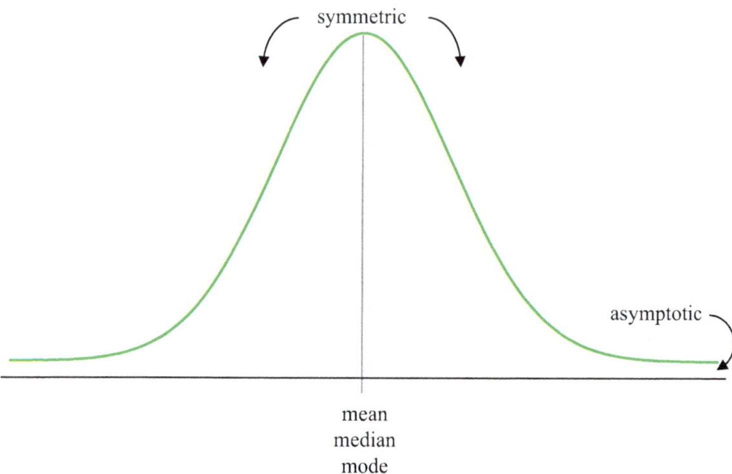

Fig. 10.1 The normal distribution

means that the tails are getting closer and closer to the x-axis, but never reach it. The area under the normal distribution curve indicates how likely it is that a certain value of a normally distributed variable occurs within an interval.

Specifically, we want to illustrate this with two variables that we can easily assume are normally distributed: the intelligence of individuals and the profit of companies in an industry. Let's look at Fig. 10.2 and start with the intelligence quotient. We keep in mind

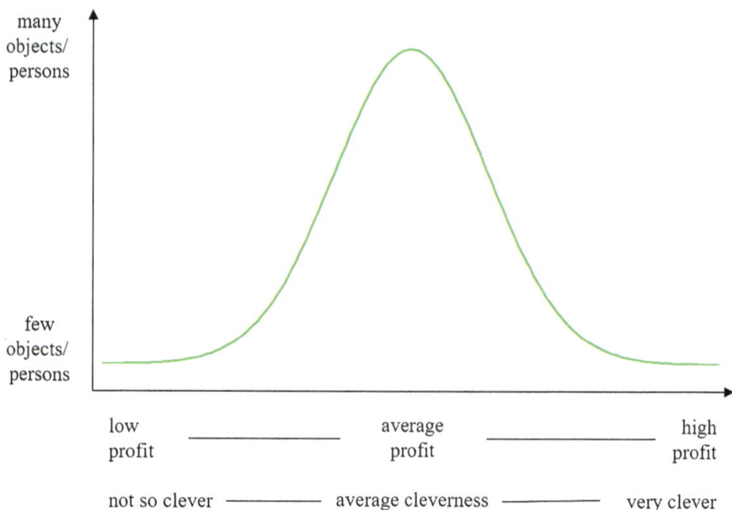

Fig. 10.2 The normal distribution and observations

that the area under the normal distribution curve represents the probability that a certain value occurs within an interval.

In the figure, the number of people is plotted on the y-axis and the intelligence quotient on the x-axis, with people with a low intelligence quotient on the left, people with a high intelligence quotient on the right, and people with a medium intelligence quotient in the middle. From experience, we know that very smart people and very stupid people are very rare. The curve reflects this. The areas far to the left and far to the right under the normal distribution curve are very small, the curve runs close to the x-axis here. People who are average in intelligence, on the other hand, are very common. The central area under the normal distribution curve is very large. The same applies to company profits. We have many companies that make an average profit, few that perform extraordinarily well, and also few with extraordinarily poor performances. Accordingly, the areas to the right and left under the normal distribution curve are small, the central area in the middle is large.

Let's look at this in more detail. We can easily divide the normal distribution into intervals. In the following figure, we have done this for a normal distribution with the average value of 100 and the standard deviation of 20. 100 is the average value, 120 or 80 are 1 standard deviation (1×20) away from the average value, 140 or 60 are 2 standard deviations (2×20) away from the average value, and 160 or 40 are 3 standard deviations (3×20) away from the average value (Fig. 10.3).

Looking at the figure, we see that the two central intervals from 80 to 100 and from 100 to 120 each have an area of 34.13%, or together, double the area of 68.26% of the total area under the normal distribution curve. In other words, the area from minus one standard deviation to one standard deviation is 68.26% of the total area. What does this mean specifically? It means that values between a standard deviation of -1 and 1 have a probability of 68.26% to occur. Looking at our example with the intelligence quotient, we can say that 68.26% of people have an intelligence quotient between 80 and 120 points. If we apply this to our sample selection and pick a random person from all

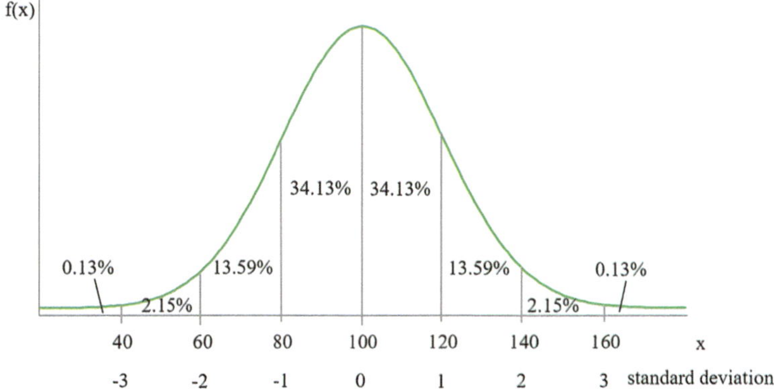

Fig. 10.3 Areas of the normal distribution and standard deviation

people, we find with a 68.26% probability someone whose intelligence quotient falls within this interval. We can continue in the same way. The probability of finding a person whose intelligence quotient falls in the interval from 120 to 140 points, or between one and two standard deviations, is 13.59%, for an intelligence quotient between 140 and 160 or between two and three standard deviations it is 2.15%, and for more than 160 points or over three standard deviations from the average value it is 0.13%. Of course, we can proceed in the same way with the deviations to the left of the average value.

So far, we have looked at a special normal distribution, the normal distribution with an average value of 100 and a standard deviation of 20. A small problem here is that we do not have just one normal distribution, but a multitude of normal distributions. So we need to go further and look at the normal distribution in detail again.

In Fig. 10.4 we have plotted the function of the normal distribution curve.

This is

$$f(x) = \frac{1}{\sigma\sqrt{2\pi}} e^{-\frac{1}{2}\left(\frac{x-\mu}{\sigma}\right)^2}$$

where

$f(x)$ is the function value at the point x, i.e., the value on the y-axis at the point of the respective x-value,
μ is the average value of the normal distribution,
σ is the standard deviation of the normal distribution,
π the circle constant with its value of 3.141..., and
e is Euler's number with its value of 2.718....

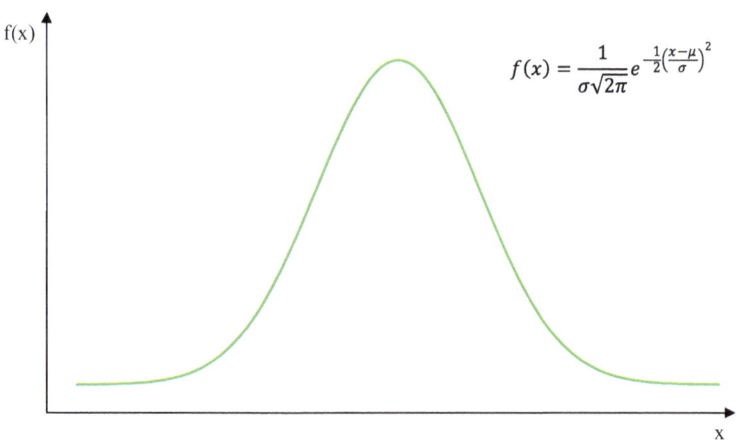

Fig. 10.4 The function of the normal distribution

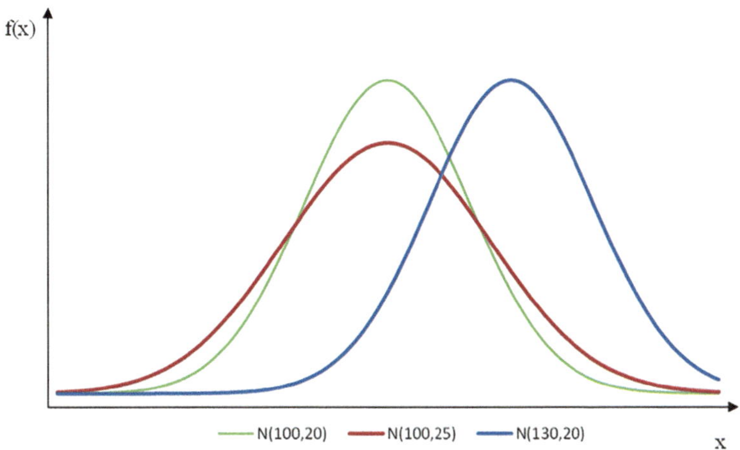

Fig. 10.5 Comparison of the normal distributions $N(100, 20)$, $N(100, 25)$ and $N(130, 20)$

From this it follows that the shape of the normal distribution depends only on two parameters, the average value and the standard deviation. With these two values, we can clearly define a normal distribution. Therefore, a normally distributed variable can be characterized as follows:

$$X \sim N(\mu, \sigma)$$

The variable X, e.g. our intelligence quotient or our company profits, is approximately normally distributed (the tilde stands for approximately) with the average value μ and the standard deviation σ.

When the average value or the standard deviation changes, the shape of the normal distribution also changes.

Specifically, an increase in the mean leads to a rightward shift of the normal distribution curve and an increase in the standard deviation leads to a broadening. This can be easily seen in Fig. 10.5. The normal distribution curve $N(130, 20)$ is further to the right than $N(100, 20)$, the normal distribution curve $N(100, 25)$ is wider than $N(100, 20)$ and $N(130, 20)$. Since both the average value and the standard deviation can theoretically take any possible value, there is not just one normal distribution, but infinitely many.

10.2 The Z-Value and the Standard Normal Distribution

We have seen that we do not only know one normal distribution curve, but that there are many normal distribution curves, the appearance and location of which depend on the average value and the standard deviation. However, we can standardize each of these normal distributions with a simple transformation so that they have an average value of 0 and a standard deviation of 1. The transformation rule is as follows:

$$z = \frac{x - \mu}{\sigma}$$

with

z is equal to the standardized value (short z-value),
x is the value of the distribution to be standardized,
μ is equal to the average value of the distribution,
σ is equal to the standard deviation of the distribution.

In Table 10.1 we apply the transformation rule to the normal distribution $N(100, 30)$ with an average value of 100 and a standard deviation of 30. Instead of this, we could also use any other normal distribution. We see that with the help of the z-transformation we convert the average value of 100 to 0 and that the new standard deviation is 1.

With the transformation, we have converted the normal distribution $N(100, 30)$ into the normal distribution $N(0, 1)$. We also call the latter standard normal distribution, where on the x-axis now not the x values, but the standardized z values are plotted. Otherwise, the interpretation remains the same. In the interval from -1 standard deviation (z-value of -1) and +1 standard deviation (z-value of +1) lie 68.26% of the possible cases. This means that the probability of finding a standardized z-value in the interval from -1 to +1 is 68.26%, that we get a standardized z-value greater than 3 is 0.13%, etc. Compare this to Fig. 10.6.

Table 10.1 Standardization of the normal distribution $N(100, 30)$

x-value	Transformation	z-value
10	$z = \frac{10-100}{30}$	−3
25	$z = \frac{25-100}{30}$	−2.5
40	$z = \frac{40-100}{30}$	−2
55	$z = \frac{55-100}{30}$	−1.5
70	$z = \frac{70-100}{30}$	−1
85	$z = \frac{85-100}{30}$	−0.5
100	$z = \frac{100-100}{30}$	0
115	$z = \frac{115-100}{30}$	0.5
130	$z = \frac{130-100}{30}$	1
145	$z = \frac{145-100}{30}$	1.5
160	$z = \frac{160-100}{30}$	2
175	$z = \frac{175-100}{30}$	2.5
190	$z = \frac{190-100}{30}$	3

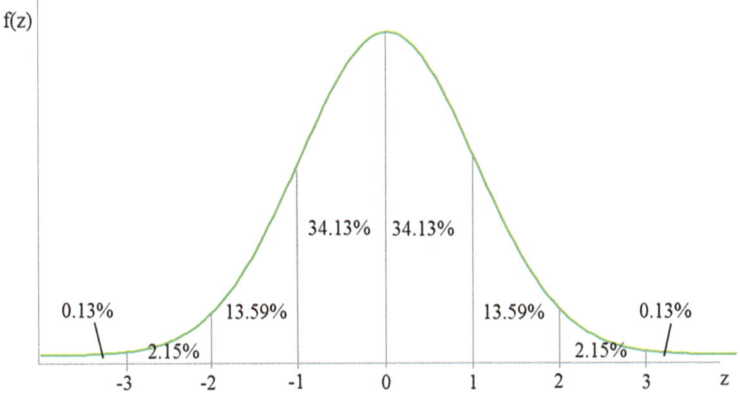

Fig. 10.6 The standard normal distribution $N(0, 1)$

But what does the z-value really mean and why do we need it? The z-value always represents an x-value of any normal distribution. With its help, we can determine for any normal distribution how likely the occurrence of a certain x-value is. An example should clarify this. We have the variable intelligence quotient and this is normally distributed with the average value 100 and the standard deviation 10. We now find a person with an intelligence quotient of 130. The z-value, which represents the 130, is $z = \frac{130-100}{10} = 3.0$.

With the help of this z value and the standard normal distribution, we can make a statement about how likely it is to find a person with an intelligence quotient of 130 and higher. The probability is 0.13% (compare Fig. 10.6). We can also ask ourselves how likely it is to find a person with an intelligence quotient between 80 and 90 points. The corresponding z values are

$$z_1 = \frac{80 - 100}{10} = -2.0 \text{ und } z_2 = \frac{90 - 100}{10} = -1.0,$$

i.e., the probability of finding such a person is 13.59%.

Each area of the standard normal distribution correspond to automatically to a specific z value. With the help of the z value, we can therefore determine the area under the standard normal distribution. As can be seen in Fig. 10.7, for example, the z value of 1.18 divides the standard normal distribution into the areas of 11.9% to the right and 88.1% to the left.

As the z- value increases, the area to the right of the z- value decreases and the area to the left increases. The reverse is also true.

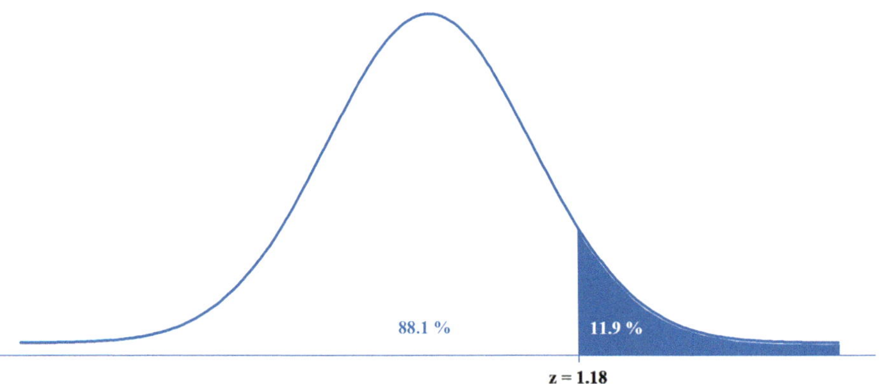

Fig. 10.7 Area under the standard normal distribution and z- value

Since the calculation of the area is is time consuming, clever minds have tabulated the standard normal distribution, its area, and its z- values for us (see Appendix 2). With this table, given a normal distribution of a variable, we can state how likely or unlikely the occurrence of a certain x- value is and (as we will see in the next chapter) reject or not reject hypotheses.

10.3 Normal Distribution, t-Distribution, χ^2-Distribution and (or rather) F-Distribution

Variables are not only normally distributed, but can also be distributed differently. At this point, we therefore want to introduce three more important distributions that we need for testing: the t-distribution, the $\chi 2$-distribution (pronounced Chi-square distribution) and the F-distribution. Basically, we use these distributions just like the standard normal distribution for testing. Here too, the areas are tabulated (see Appendix 3, 4 and 5).

The t-distribution is symmetric like the standard normal distribution with a maximum, an average value of zero and its tails running asymptotically towards the x-axis. However, it is somewhat wider and flatter, how wide and how flat depends on the number of degrees of freedom (df). If we have few degrees of freedom, the t-distribution is flatter and wider, if we have more degrees of freedom, it becomes narrower and sharper. From about 30 degrees of freedom, the t-distribution transitions into the standard normal distribution (compare Fig. 10.8).

We will discuss the meaning of the degrees of freedom later when we need the t-distribution for testing. It is important to note here that we need the t-distribution in practical work, for example, when dealing with small samples.

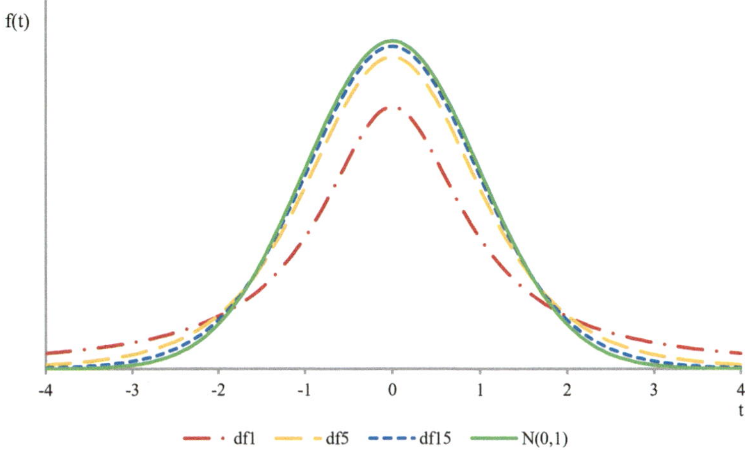

Fig. 10.8 The *t*-distributions compared to the standard normal distribution

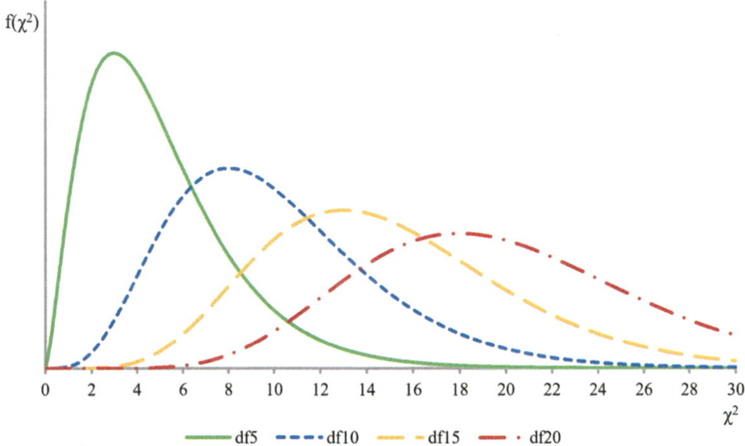

Fig. 10.9 The χ^2-distribution with 5, 10, 15 and 20 degrees of freedom

In addition to the *t*-distribution, the $\chi 2$-distribution represents another important test distribution. We need it especially when testing nominal variables. As with the *t*-distribution, there is not just one $\chi 2$-distribution, but many $\chi 2$-distributions, the shape of which depends on the number of degrees of freedom. Again we will discuss the degrees of freedom when we use the $\chi 2$-distribution for testing. Fig. 10.9 shows the $\chi 2$-distribution with 5, 10, 15 and 20 degrees of freedom.

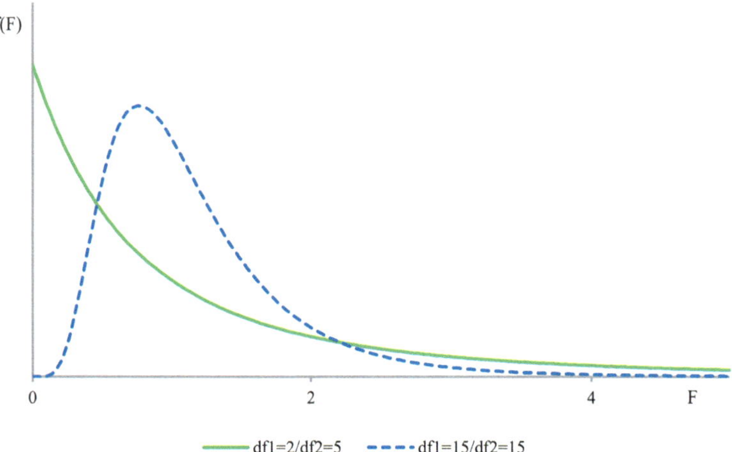

Fig. 10.10 The *F*-distribution for the degrees of freedom df_1 equal to 2 and df_2 equal to 5 as well as for the degrees of freedom df_1 equal to 15 and df_2 equal to 15

The last distribution that we want to briefly look at is the *F*-distribution. We need this, for example, when doing analysis of variance and regression. As already with the *t*-distribution and the $\chi 2$-distribution, there is not just one *F*-distribution, but many *F*-distributions, which depend on the number of degrees of freedom. The difference is only that we have not one value for the degrees of freedom, but two values. We will discuss this later in the analysis of variance and regression analysis. Fig. 10.10 plots the *F*-distribution for the degrees of freedom 2 and 5 as well as 15 and 15.

For our practical work, it is important to understand that these distributions are theoretical distributions that we need for testing. As we will see later, depending on the test procedure, one or the other distribution comes into play. Furthermore, it is important to know that the total area under the curve always amounts to 1 or 100%, and that intervals under the area indicate how likely the occurrence of a certain value of the *x*-axis is, e.g., how likely the occurrence of an *F*-value greater than 4 is. Compare this to Fig. 10.10, here the area and thus the probability is very small.

10.4 Drawing the Distributions with the R Commander

The R Commander offers us the opportunity to draw the discussed distributions. We use the menu Distributions with the submenu Continuous distributions. Afterwards, we can select the distributions. We see that there are not only the discussed distributions.

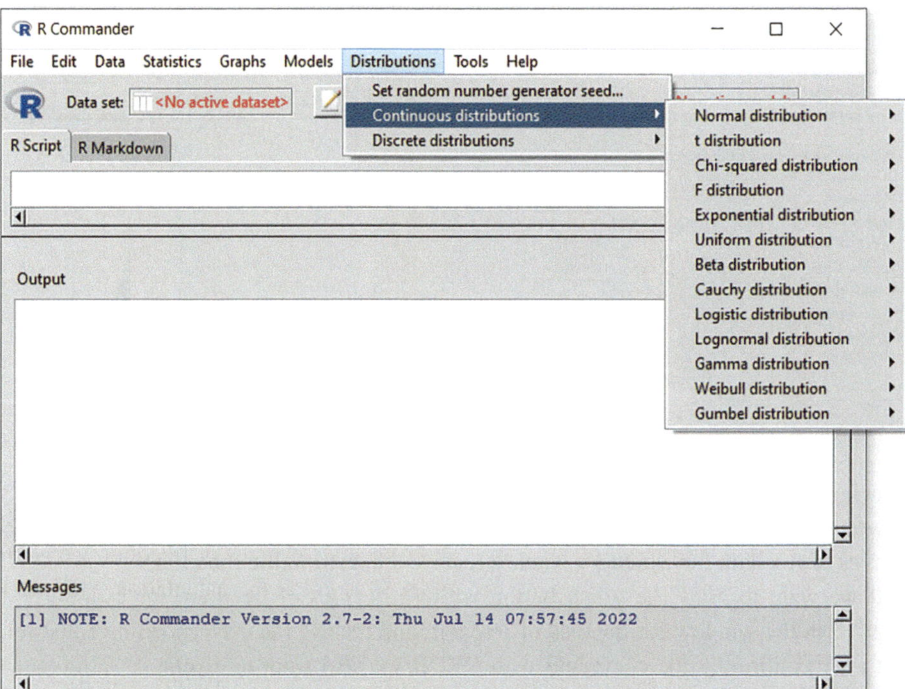

If we want to draw the normal distribution, we select the normal distribution and the command Plot normal distribution. In the window that opens, we specify the average value and the standard deviation, e.g. N(100, 20). We can also define a range that will be displayed. If we specify quantiles, for example, and then enter the range from 0.9 to 1, we get the value range that includes the 10% largest values. When we look at the result, we see that the probability of finding a value greater than 125.63 is 10%. We confirm the command with OK and receive the corresponding normal distribution.

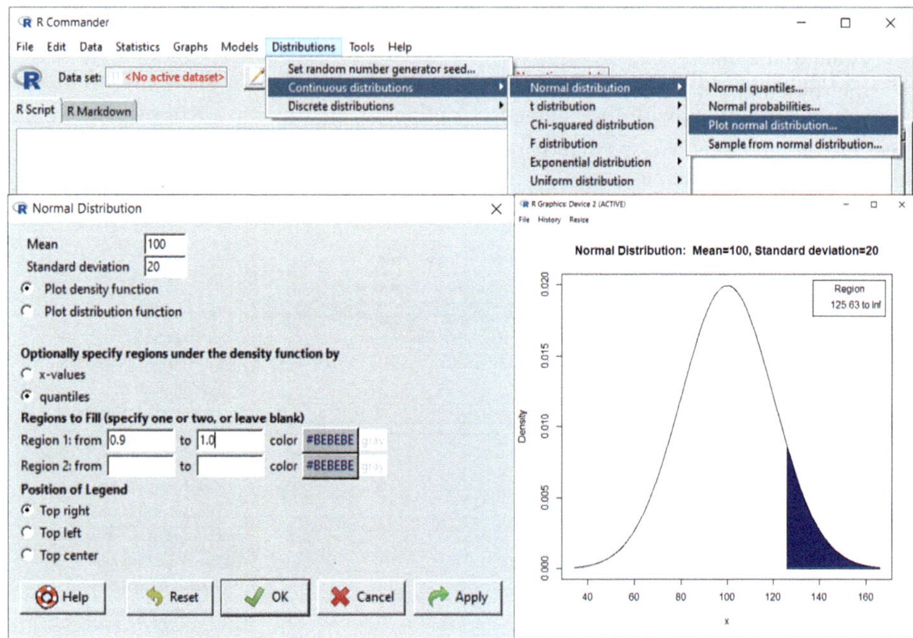

Drawing the t-distribution, the $\chi 2$-distribution, and the F-distribution works similarly. However, instead of the average value and the standard deviation, we need to define the degrees of freedom for which we want to receive the distribution functions.

If we want to draw the t-distribution with 15 degrees of freedom, for example, we enter 15 in the window for degrees of freedom and receive the corresponding t-distribution after confirming the command with OK. If we also want to display a value range, we can specify the quantile we want to display. If we enter the range from 0.95 to 1 as in this example, we get the probability of finding a value greater than 1.75, which is 5% for this input.

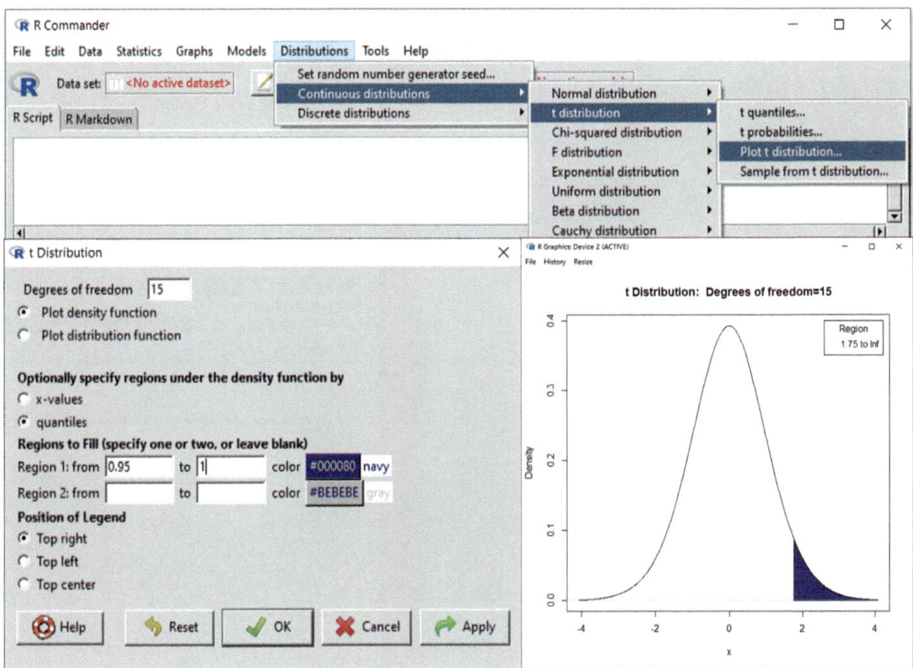

For the χ2-distribution with 10 degrees of freedom, we accordingly enter 10 for degrees of freedom. In this example, we also display the 1% quantile by specifying quantiles and entering the range from 0.99 to 1. 1% of the values are greater than 23.21.

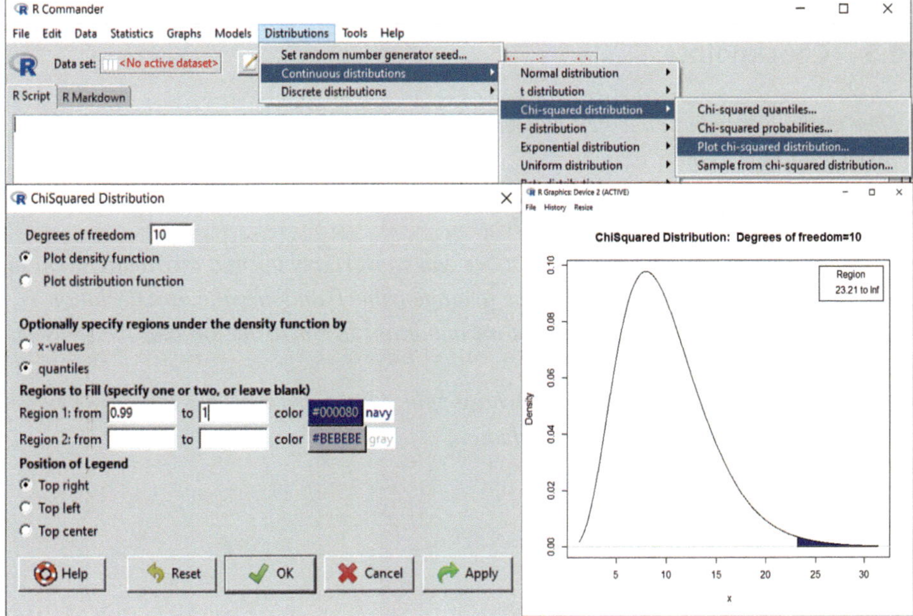

If we want to draw the F-distribution, we need to specify both the degrees of freedom for the numerator and the denominator, e.g. Fg1 = 10 and Fg2 = 15. The specified quantile range shows us that 5% of the values are greater than 2.54.

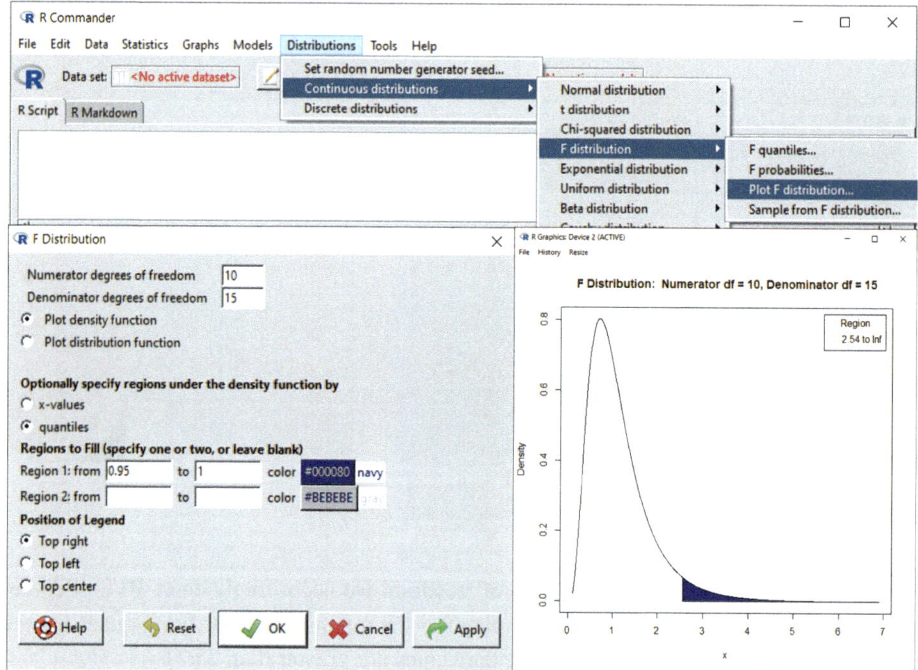

10.5 Checkpoints

- *The normal distribution is bell-shaped, symmetrical with a maximum, and its tails run asymptotically to the x-axis.*
- *The area under the normal distribution indicates how large the probability is of finding a certain value in a certain interval.*
- *With the help of the z-transformation, we can convert any normal distribution into the standard normal distribution with the average value 0 and the standard deviation 1.*
- *The standard normal distribution is an important test distribution and its areas and values are tabulated.*
- *Other important test distributions, whose values are tabulated, are the t-distribution, the $\chi 2$-distribution, and the F-distribution.*

10.6 Application

10.1 Draw the normal distribution curves $N(-20, 15)$, $N(20, 15)$, $N(0, 25)$ using the R Commander, mark the area in the interval from 0 to 20 and compare them.

10.2 Perform the z-transformation to convert the normal distribution $N(20, 15)$ into the standard normal distribution $N(0, 1)$. What are the z-values for the following x-values: $-25, -20, -10, -7, 5, 20, 35, 47, 50, 60, 65$?

10.3 What is the area under the standard normal distribution curve to the right of $z = 0.5, 0.75, 1.0, 1.26$?

10.4 What is the area under the standard normal distribution curve to the left of $z = -0.5, -0.75, -1.0, -1.26$?

10.5 Compare the solutions from task 3 and task 4. What can we observe?

10.6 Assume that the age of our company founders is normally distributed with an average value of 35 and a standard deviation of 7.
- What is the probability that we will find a company founder who is older than 42 years?
- What is the probability that we will find a company founder younger than 21 years?
- What is the probability that we will find a company founder aged between 42 and 49 years?

10.7 Assume that one of our company founders has developed a novel method to cut panel sheets. The precision of the cut width is particularly important. He regularly checks the cut width and finds that the cut width averages 0.1 mm, with a standard deviation of 0.005 mm. What is the probability that the cut width will not meet a customer's required tolerance range of 0.0908 mm to 0.1092 mm?

10.8 We need to be in the top 5% of students in a selection process to receive a scholarship. We know that the point distribution in the test is normally distributed and that on average 80 points are achieved with a standard deviation of 10.0 points. How many points do we need to score to be among the top 5% of students?

11 Hypothesis Testing: What Applies?

We now know what a hypothesis is, we also know about important test distributions. What we do not yet know is how to test hypotheses and generate statements about our population using the test distributions. The aim of this chapter is to learn exactly that. We clarify the term statistical significance. Once we have understood this, we can read scientific studies. We explain the significance level and the error that is allowed in testing. We also point out that statistical significance is not synonymous with practical relevance. Finally, we discuss the steps we must go through when conducting any hypothesis test.

11.1 What does Statistical Significance Mean?

Statistical significance is a central concept for testing hypotheses. With the help of this concept, hypotheses are rejected or not, and statistically verified knowledge is generated. But how does it work? We take a step back and start with our hypotheses. As we already know, we formulate a research hypothesis from our question using existing studies and theories, and from this our testable null hypothesis H_0 and our alternative hypothesis H_A.

Let's revisit the example of the age of our company founders:

Research question: How old are the company founders on average?

From this, we have derived the following null and alternative hypothesis using existing studies and theories:

H_0: *Company founders are on average 40 years old.*
H_A: *Company founders are on average younger or older than 40 years.*

F. Kronthaler, *Statistics Applied with the R Commander*,
https://doi.org/10.1007/978-3-662-69107-6_11

We can formulate the null and alternative hypothesis a little shorter, in mathematical terms:

$$H_0\colon \ \mu = 40$$
$$H_A\colon \ \mu \neq 40$$

Let's take a closer look at the short version. We described H_0 with $\mu = 40$, which is the short version of the following statement: Company founders are on average 40 years old in the population. μ stands for the average value of the population. For H_A we wrote $\mu \neq 40$. In detail, this means that the company founders are not 40 years old in the population.

We test our hypotheses with the average value of the sample $\bar{x}$. If our sample result $\bar{x}$ deviates much from the value we assumed for the null hypothesis and exceeds a critical value, we assume that our statement about the population is not correct. We reject this and then proceed with the alternative hypothesis.

But what is a sufficient large deviation or a critical value? We best illustrate this with a diagram of the concept of statistical significance using the one-sided hypothesis test. We use this when we have directional hypotheses (see Fig. 11.1 and Chap. 9).

In the figure, we see the distribution of our possible sample means, in short, the sample distribution. We already know that there is not just one sample, but many possible samples and thus many possible sample means. On the x-axis, we have plotted the possible sample means. On the y-axis, we see how often certain sample means occur. Furthermore, we already know that the expected value of the sample is equal to the average

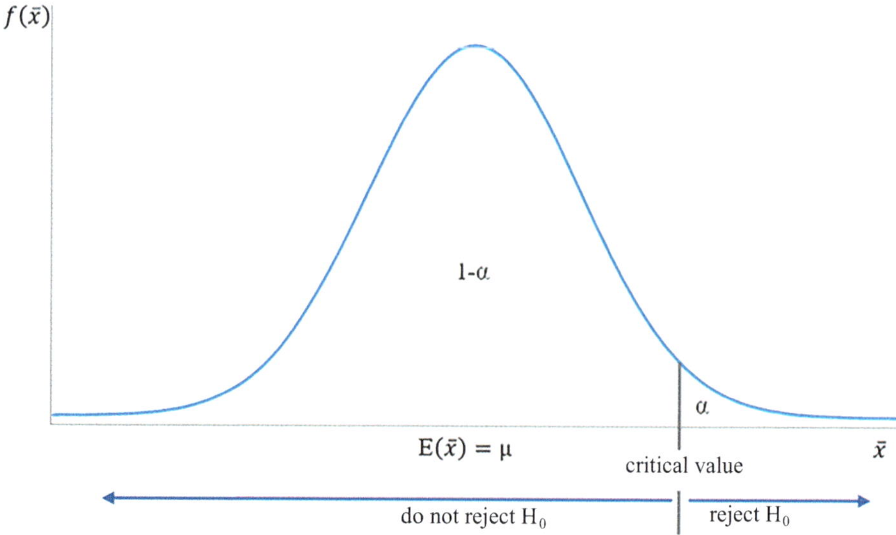

Fig. 11.1 The concept of statistical significance in the one-sided hypothesis test (rejection area on the right)

value of the population: $E(\bar{x}) = \mu$ (compare this to Chap. 8). We can see from the figure that values close to the expected value have a high probability of occurring, given our null hypothesis is correct. Values far away from the expected value have a very small probability. If the probability is too small, i.e., if our sample mean exceeds a critical size, then we assume that our null hypothesis cannot be correct and we reject it. α is the area of the sample distribution in the rejection area, i.e., the probability that a value is found that is greater than the critical value.

Of course, it can also be that the rejection area is on the left side and that we reject the null hypothesis when a critical value is undershot (compare Fig. 11.2).

For non-directional hypotheses, we have to test on both sides and place our rejection area on both sides, as can be seen in the following figure (Fig. 11.3).

We divide our area α into two halves, thus having $\alpha/2$ on the left and right respectively, and thus a critical value on the left that we can undershoot and a critical value on the right that we can exceed. At this point, it makes sense to take a closer look at the area α .

11.2 The Significance Level α

We have already said that α is the area under the sample distribution in the rejection area and represents the probability of finding in absolute terms a value larger than the critical value. But how small must the probability be for us to reject our null hypothesis?

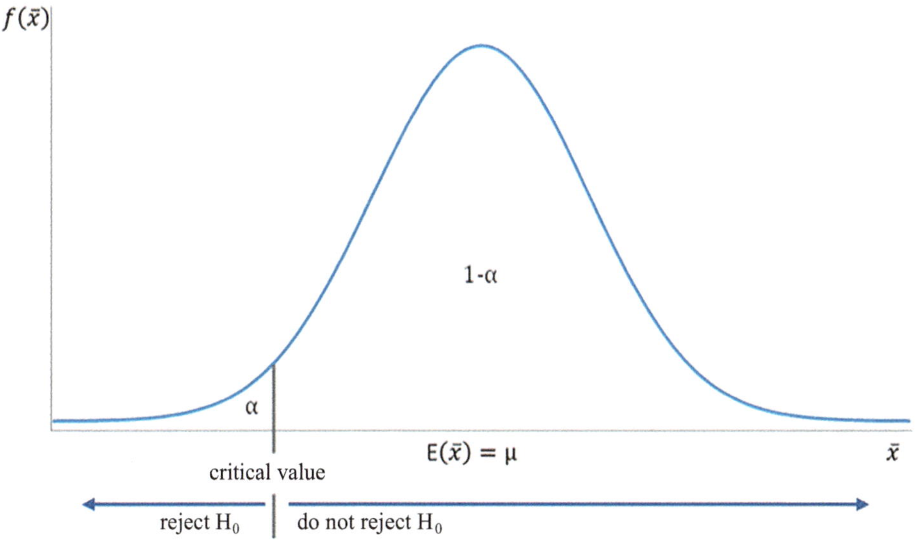

Fig. 11.2 The concept of statistical significance in one-sided hypothesis testing (rejection area on the left)

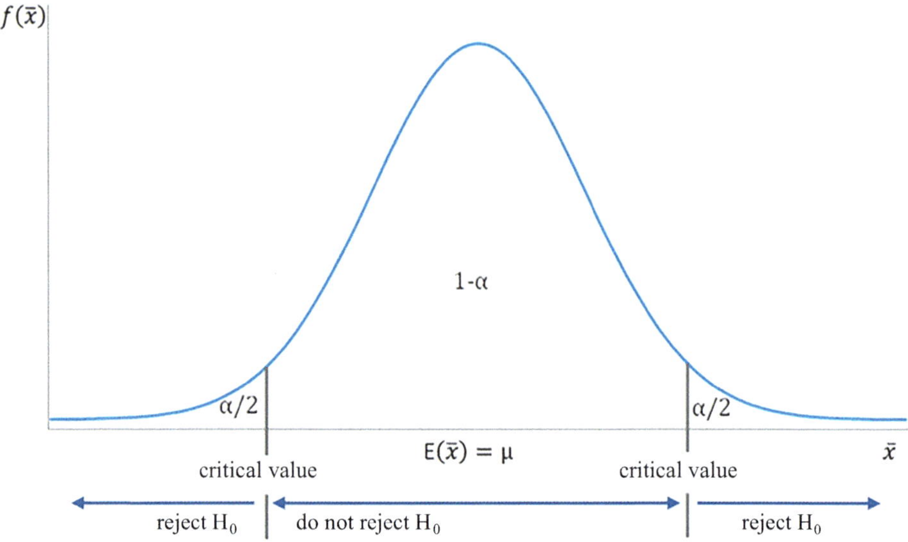

Fig. 11.3 The concept of statistical significance in two-sided hypothesis testing

In science, three different-sized areas have become established: $\alpha = 10\ \%\ or\ \alpha = 0.1$, $\alpha = 5\ \%\ or\ \alpha = 0.05$ and $\alpha = 1\ \%\ or\ \alpha = 0.01$. This means that we reject the null hypothesis if the probability of finding a certain value is less than 10%, 5%, or 1%. Instead of 10%, 5%, or 1%, we can also write 0.1, 0.05, or 0.01, we then do not use percentage values, but the decimal notation. We refer to these areas as the significance level α. If our sample yields a value whose probability is less than 10%, 5%, or 1%, then we say that we reject our null hypothesis at a significance level of 10%, 5%, or 1%. The result is statistically significant. In other words, the probability of finding a sample value of this magnitude is less than 10%, 5%, or 1% if the null hypothesis is correct. α is thus the probability at which we reject our null hypothesis.

α is more than that. α is also the error probability with which we falsely reject the null hypothesis. Let's look again at Fig. 11.3. On the x-axis, all possible sample mean values are plotted under the assumption that our null hypothesis is correct. Some are more likely, some are less likely. Values to the left and right are unlikely, but not impossible. However, if we find a value far to the left or far to the right, we reject the null hypothesis, knowing that the value is possible but unlikely given the null hypothesis is correct. This means that there is a possibility that we make a mistake in our test decision. This error is our chosen significance level α.

This possible error has a big influence on how large we choose our α. If the consequences of a wrong rejection are severe, if life or death depends on it, then we choose a small α, α equal to 1% or even smaller, e.g. α equal to 0.1%. If the consequences are less severe, we allow a larger α, perhaps α equal to 5% or 10%. Let's illustrate this with an example. A bridge builder tells us that his bridge over a 100 m deep canyon is 90% safe to cross, or that the probability of it collapsing is only 10%. Do we cross this bridge? Probably not, the probability of the bridge collapsing would have to be significantly lower. Let's take a second example. In the weather forecast, we learn that there is a 5% chance of rain tomorrow. Will we leave the umbrella at home? Very likely, if it really rains, we will get wet, but nothing else will happen. It is important at this point that we consider the consequences of a wrong rejection of the null hypothesis in advance of the hypothesis test and choose our α according to these considerations.

But why don't we simply always choose our α to be very small? There are two reasons for this. First, the rejection of the null hypothesis becomes increasingly unlikely the smaller our α becomes. Second, we can not only make the mistake of falsely rejecting our null hypothesis, but it could also happen that we do not reject the null hypothesis even though it is false. This is the second possible error in hypothesis testing, the β error. Both errors are shown in Table 11.1.

This table is so important that we want to discuss it in more detail. The first row indicates whether our null hypothesis actually reflects the state of the population, that is, whether H_0 is true or not. We do not know this. Our null hypothesis is merely a conjecture about the behavior in the population. Accordingly, our null hypothesis can be both correct and incorrect. The second and third rows show the result of the hypothesis test. Let's first look at the second row. The result of the hypothesis test is that we reject our null hypothesis. This is correct if our null hypothesis is indeed false. However, if our null hypothesis correctly reflects the behavior in the population, we have made a wrong test decision, the α error. We have falsely rejected the null hypothesis. Now let's look at the third row. The sample result leads us to not reject our null hypothesis. This is correct if our null hypothesis is correct. However, if our null hypothesis does not correctly reflect the behavior in the population, we have made a mistake, the so-called β error.

It is now important that the α error and the β error are not independent of each other. If we reduce the α error, we increase the β error and vice versa. In Fig. 11.4, this relationship is graphically represented. If we conceptually reduce the area of the α error, the area of the β error increases and vice versa.

For practical work, it is also important that the β error is relatively difficult to determine. To determine the β error, we need knowledge about how to correctly formulate the null hypothesis so that it represents the population. However, we do not have this knowledge, otherwise we would not have formulated a false null hypothesis. For this reason, in practice, we usually work with the α error.

Table 11.1 α error and β error in hypothesis testing

	H_0 is true	H_0 is not true
H_0 is rejected	α-error	correct decision
H_0 is not rejected	correct decision	β-error

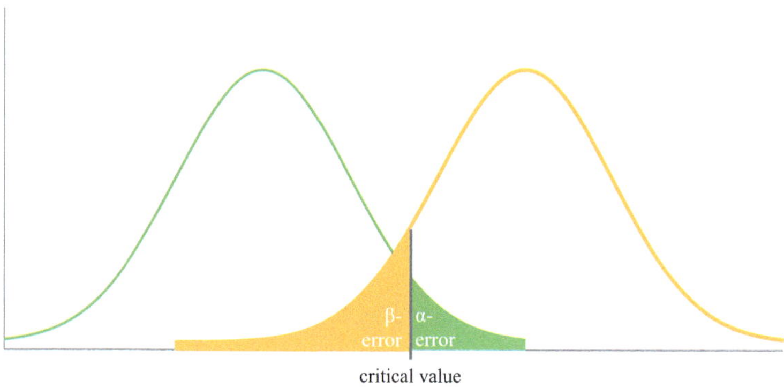

Fig. 11.4 Relationship between α-error and β-error

From what has been said, we can draw an important insight. When we perform the hypothesis test and reject or do not reject our null hypothesis, there is a probability that we have made a mistake, even if we have worked according to all the rules of the art. This implies that a test decision is always only an indication of the behavior of the population, but never a real proof.

11.3 Statistically Significant, but also Practically Relevant?

We have seen how we come to the decision whether a result is statistically significant and what this means. We can say with relatively high certainty that a result is not random. But is such a statistically significant result always practically relevant or practically significant? This can be clearly denied. Statistically significant does not automatically mean practically relevant. The reason is that the sample size has an influence on whether a result becomes statistically significant. If the sample is large enough, every result will be statistically significant, even very small differences, which may be completely irrelevant.

We want to illustrate this with a completely fictitious example. Let's imagine we are interested in beer consumption at the Oktoberfest in Munich (or at another folk festival). Specifically, we analyze whether there is a difference in beer consumption between female and male festival visitors. Let's assume we collect data and find that during a two-hour festival tent visit, men drink an average of 1.98 l of beer (almost two liters of beer) and women 1.94 l (also almost two liters of beer). If the sample is large enough, this result will be statistically significant. But is this result also practically relevant? Probably not, a difference of 0.04 liters is irrelevant at these quantities, it is in any case too much beer.

We see that a statistically significant result must always be analyzed for its practical significance.

We can measure this using the absolute numbers and effect size measures. Effect size measures are standardized numbers that provide information about whether a result has a strong the effect. A major advantage of calculating the effect size is that due to the standardization, results of different studies can be compared with each other. A major advantage of absolute numbers is that we do not loose the original unit, e.g. liter.

Commonly used measures for effect size are the correlation coefficient of Bravais-Pearson r and *Cohen's d*. We will see the calculation in the next chapters when we perform the test procedures. But here a note on the interpretation of the effect size should be given. It is generally accepted that the size of the effect size r can be interpreted as follows:

An effect size of $r = 0.1$ indicates a small effect,
an effect size of $r = 0.3$ is a medium effect,
an effect size of $r = 0.5$ and larger measures a large effect.

In summary, it can be said that in addition to statistical significance, the practical relevance should be analyzed using the effect size and the absolute numbers.

11.4 Steps in Performing the Hypothesis Test

Before we now turn to the hypothesis test, let's briefly go through the steps that need to be practically carried out. The hypothesis test consists of the following five steps:

1. Formulation of the null hypothesis, the alternative hypothesis, and the significance level
2. Specify the test distribution and the test statistic
3. Determine the rejection area and the critical value
4. Calculation of the test value
5. Decision and interpretation

1. *Formulation of the null hypothesis, the alternative hypothesis, and the significance level* The first step in the hypothesis test always involves formulating the research hypothesis from the research question, and from this the null hypothesis and the alternative hypothesis. It is important, and cannot be emphasized enough, that the hypotheses cannot be formulated off the cuff, but must reflect the knowledge, theories, and studies available on the subject. In addition, we must determine at what significance level we reject the null hypothesis. The choice of the significance level depends on the consequences of a false rejection of the null hypothesis.
2. *Specify the test distribution and the test statistic* We had learned about various test distributions in Chap. 10. It is important that we have different test situations depending on the null hypothesis, which are associated with different test distributions and test statistics. Once we have discussed the knowledge about the different

test situations, we can automatically associate the null hypothesis with a certain test distribution and test statistic.

3. *Determine rejection area and the critical value* In Chap. 10 we saw that the area under the test distributions indicates how likely the occurrence of a certain value is. After we have set the significance level and know the test distribution (standard normal distribution, *t* distribution, $\chi 2$ distribution or *F* distribution), we are able to determine the critical value and thus the rejection area.
4. *Calculation of the test value* The next step is the calculation of the test statistic and thus the test value. For this, we e.g. compare the sample value found with the value expected under the null hypothesis. As already mentioned the test statistic depends on the test situation.
5. *Decision and interpretation* The last step involves the test decision and interpretation. For this, we compare the test value with the critical value. If our test value falls into the rejection area, we reject the null hypothesis at our significance level and we continue with the alternative hypothesis. It is important here that our test decision is not a proof of the null hypothesis or the alternative hypothesis. However, we get a good indication of how the population is likely to behave.

11.5 How Do I Choose My Test Procedure?

Above we already mentioned that there are different test situations. These depend on what is to be tested and what scale level the data have. In Fig. 11.5, important test situations are shown and it is depicted which test procedure comes into play in which situation.

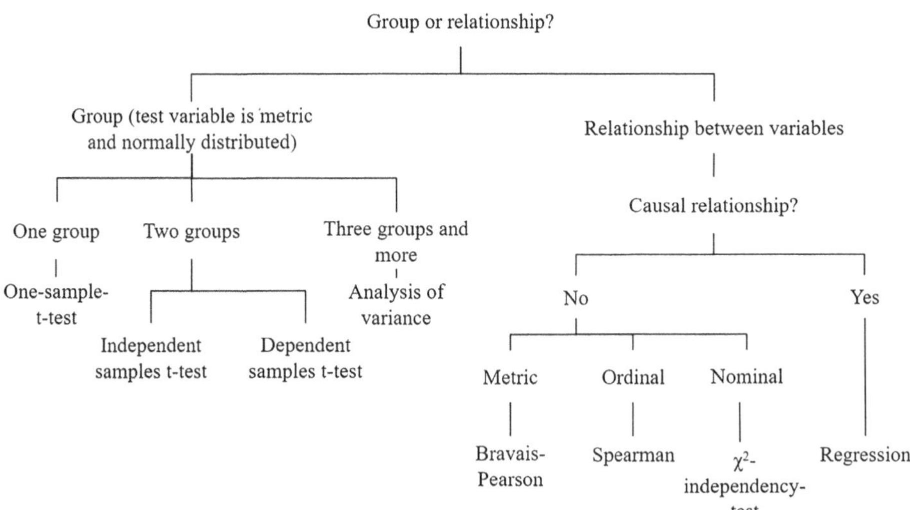

Fig. 11.5 Selection of a test procedure

It is important to first determine whether we are examining groups or a relationship between variables. In the former case, we go along the left axis and come to the one-sample test, the test for independent samples, and the test for dependent samples. If we are investigating a relationship between variables, we go along the right axis and, depending on the data level, come to the various correlation coefficients or to regression. We are now ready and can start. In the following chapter, we start with our first test situation, the one-sample t-test.

11.6 Checkpoints

- *The significance level α is the probability of finding a value greater than a critical value and thus the probability or certainty with which the null hypothesis is rejected.*
- *α is usually chosen to be 10%, 5% or 1%. Alternatively, $\alpha = 0.1$, $\alpha = 0.05$ and $\alpha = 0.01$ are written.*
- *The choice of α depends on what happens if we falsely reject the null hypothesis. If the consequences are severe, we choose a smaller α. If the consequences are less severe, we choose a larger α.*
- *In hypothesis testing, two errors are possible. The α-error is the probability of rejecting the null hypothesis even though it is true. The β-error is the probability of not rejecting the null hypothesis even though it is false.*
- *α-error and β-error are interconnected. If one error is reduced, the other increases and vice versa.*
- *A statistically significant result must always be analyzed for its practical significance.*
- *Hypothesis testing involves five steps: 1) formulating the null hypothesis, the alternative hypothesis, and the significance level, 2) specifying the test distribution and the test statistic, 3) determining the critical value and the rejection region, 4) calculating the test value, and 5) making the decision and interpretation.*
- *The choice of test procedure depends on the test situation and the scale level of the data.*

11.7 Application

11.1 What are the two meanings of the significance level α?

11.2 Which of the following statements is correct:

- Exceeding the critical value leads to the rejection of the null hypothesis.
- It is possible to set the α error to zero.
- The smaller the α error, the better the result.
- The selection of the α error depends on the consequences of a false rejection of the null hypothesis.

11.3 When is testing done on both sides, left side, or right side?

11.4 Formulate both the null hypothesis and the alternative hypothesis for the two-sided test, the left-sided test, and the right-sided test for the following research question: How old are company founders?

11.5 What is the relationship between the α error and the β error?

11.6 Go to the library and find a scholarly article of interest that generates knowledge about the population through a sample. Explain how the author has generated knowledge about the population using the concept of statistical significance.

Part IV
Method for Testing Hypotheses

Time for the Application of the Hypothesis Test

Now we start with the hypothesis test and generate information about the entire population. If our information about the 100 newly founded companies were data from a real sample, we would now generate information about all newly founded companies through this data.

We begin with the test for a group mean and the tests for difference of means for metric data. These test procedures assume that the test variable is metric and normally distributed in the entire population. The assumption of normal distribution is particularly important for small samples. For large samples, the procedures are relatively robust against a violation of this assumption. Subsequently, we discuss the analysis of variance, which can be considered as an extension of the test for difference of means for multiple groups, test procedures for correlation in metric, ordinal, and nominal variables. Finally, we address additional test procedures for nominal data.

12 The Test for a Group Mean or One Sample t-Test

12.1 Introduction to the Test for a Group Mean

Let's recall, in Chap. 3 we learned and calculated the arithmetic mean. For example, we found out for our companies that they have grown by an average of 7.1% over the last 5 years, that they spend an average of 19.81% of their revenue on marketing and an average of 4.65% on innovation. Furthermore, we know that the company founders are on average 34.25 years old and that they have an industry professional experience of 7.42 years. We have learned that this statement only applies to the sample, not to the population. But we are not satisfied with that. We want to generate a statement about the population using the hypothesis test. Specifically, with the test for a group mean, we analyze whether our assumption about the mean in the population is compatible with the sample finding.

12.2 The Research Question and Hypotheses in the Test for a Group Mean: Are Company Founders on Average 40 Years Old?

In order to test our assumption about the mean in the population, we first need to formulate this assumption. As shown above, we specify the research question and formulate our hypothesis with the help of existing studies and theories. Let's take the example of the age of company founders. The research question is: How old are company founders on average? We generate our null hypothesis and our alternative hypothesis through theories and studies:

F. Kronthaler, *Statistics Applied with the R Commander*,
https://doi.org/10.1007/978-3-662-69107-6_12

H_0: *Company founders are on average 40 years old.*
H_A: *Company founders are on average younger or older than 40 years.*

or

$$H_0: \quad \mu = 40$$
$$H_A: \quad \mu \neq 40$$

In addition, we set the significance level at which we reject our null hypothesis. A false rejection of the null hypothesis does not seem to be a matter of life or death. At the same time, we want to be reasonably sure. Accordingly, we set the significance level at 5%:

$$\alpha = 5\% \quad \text{or} \quad \alpha = 0.05$$

Just a reminder. A significance level of 5% means that we reject the null hypothesis if we receive a sample finding that has a probability of less than 5% when the null hypothesis is correct.

12.3 The Test Distribution and Test Statistic in the Test for a Group Mean

With this, we can deal with the second step in the hypothesis test. As we have already heard, every test is associated with a certain test distribution and test statistic. In the test for a group mean with a large sample $n > 30$ and known standard deviation in the population one chooses the standard normal distribution as the test distribution. The test statistic compares the sample mean with the average value that we have assumed under the null hypothesis:

$$z = \frac{\bar{x} - \mu}{\sigma_{\bar{x}}}$$

with

z is the z-value of the standard normal distribution,
$\bar{x}$ is the sample mean,
μ is the mean of the population assumed under the null hypothesis,
$\sigma_{\bar{x}}$ is the standard deviation of the sample distribution with $\sigma_{\bar{x}} = \frac{\sigma}{\sqrt{n}}$ is, where
σ is equal to the standard deviation in the population and
n is equal to the number of observations.

12.4 The Critical Value in the Test for a Group Mean

We determine the critical value and thus the rejection area with the help of the standard normal distribution table. For this, we need to know whether we are testing one-sided or two-sided and how large our significance level is. If we test one-sided, we place the significance level either on the left or on the right side. If we test two-sided, we divide the significance level on both sides. In our example, we test on both sides, as we have an undirectional null hypothesis, with a significance level of 5%. This means that we have in the rejection area both on the left and on the right a n area of 2.5% with the critical values −1.96 and 1.96 (compare Appendix 2).

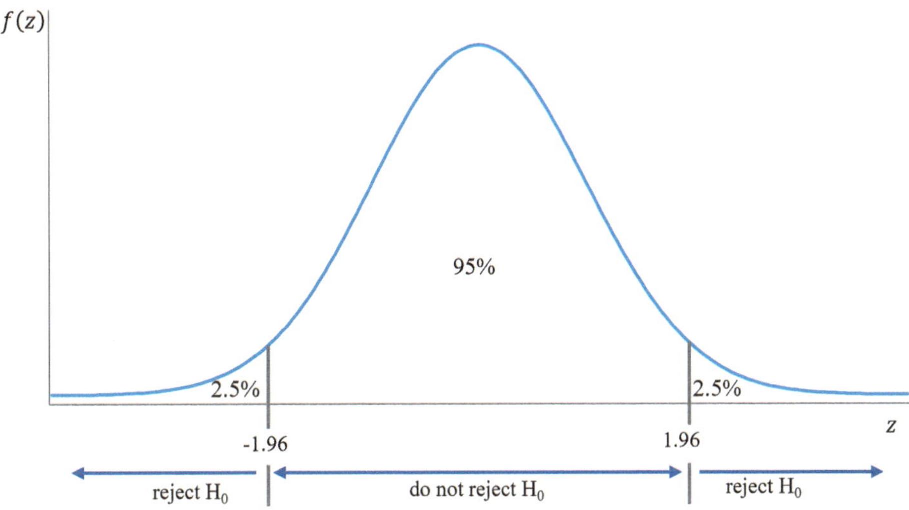

12.5 The z-Value

Now we just need to calculate our z value. We do this with the help of the test statistic. For this, we use the mean of our sample, the mean from the null hypothesis, and the standard deviation of the sample distribution. As we have already calculated earlier, the mean of the sample is $\bar{x} = 34.25$ years. For the mean of the null hypothesis, we use the value assumed under H_0 of $\mu = 40$ years. To determine the standard deviation of the sample distribution, we still need the standard deviation in the population. We know that this is 8 years and accordingly we get the following value for the standard deviation of the sample distribution: $\sigma_{\bar{x}} = \frac{\sigma}{\sqrt{n}} = \frac{0.8}{\sqrt{100}} = 0.80$ years. Our z value is thus:

$$z = \frac{\bar{x} - \mu}{\sigma_{\bar{x}}} = \frac{34.25 - 40}{0.80} = -7.19$$

12.6 The Decision

Now all that remains is the decision. For this, we compare our *z* value with the critical values. We see that our *z* value is to the left of the critical value -1.96, thus in the rejection area. We reject the null hypothesis and proceed with the interpretation. We can now say that the company founders in the population are not equal to 40 years old. Our sample mean also suggests that the company founders are on average younger than 40 years old. We could now calculate the confidence interval, as we did in Chap. 8. If we determine the 95% confidence interval, for example, it shows that the true value is between 32.68 and 35.82 years with a probability of 95%.

12.7 The Test for a Group Mean With Unknown Standard Deviation in the Population or With a Small Sample n ≤ 30

We generally have the problem that we do not know the standard deviation in the population. In this case, we replace in the formula of the standard deviation of the sample distribution the standard deviation of the population σ with the standard deviation of the sample *sd*. Thus, the standard deviation of the sample distribution is calculated as follows:

$$\hat{\sigma}_{\bar{x}} = \frac{sd}{\sqrt{n}}$$

with

$\hat{\sigma}_{\bar{x}}$ is equal to the estimated standard deviation of the sample distribution,
sd is equal to the standard deviation of the sample and
n is equal to the number of observations.

The test distribution is then no longer the standard normal distribution, but the *t*-distribution with $df = n - 1$ degrees of freedom. The same is the case when we have a small sample. We speak of a small sample in the test for a group mean when the number of observations n is less than or equal to 30 ($n \leq 30$). The critical values are taken in both cases not from the standard normal distribution table, but from the *t*-distribution table (Appendix 3). Consequently, the test value is *t*-distributed and the test statistic looks as follows:

$$t = \frac{\bar{x} - \mu}{\hat{\sigma}_{\bar{x}}}$$

Apart from that, the procedure is identical. We want to demonstrate this using the example of a small sample. We assume that we only have 20 observations (n = 20) and take the average value of the sample and the standard deviation from our data set. The values are $\bar{x} = 31.55$ and $sd = 7.96$. In addition, we want to test one-sided. To test one-sided, we have to reformulate the hypotheses. The null hypothesis and the alternative hypothesis should be as follows.

H_0: *Company founders are on average older or equal to 40 years old.*
H_A: *Company founders are on average younger than 40 years old.*

or

$$H_0: \quad \mu \geq 40$$
$$H_A: \quad \mu < 40$$

We leave the significance level at $\alpha = 5\ \%$ or $\alpha = 0.05$.

The test distribution is t-distributed with $df = n - 1 = 20 - 1 = 19$ degrees of freedom and the test statistic is:

$$t = \frac{\bar{x} - \mu}{\hat{\sigma}_{\bar{x}}}$$

The rejection area at the significance level of 5% is on the left side. We reject the null hypothesis if we fall below a certain value. We take the critical value from the t-distribution table, it is at -1.729 (see Appendix 3).

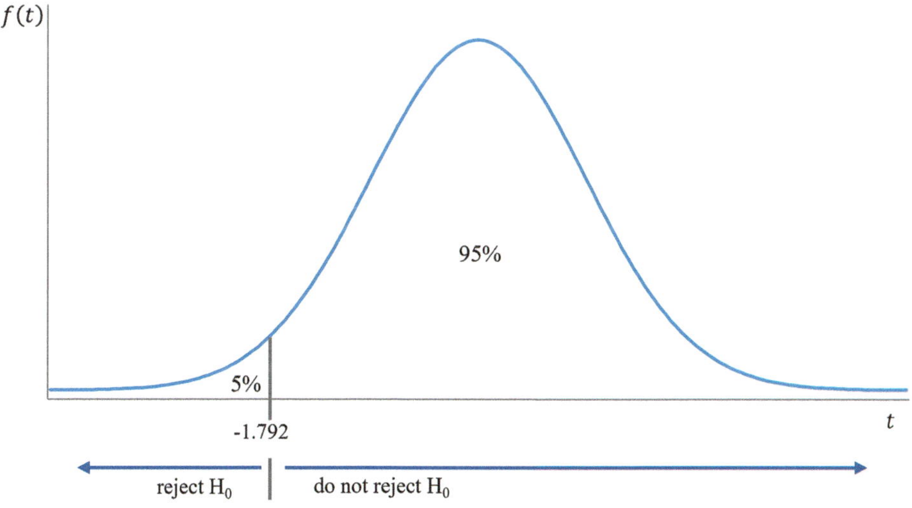

The test statistic is calculated as follows:

$$t = \frac{\bar{x} - \mu}{\hat{\sigma}_{\bar{x}}} = \frac{31.55 - 40}{1.78} = -4.75$$

Now, if we compare our calculated t-value with the critical value, we see that we fall below it and that we are in the rejection area. This means we reject the null hypothesis and continue with the alternative hypothesis. Therefore, our company founders are on average younger than 40 years.

Freak Knowledge

We have just seen that with a small sample, the test value is t-distributed and the test distribution is the t-distribution. In Chap. 10 we saw that with a large number of observations, the t-distribution converges to the standard normal distribution. Therefore, in statistical programs, the t-distribution is usually used as the test distribution and the test for a group mean is often also referred to as a one sample t-test.

12.8 The Effect Size in the Test for a Group Mean

Before we now turn to the final interpretation, we want to calculate the effect size. The measure usually used in the test for a group mean is *Cohen's d. Cohen's d* is calculated in the test for a group mean as follows:

$$Cohen's\ d = \frac{\bar{x} - \mu}{sd}$$

with

$\bar{x}$ is the arithmetic mean of the sample,
μ is the assumed mean of the population under the null hypothesis and
sd is the standard deviation of the sample.

Looking at the formula, the only difference to the test statistic is that we do not divide by the standard deviation of the sample distribution (standard error), but by the standard deviation of the sample. What does this mean? When we divide by the standard deviation, we consider the deviation of the sample mean from the assumed mean of the population in relation to the standard deviation. A Cohen's d of 1 thus means that the deviation is 1 standard deviation. The literature suggests interpreting Cohen's d in absolute terms as follows:

- small effect: Cohen's d is in the range between 0.2 to <0.5,
- medium effect: Cohen's d is in the range between 0.5 to <0.8,
- large effect: Cohen's d is ≥ 0.8.

It should be noted that Cohen's d can also take negative values, namely when the arithmetic average value is smaller than the assumed average value in the population.

The calculation of the effect size for our example thus results in a large effect:

$$Cohen's\ d = \frac{\bar{x} - \mu}{sd} = \frac{31.55 - 40}{7.96} = -1.06$$

With this, we can make the final interpretation roughly as follows. The test shows that the sample mean is statistically significantly different from the assumed mean in the null hypothesis. We reject the null hypothesis at the 5% significance level and assume the alternative hypothesis. The effect is large and in absolute terms more than eight years. We could now also specify the confidence interval, which would be even better.

Freak Knowledge

The discussed test for a group mean assumes that the data are metric and normally distributed. But what do we do if the data are metric but not normally distributed or if our data are ordinal. In such a case, the Wilcoxon test is available to us. We resort to this especially when we have a small sample and metric data that are not normally distributed. The problem of testing a test for a group average with ordinal data (we take the median) is rather rare.

12.9 Calculation with the R Commander

To calculate the one sample t-test, we use the Statistics menu in the R Commander. In the submenu Means, we select the command Single-sample t-test. In the window that opens, we select the variable for which we want to perform the test. We also specify under alternative hypothesis whether we are testing two-sided, left-sided or right-sided and under null hypothesis we state the assumption formulated in the null hypothesis. Furthermore, we could also define the confidence interval. By default, the 95% confidence interval is defined.

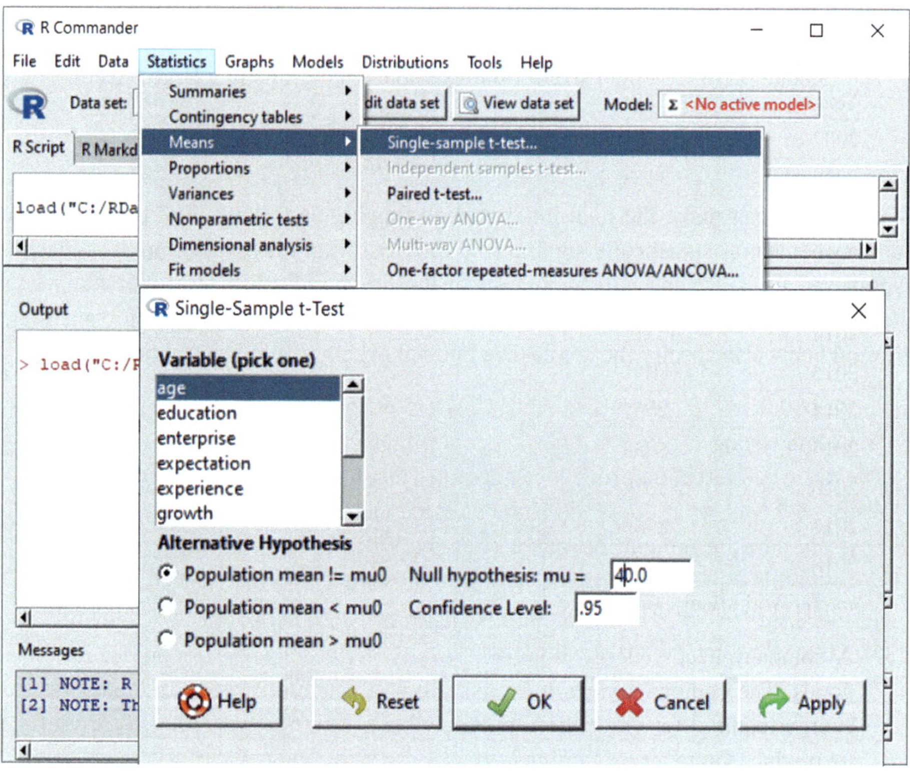

We confirm the command with OK and receive the result of the one sample t-test.

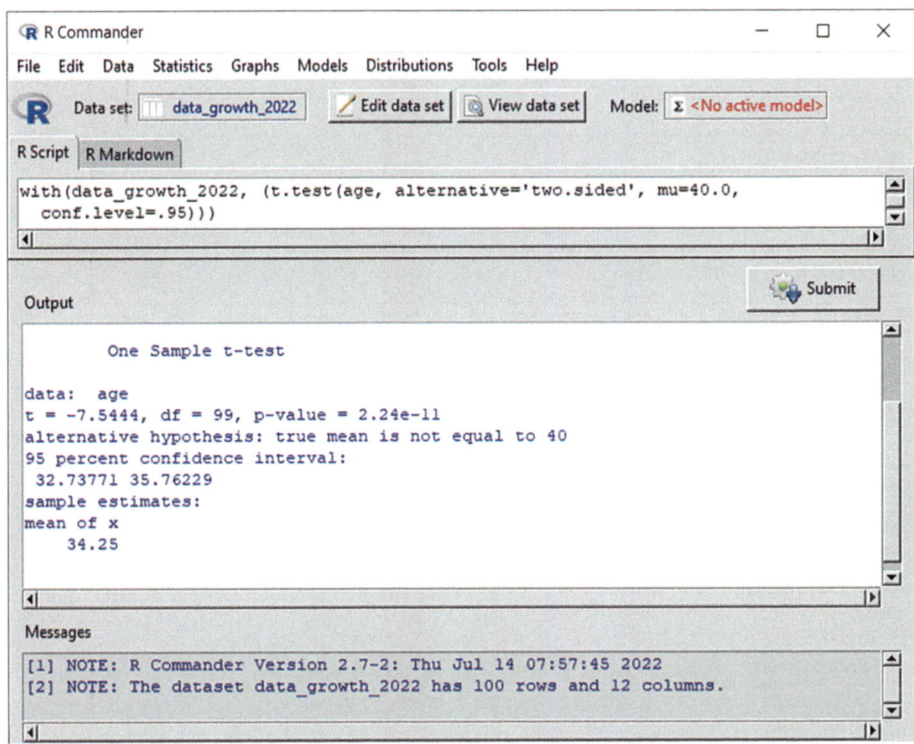

We can interpret the result as follows. In the first line of the result, we see that we have performed the one sample t-test. The test variable is listed in the second line. Next, we get the calculated t-value, the number of degrees of freedom and the p-value. The t-value is the value calculated using the test statistic, which is −7.54. We have 100 observations, i.e., 99 degrees of freedom. Behind this, we get the p-value, p for probability, the probability that the null hypothesis is correct. In our case, the probability is 0.00000000224%. Here are two notes: e-11 means that the number starts at the eleventh place after the decimal point. Furthermore, the probability is given in decimal notation, to get to percentages, we have to multiply by 100. The probability that the null hypothesis is correct is very small. It follows that we reject the null hypothesis at any specified significance level (e.g., $\alpha = 0.05$ or 5%). We therefore proceed with the alternative hypothesis stated in the fourth line. The average value in the total population is not equal to 40 years. In the lines below, we also get the mean and the confidence interval from the sample.

Alternatively, we can also enter the command directly in the command line. The command is t.test(data_growth_2022$age, alternative = 'two.sided', mu = 40.0, conf.level = .95).

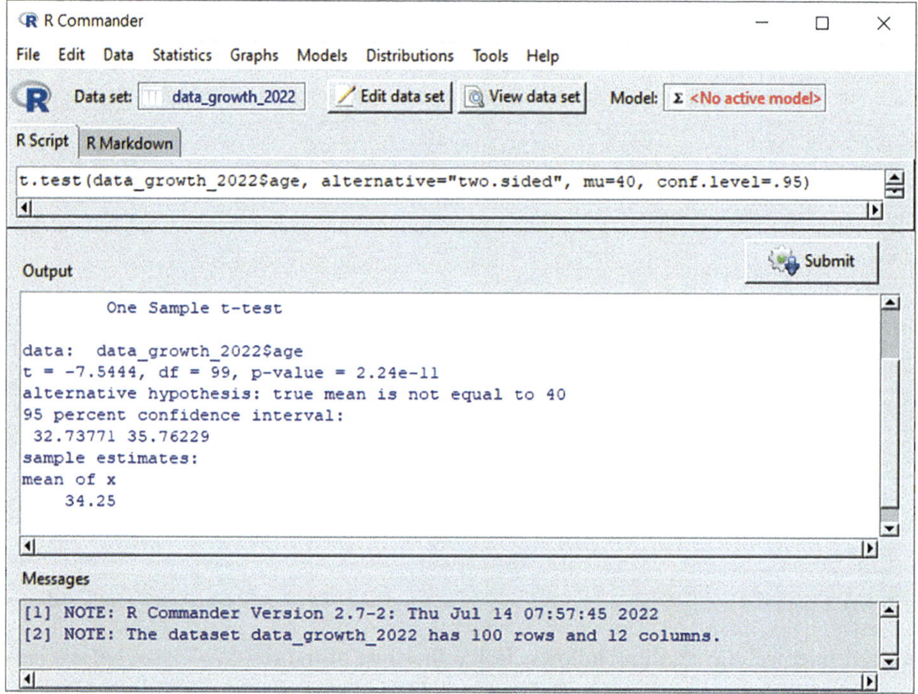

12.10 Checkpoints

- *In the test for a group mean, we analyze whether the assumed mean for the population is in line with our sample findings.*
- *With a large sample and knowledge about the standard deviation in the population, the test distribution is the standard normal distribution and our test value is standard normally distributed. We take the critical value from the standard normal distribution table accordingly.*
- *With a small sample or unknown the standard deviation in the population, the test distribution is the t-distribution and our test value is t-distributed. We take the critical value from the t-distribution table accordingly.*
- *Cohen's d is a commonly used measure for calculating the effect size.*
- *The test for a group mean is often also referred to as one sample t-test.*

12.11 Application

12.1. We are interested in the industry professional experience that company founders have on average. Based on preliminary considerations and studies read, we suspect that they have typically accumulated ten years of industry professional experience before founding a company. We therefore formulate the following null hypothesis: The average industry professional experience of founders is 10 years. Test the null hypothesis at the 10% significance level, taking into account all relevant steps.

12.2. One of our companies has grown by an average of 5% in recent years. We are now interested in whether the company has grown below or above average compared to the population. To test this, we formulate the following null hypothesis: The average growth of start-ups in the population in recent years was 5%. Test the null hypothesis at the 5% significance level, taking into account all relevant steps.

12.3. Let's assume that our sample only includes the first 25 companies in our dataset. Repeat task 2 for this problem.

12.4. We are interested in whether the proportion that companies spend on marketing has decreased in recent years. We know that ten years ago the effort was on average 15%. To test this, we formulate the null hypothesis: The average effort for marketing in the population of start-ups is greater than or equal to 15%. Test the null hypothesis at the 10% significance level, taking into account all relevant steps with the R Commander.

12.5. We know that ten years ago, start-ups spent an average of 5% of their turnover on innovation. We suspect that this effort has increased, as the increasing importance of innovations is discussed in the literature. We want to verify this using the following null hypothesis: The average effort for innovation in the population of start-ups is less than or equal to 5% of turnover. Test the null hypothesis at the 5% significance level, taking into account all relevant steps with the R Commander.

13 The Test for a Difference Between Means with Independent Samples or Independent Samples t-Test

13.1 Introduction to the Test for a Difference Between Means with Independent Samples

In the last chapter, we pursued the question of how old our company founders are on average and we conducted a test for a group mean. We can expand this question. For example, we might be interested in whether women are older than men when founding a company. In this case, we compare the average founding age of women with the average founding age of men. We compare two means and test whether they differ or whether they are the same. Whenever we compare two means from two different groups, we perform the test for a difference between means with independent samples. Other examples from our data set would be the questions of whether industrial companies or service companies achieve a higher average growth, or whether women and men differ in terms of industry professional experience.

13.2 The Research Question and Hypotheses in the Test: Are Women and Men the Same Age at the Time of Founding?

First, as always, the research question needs to be specified. This could be as follows. Are men and women of the same age when founding a company? To formulate our null hypothesis and our alternative hypothesis, we first read the literature. With the help of this, we may come to the conclusion that women and men differ in average founding age. Our null hypothesis is therefore as follows:

H_0: *Women and men are on average the same age when founding a company.*
H_A: *Women and men are not the same age on average when founding a company.*

F. Kronthaler, *Statistics Applied with the R Commander*,
https://doi.org/10.1007/978-3-662-69107-6_13

We can also write this more briefly as follows:

$$H_0: \quad \mu_1 - \mu_2 = 0$$
$$H_A: \quad \mu_1 - \mu_2 \neq 0$$

Translated, this means for the null hypothesis, in the population the difference between the mean of group one μ_1 (e.g., women) and the men of group two μ_2 (e.g., men) is zero. That is, the means are the same in the population. The alternative hypothesis states that the difference between the means is not zero. We also specify the significance level at 10%.

$$\alpha = 10\ \% \quad \text{or} \quad \alpha = 0.1$$

Now we can perform our test.

13.3 The Test Distribution and the Test Statistic

The test distribution is the t-distribution with $df = n_1 + n_2 - 2$ degrees of freedom, where n_1 is the number of observations of group one and n_2 is the number of observations of group two. The test statistic is as follows:

$$t = \frac{\bar{x}_1 - \bar{x}_2}{\hat{\sigma}_{\bar{x}_1, \bar{x}_2}}$$

where

t	is the t-value of the t-distribution,
$\bar{x}_1$ and $\bar{x}_2$	are the sample means of the respective groups,
$\hat{\sigma}_{\bar{x}_1, \bar{x}_2}$	is the estimated standard deviation of the sampling distribution.

As we can see, the test statistic compares the two means. If both means are equal, the upper part of the equation, the numerator, is zero and the t-value is also zero. The larger the difference between the means, the larger the t-value becomes. We now need to take a closer look at the standard deviation of the sampling distribution. This is estimated as follows:

$$\hat{\sigma}_{\bar{x}_1, \bar{x}_2} = \sqrt{\left[\frac{(n_1 - 1)sd_1^2 + (n_2 - 1)sd_2^2}{n_1 + n_2 - 2}\right]\left[\frac{n_1 + n_2}{n_1 \times n_2}\right]}$$

with

n_1 and n_2	as the number of observations of the respective group,
sd_1	as the standard deviation of the sample of the first group and
sd_2	as the standard deviation of the sample of the second group.

Admittedly, the formula looks a bit complicated, but like the test for a group mean, it only includes the standard deviation and the number of observations of the sample. We now know the test statistic and the test distribution and can therefore turn to the critical value and the rejection area.

13.4 The Critical t-value

We determine the critical value with the help of the *t*-distribution table and the degrees of freedom. As with the test for a group mean, it is important to decide whether we want to test one-sided or two-sided and at what significance level. In our example, we test two-sided and have a significance level of 10 %. This means we have both left and right in the rejection area an area of 5 %. We take the critical values from the *t*-distribution table with the help of our degrees of freedom. Let our founders be group 1 and our female founders be group 2, then $n_1 = 65$ and $n_2 = 35$, we have 65 men and 35 women in the sample. Thus, we have $df = n_1 + n_2 = 65 + 32 - 2 = 98$ degrees of freedom. With this knowledge, we go into the *t*-distribution table and determine our critical values of -1.66 and 1.66 (compare Appendix 3).

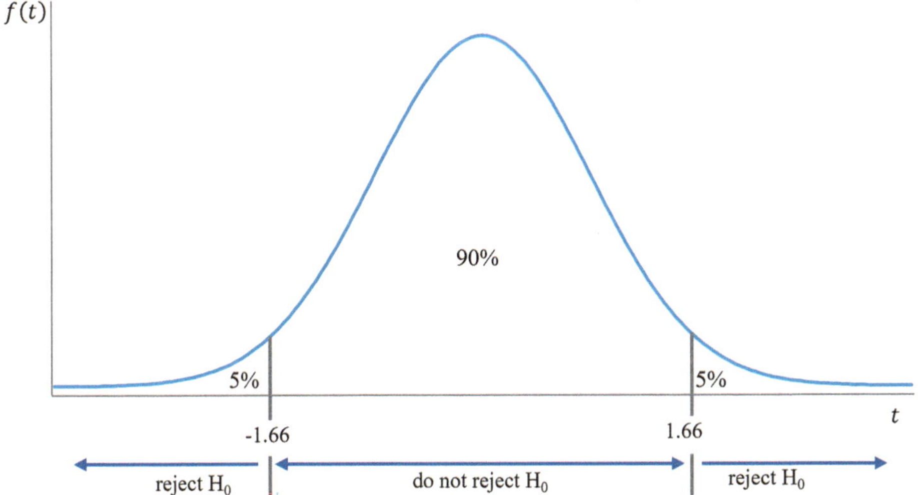

13.5 The t-value and the Decision

Now we just need to calculate our *t*-value, for this we fill our test statistic with the required values. The calculation of the means and standard deviations for our founders and female founders should not be a problem. We insert the values and get a *t*-value of 0.93.

$$t = \frac{\bar{x}_1 - \bar{x}_2}{\hat{\sigma}_{\bar{x}_1,\bar{x}_2}} = \frac{\bar{x}_1 - \bar{x}_2}{\sqrt{\left[\frac{(n_1 - 1)sd_1^2 + (n_2 - 1)sd_2^2}{n_1 + n_2 - 2}\right]\left[\frac{n_1 + n_2}{n_1 \times n_2}\right]}}$$

$$= \frac{34.77 - 33.29}{\sqrt{\left[\frac{(65 - 1) \times 7.55^2 + (35 - 1) \times 7.76^2}{65 + 35 - 2}\right]\left[\frac{65 + 35}{65 \times 35}\right]}} = \frac{1.48}{1.598} = 0.93$$

With the help of our *t*-value, we can now make our test decision. A comparison with the critical values shows us that we do not reject the null hypothesis. So we go on with the null hypothesis.

13.6 The Effect Size

Before we come to the final interpretation, we want to see how we can calculate and interpret the effect size. In this case, this would not be necessary, as we already assume that there is no difference between women and men in the average founding age in the population. However, if we have a significant result, we should address it.

A commonly used measure of effect size in the independent samples t-test is the Bravais-Pearson correlation coefficient r. Compared to Cohen's d, this has the advantage that it has a fixed defined range of values from 0 to 1. To calculate the effect size in this way, we can convert the found t-value into the correlation coefficient. The formula for this is:

$$r = \sqrt{\frac{t^2}{t^2 + df}}$$

with

t is the t-value of the test statistic and
df are the degrees of freedom.

We can interpret the value as follows:

- a r less than 0.3 indicates a small effect,
- a r in the range between 0.3 to less than 0.5 is a medium effect,
- a r in the range of 0.5 to 1.0 means a large effect.

Since r cannot be greater than 1.0, 1.0 is of course also the theoretically possible maximum effect.

If we insert the values from our example into the formula, we get a value that indicates no effect:

$$r = \sqrt{\frac{t^2}{t^2 + df}} = \sqrt{\frac{0.93^2}{0.93^2 + 98}} = 0.09$$

With this, we can conclude the interpretation. The result of the independent samples t-test is not significant, i.e., we do not expect any difference in the average age between women and men in the population based on the test result. The effect size also indicates no effect, and the difference in the sample is relatively small at about 1.5 years (compared to the average founding age).

Freak Knowledge
In the test procedures just described, the prerequisite, as mentioned, is a metric and normally distributed test variable. If our test variable is metric but not normally distributed, we can also apply the described test procedure with a large sample. This is then relatively robust against a violation of the assumption of normally distributed data. For metric, but not normally distributed data and a small sample as well as for ordinal data, the Mann-Whitney test is available.

13.7 Equal or Unequal Variances

In the example just shown, we simply assumed that the variation in the variable Age is the same for both groups, founders and female founders. Of course, we can look at this, for example with the help of box plots (see Chap. 4).

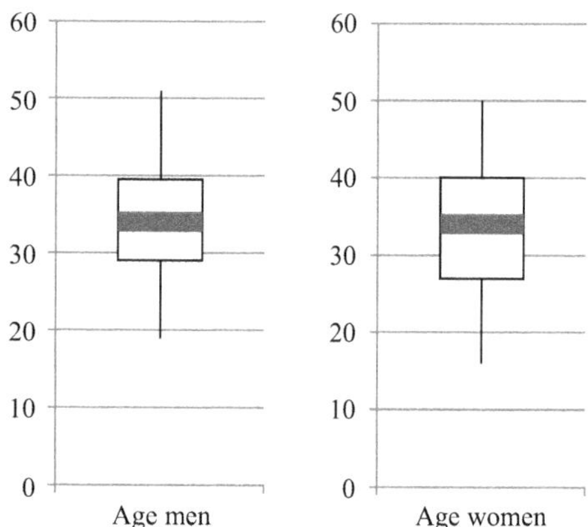

According to the box plots, the variation among women seems to be slightly higher, both the interquartile range and the range are somewhat larger, but not much. The same is shown when we compare the variance, which is 57.06 for the founders and 60.27 for the female founders. With some experience, it is easy to see that in this case the variances are roughly the same.

Freak Knowledge
Of course, we can also test whether the variances are equal, using the Levene's test for equality of variances. This works with the null hypothesis that the variances of both groups are equal and the alternative hypothesis that the variances of both groups are not equal.

The variation in the groups studied does not always have to be the same. It is important for us that the test procedures differ slightly for equal or unequal variances. As we will see, the R Commander offers us the opportunity to calculate the test for difference between means for equal or unequal variances. This said, before we carry out our test with the R Commander, we should look at the variation in the samples. If we are unsure about the equality or inequality of the variances, we simply let the R Commander calculate the test for both situations and hope that the results do not differ. If they do, we have to deal more intensively with the question.

13.8 Calculation with the R Commander

The calculation of the independent samples t-test using the R Commander requires, after reading the data, as already shown in section 6.8, first a manipulation of our data set. We need to tell R that the variable sex is a grouping variable. For this, we convert the variable into a factor. We use the menu Data with the submenu Manage variables in active data set and the command Convert numeric variables to factors.

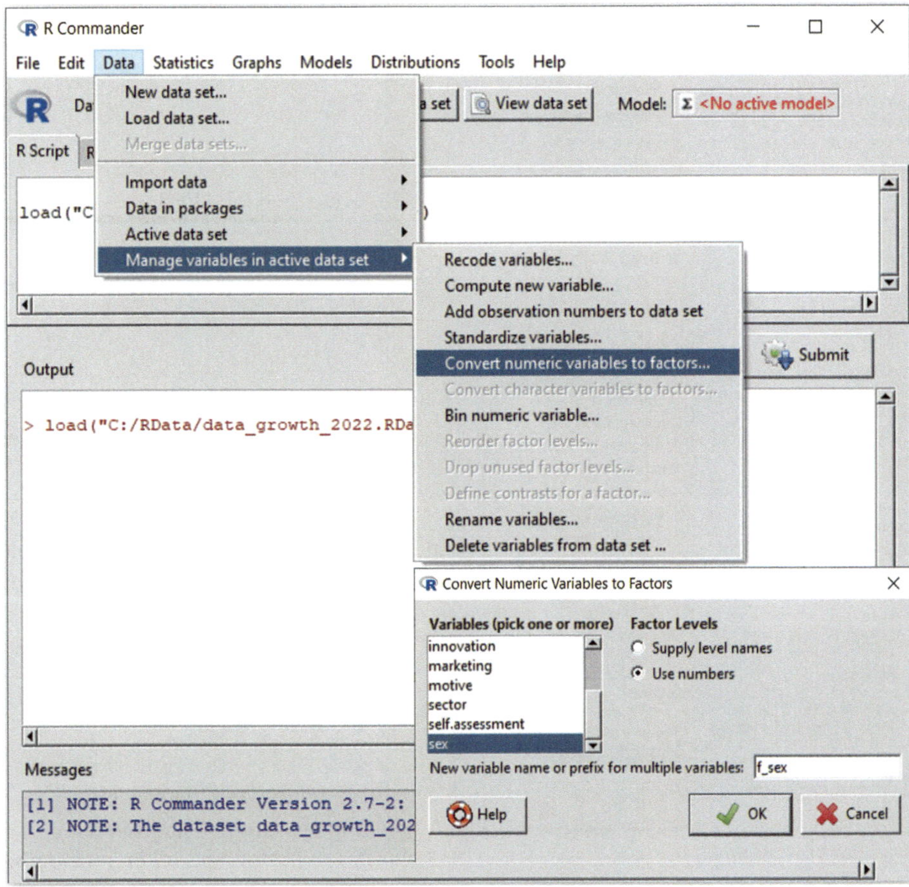

Subsequently, we can, analyze whether the variances for the two groups (female and male founders) are approximately equal using the boxplot. In the menu Graphs, we select the Boxplot command. In the next window, we specify the test variable and with using the button Plot by groups, we define that we want to create the boxplots for men and women.

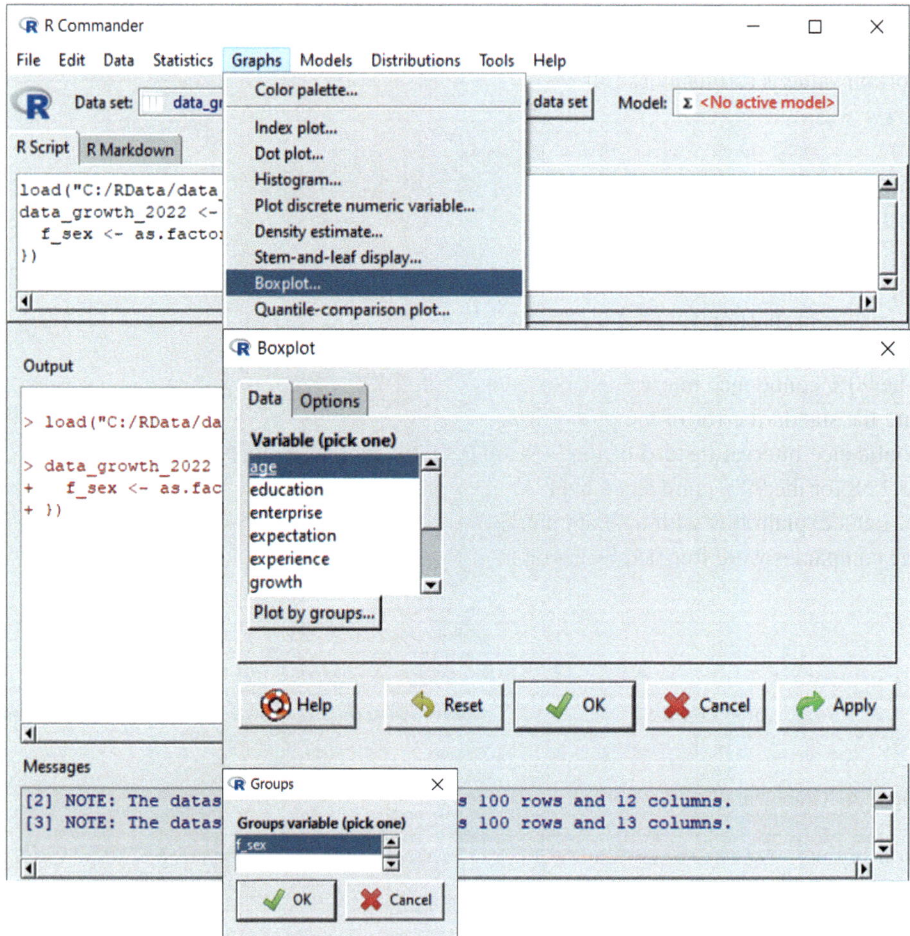

We confirm everything with OK and receive the boxplots for the female and male founders. As we know from our legend, 0 stands for the group of male founders and 1 for the group of female founders.

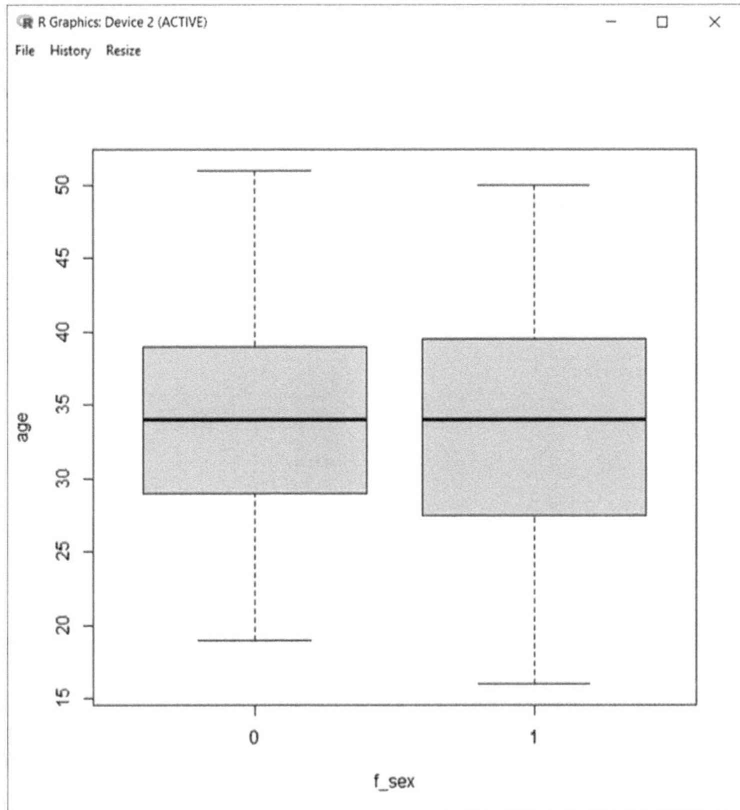

After we have decided whether we are conducting the test with equal or unequal variances, we can start with conducting the independent samples t-test. As already decided in the chapter above, we conduct the test with equal variances. In the Statistics menu, we select the Means submenu and go to the Independent samples t-test command. A small hint at this point: if we have forgotten to convert our grouping variable into a factor, the command is not active. Subsequently, we mark the grouping and test variable and select in the Options tab whether we test two-sided, left-sided or right-sided and whether we assume equal or unequal variances. We can also define the confidence interval. By default, this is again the 95% confidence interval.

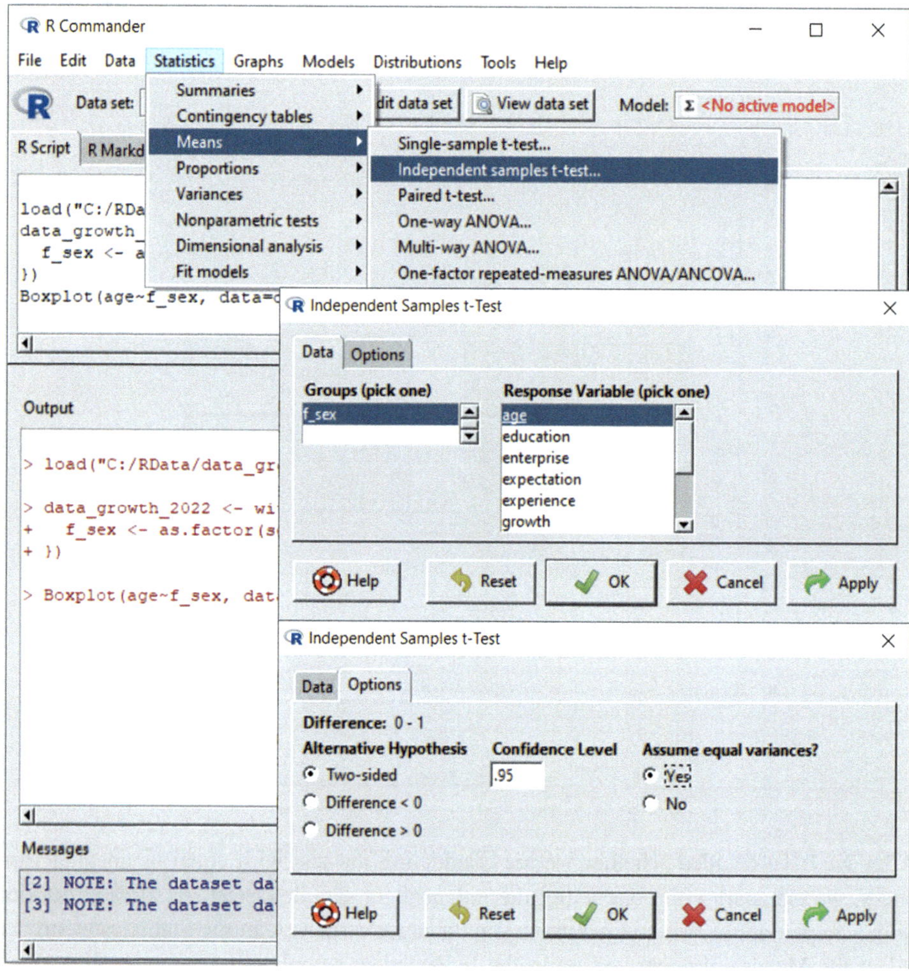

We confirm everything with OK and receive the result of our test.

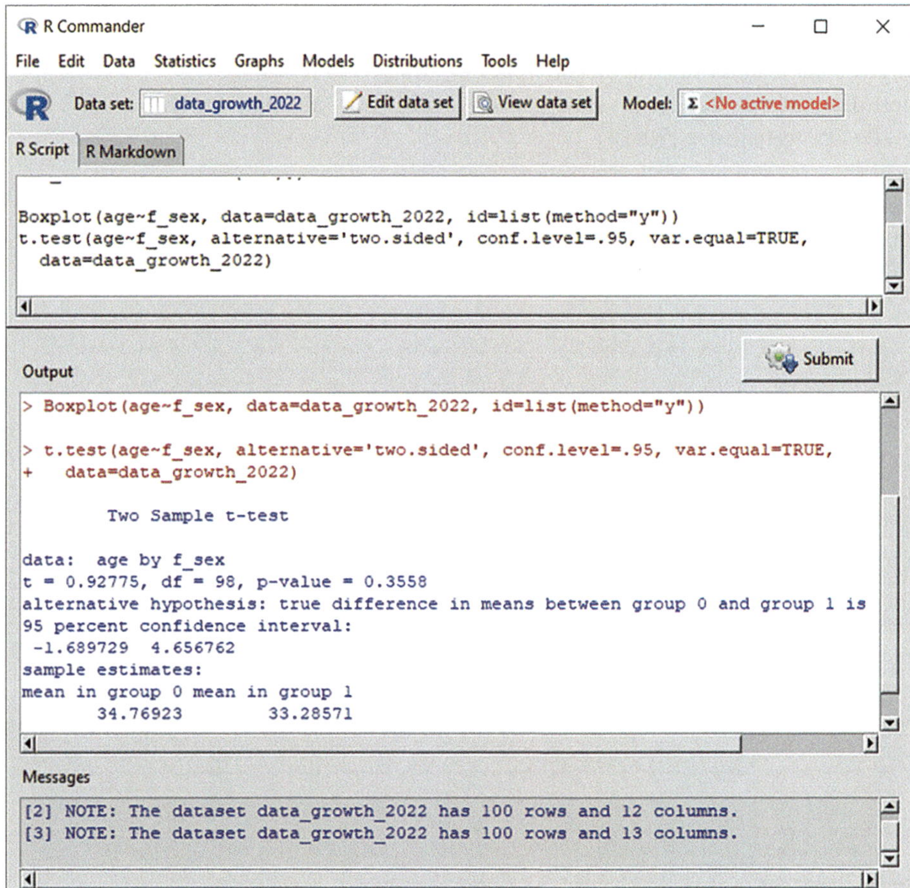

If we compare the output with our example from above, most of the numbers should be familiar. In the first line, we see that we have conducted the t-test for independent samples (Two Sample t-test). The test and grouping variable are given in the second line. In the third line, we see the t-value of 0.92775, the number of degrees of freedom of 98 and the p-value. The p-value shows that the probability that the null hypothesis is correct is 35.58%. This implies that we do not reject the null hypothesis at any specified significance level (e.g., $\alpha = 0.10$ or 10%). So we do not proceed with the alternative hypothesis indicated in the fourth line. In the population, there is no difference in founding age between women and men. We would also come to this conclusion if we compared the calculated t-value, as done above, with the critical t-value. Furthermore, we see in lines 5 to 9 the calculated sample means for the founders (34.77 years) and female founders (33.29 years) and the 95% confidence interval. The latter gives the range within which the true difference in the population is. The 95% confidence interval ranges from -1.69

to 4.66 years. The age difference between founders and female founders can therefore be either negative or positive.

The command for the test for difference of means is as follows: t.test(data_growth_2022$age ~ data_growth_2022$f_sex, alternative = "two.sided", conf.level = .95, var.equal = TRUE).

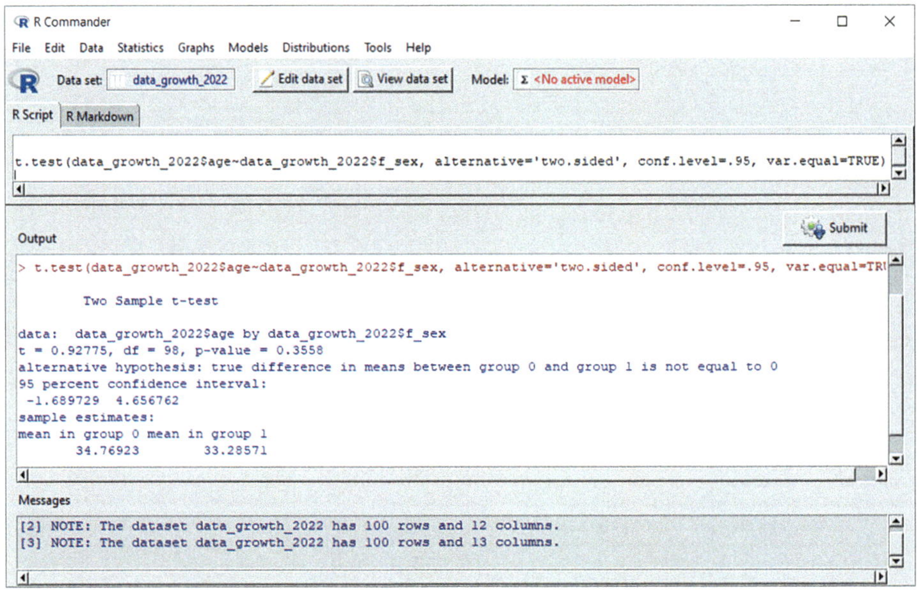

13.9 Checkpoints

- *In the test for difference between means with independent samples, we analyze whether two groups differ in their population in terms of a mean value.*
- *The test assumes that the test variable is metric and normally distributed.*
- *In the case of a small sample and a metric but not normally distributed test variable or an ordinal test variable, we resort to the Mann–Whitney test.*
- *The test for difference of means in independent samples is often also referred to as the independent samples t-test.*
- *Depending on whether the variances are equal or unequal in both groups, we perform the test for equal variances or unequal variances.*

13.10 Application

13.1. We wonder whether men and women are the same age when starting a business. Theoretical considerations and previous studies lead us to assume that men are on average older than women. Perform the test manually at the 10% significance level, taking into account all relevant steps. The following values are available: $\bar{x}_m = 34.77$, $sd_m = 7.55$, $n_m = 65$, $\bar{x}_f = 33.29$, $sd_f = 7.76$, $n_f = 35$.

13.2. We are interested in whether industrial companies or service companies have experienced higher growth in the last five years. Studies and theoretical considerations suggest that there should be a difference. Perform the test, taking into account all relevant steps, at the 5% significance level. The following values are available: $\bar{x}_i = 6.35$, $sd_i = 5.03$, $n_i = 34$, $\bar{x}_s = 7.49$, $sd_s = 5.77$, $n_s = 66$.

13.3. Perform the test from task 13.2 again using the R Commander and the data_growth dataset at the 5% significance level. Compare the results. Take into account any rounding errors.

13.4. We are interested in whether industrial companies or service companies spend relatively more money on innovation. Studies and theoretical considerations suggest that industrial companies spend a higher proportion of their turnover on innovation. Perform the test, taking into account all relevant steps, at the 1% significance level. The following values are available: $\bar{x}_i = 4.71$, $sd_i = 3.47$, $n_i = 34$, $\bar{x}_s = 4.62$, $sd_s = 3.38$, $n_s = 66$.

13.5. Perform the test from task 13.4 again using the R Commander and the data_growth dataset at the 1% significance level.

13.6. We wonder whether men and women bring more industry professional experience when starting a business. Existing studies suggest that there is a difference. We perform the corresponding hypothesis test at the 10% significance level, taking into account all relevant steps. The following values are available: $\bar{x}_m = 7.72$, $\text{sd}_m = 3.42$, $\text{n}_m = 65$, $\bar{x}_f = 6.86$, $\text{sd}_f = 3.67$, $\text{n}_f = 35$.

13.7. Perform the test from task 13.6 again using the R Commander and the data_growth dataset at the 10% significance level.

13.8. We are interested in whether founders or female founders spend more on marketing. Theoretical considerations lead us to assume that men spend more money on marketing than women. We perform the corresponding hypothesis test, taking into account all relevant steps, using the R Commander.

13.9. What values do we need to perform a test for difference between means?

13.10. If the test variable is metric, but not normally distributed or ordinal, which test for group differences is used?

14 The Test for a Difference Between Means with Dependent Samples or Dependent Samples t-Test

14.1 Introduction to the Test for a Difference Between Means with Dependent Samples

In the test for a difference between means with independent samples, we examined whether two independent groups differ in terms of a matter. We studied two different groups and then compared them. In the test for difference between means with dependent samples, we examine the same group twice and ask whether a certain measure has an influence on the group. For example, we can ask whether alcohol has an influence on the driving ability of individuals, or whether a medication improves the health status of individuals. We examine the person before and after the measure, then compare the results and thus ask about the effect of the measure. In a business context, the method is used, for example, to evaluate the effect of an advertising campaign, e.g., to ask what influence an advertising measure has on people's attitude towards a particular product or company. But one could also examine how energy drinks affect the performance of athletes.

14.2 The Example: Training of Company Founders in the Pre-Startup Phase

The question that the test requires involves one of the few situations that our data set does not cover. Therefore, we have to introduce a new data set, data_further_education. Let's imagine that the Association of Young Entrepreneurs e.V. conducts trainings. These trainings are intended to support potential founders in the phase before the founding of their company in correctly assessing the market potential of their company. To check

F. Kronthaler, *Statistics Applied with the R Commander*,
https://doi.org/10.1007/978-3-662-69107-6_14

whether the training has an influence on the perception of the course participants, they are asked about the expected turnover of their yet to be founded company before and after the course. This survey results in the presented data set (Table 14.1).

The legend for the data set can be found in Table 14.2.

As we can deduce from the data set and the legend, column two contains the revenue that the potential founders expected with the help of their business idea before they conducted the training. Column three contains the revenue they expect after the training. The difference is plotted in column four. In column five, we have already calculated the squared difference, as we will need it shortly.

Table 14.1 Data expected turnover before and after training

Participant	Pre	Post	Diff	d_i^2
1	165	155	10	100
2	210	185	25	625
3	150	150	0	0
4	130	145	−15	225
5	150	165	– 15	225
6	110	115	– 5	25
7	330	310	20	400
8	220	215	5	25
9	165	175	−10	100
10	200	175	25	625
11	230	220	10	100
12	275	255	20	400
13	240	215	25	625
14	200	180	20	400
15	155	140	15	225
16	210	175	35	1225
17	195	180	15	225
18	195	200	−5	25
19	140	155	−15	225
20	145	135	10	100
21	160	160	0	0
22	175	165	10	100
23	185	170	15	225
24	110	100	10	100
25	140	125	15	225

Table 14.2 Legend for the data expected turnover before and after training

Variable name	Variable description	Values	Missing values	Scale
Participant	Participant of the training			
Pre	Expected revenue before training	In 1000 Fr.	n. d.	Metric
Post	Expected revenue after training	In 1000 Fr.	n. d.	Metric
Diff	Difference in revenue before and after training	In 1000 Fr.	n. d.	Metric

Source Own survey 2013

14.3 The Research Question and the Hypotheses in the Test: Does the Training Have an Influence on the Assessment of Market Potential?

The research question in our example is as follows: Does the training have an impact on the assessment of business founders regarding the market potential of their planned companies? The null hypothesis and the alternative hypothesis are as follows:

H_0: *The training has no impact on the assessment of business founders regarding the market potential of their planned companies.*

H_A: *The training has an impact on the assessment of business founders regarding the market potential of their planned companies.*

Or we simply write shorter:

$$H_0\colon \mu_{Pre} - \mu_{Post} \neq 0$$

$$H_A\colon \mu_{Pre} - \mu_{Post} = 0$$

Here are two more notes: First, we have formulated the null and alternative hypothesis without direction, as we have not made any statement regarding the direction of the impact. Second, we recognize that we compare the means before and after the training/measurement. μ_{Pre} is the mean before the test and μ_{Post} is the mean after the test.

Let's also set the significance level. In this case, we want to test at the 10% significance level:

$$\alpha = 10\ \%\ or\ \alpha = 0.1$$

14.4 The Test Statistic

The test statistic in the test for a difference between means with dependent samples is as follows:

$$t = \frac{\sum d_i}{\sqrt{\frac{n\sum d_i^2 - \left(\sum d_i\right)^2}{n-1}}}$$

with

d_i is equal to the difference between the value before and after the measure and
n is equal to the size of the sample or the number of observed persons or objects.

We see that the test value is t-distributed and that we compare the difference between the values before and after the measure using the test statistic. If the measure had no influence, the values before and after the measure would be the same, the difference would therefore be zero and thus the numerator of the equation in the test statistic would also be zero. This implies that the t value would also be zero and we would therefore not reject our null hypothesis.

14.5 The Critical t-value

Now we need to determine the critical t-values. We already know that these depend on whether we are testing two-sided or one-sided, how high the specified significance level is, and how many degrees of freedom we have. In our case, we have 25 observations. The number of degrees of freedom in the test for a difference between means with dependent samples is $df = n - 1$, i.e., we have $df = 25 - 1 = 24$ degrees of freedom. Thus, the critical t values are -1.711 and 1.711.

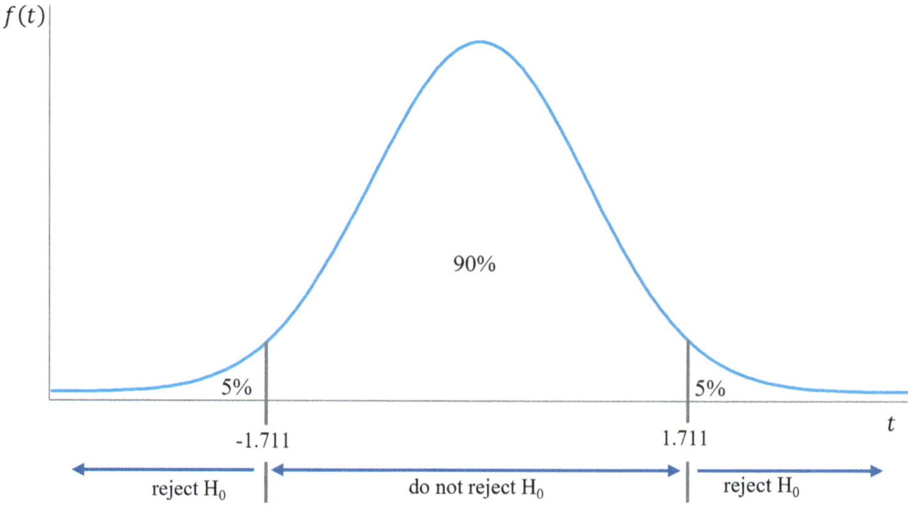

14.6 The t-value and the Decision

All that remains for us to do is to calculate the test statistic and make the decision. For this, we insert the sum of the differences and the sum of the squared differences as well as the number of observed pair values into the test statistic. We get:

$$t = \frac{\sum d_i}{\sqrt{\frac{n\sum d_i^2 - \left(\sum d_i\right)^2}{n-1}}} = \frac{220}{\sqrt{\frac{25\times 6'550-220^2}{25-1}}} = 3.17$$

If we compare the calculated t value with our critical values, we see that we reject the null hypothesis and proceed with the alternative hypothesis. This means that the training has an impact on the assessment of the market potential of potential company founders.

14.7 The Effect Size

Before we come to the final interpretation, we again calculate the effect size. For this, we use the already known formula, with which we convert the t-value into the correlation coefficient r (see independent samples t-test):

$$r = \sqrt{\frac{t^2}{t^2 + df}}$$

with

t is the t-value of the test statistic and
df are the degrees of freedom.

The interpretation remains the same, of course:

- a r less than <0.3 indicates a small effect,
- a r in the range between 0.3 to <0.5 is a medium effect,
- a r in the range of 0.5 to 1.0 means a large effect.

We insert the values of our example into the formula and see that we have a medium to large effect:

$$r = \sqrt{\frac{t^2}{t^2 + df}} = \sqrt{\frac{3.17^2}{3.17^2 + 24}} = 0.54$$

With this, we are able to finally evaluate our training measure. We reject the null hypothesis at the significance level of 10% and thus assume that the training measure has an influence on how the new founders assess their market potential. The effect of the training measure is medium to large. The mean value before the training was 183.4 or CHF 183,400, after the training it was 174.6 or CHF 174,600. This means that the training causes the potential founders to assess the market potential about 5% lower after the training than it was the case before the training.

14.8 Calculation with the R Commander

The calculation of the test for a difference between means with dependent samples with the R Commander is done with the command Paired t-test. We find this in the Statistics menu within the submenu Means. After we have called up the command, we specify the variables. As a rule, we take the variable before the measure as the first variable and the variable after the measure as the second variable. In the Options tab, we also specify whether we are testing two-sided, left-sided or right-sided and indicate the confidence interval.

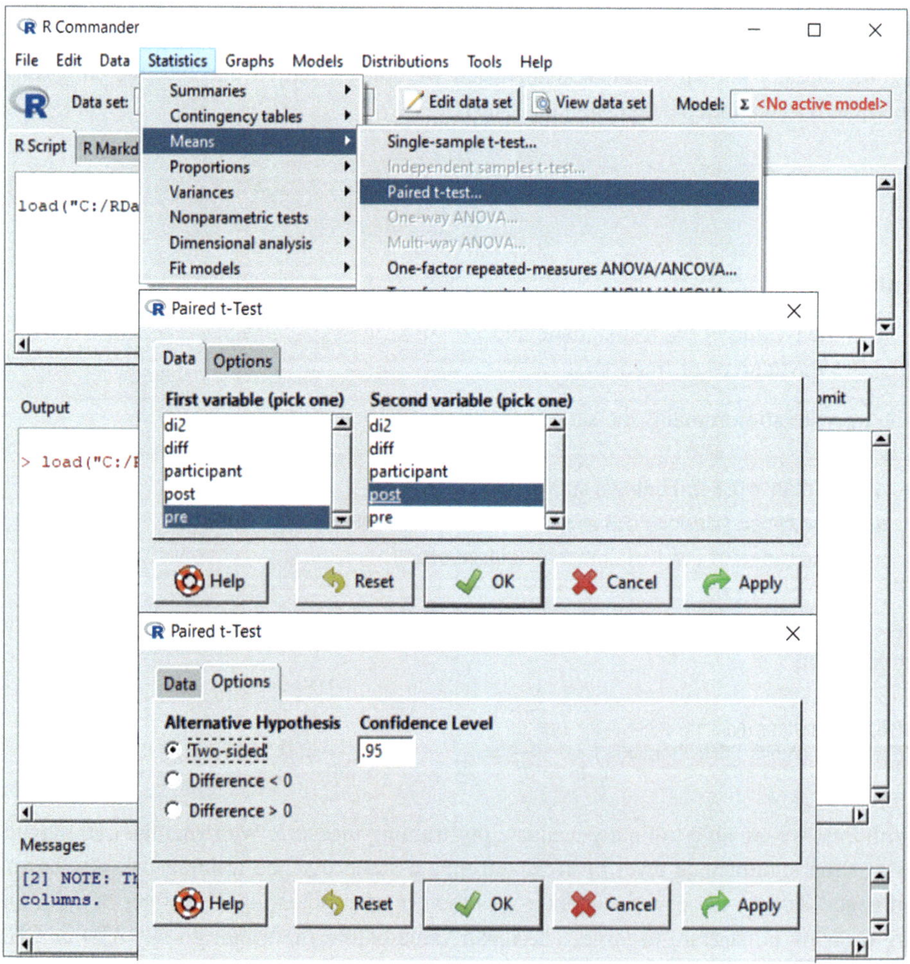

We then confirm everything with OK and receive the result of the test.

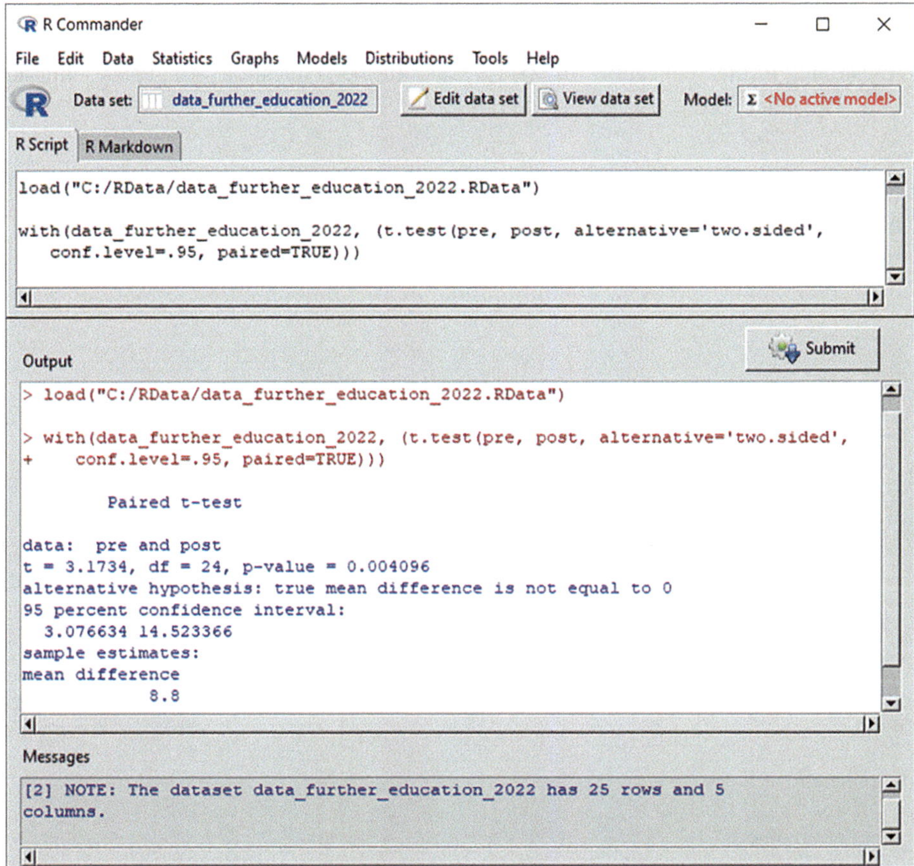

Most of the numbers should be familiar. The first line states that the t-test with dependent samples was performed (Paired t-test). The variables are then given in the second line. In the third line, we find the t-value, the number of degrees of freedom, and the p-value. The t-value is 3.1734. We had calculated 3.17 by hand. Except for rounding errors, the values are the same. We have 24 degrees of freedom and the p-value shows the probability of the null hypothesis to be correct. The probability is 0.004096 or 0.41%. This implies that we reject the null hypothesis at the specified significance level of $\alpha = 0.1$ or $\alpha = 10\%$. We therefore proceed with the alternative hypothesis indicated in the fourth line. We also arrive at this result if we compare the calculated t-value with the critical t-value. Furthermore, in lines 5 to 9, we see the average difference of 8.8 (CHF 8.800) estimated from the sample as well as the 95% confidence interval. This shows us that the true difference in the population is likely to be in the interval between 3.077 (CHF 3.077) and 14.523 (CHF 14.523) with a 95% probability. Thus, we know both the estimated difference for the population and the confidence interval.

Of course, we can also enter the command for the t-test with dependent samples by hand. The command is as follows:

t.test(data_further_education_2022$pre, data_further_education_2022$post, alternative = “two.sided”, conf.level = .95, paired = TRUE)

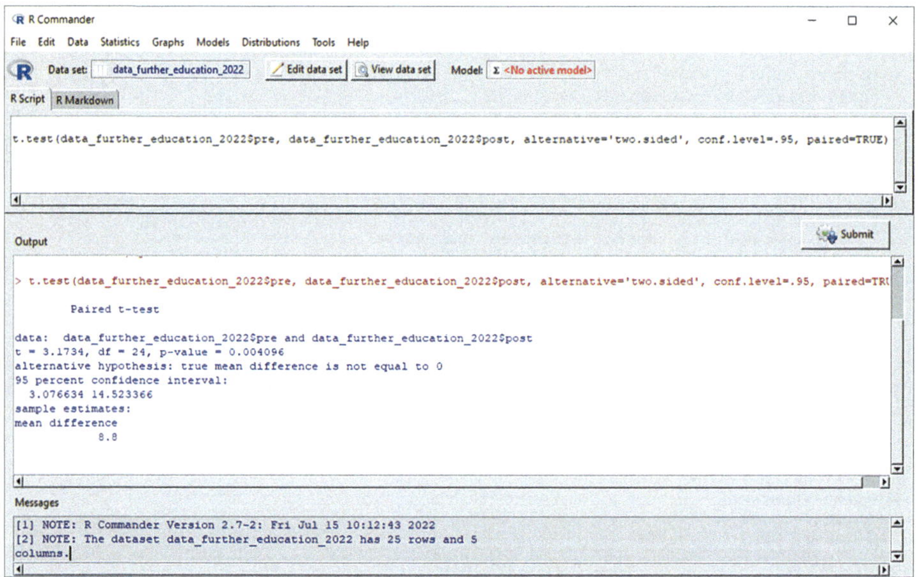

Freak Knowledge

If our data is metric but not normally distributed, we can also apply the described test procedure with a large sample. This is relatively robust against a violation of the assumption of normally distributed data. For metric but not normally distributed data and a small sample, as well as for ordinal data, the Wilcoxon test for dependent samples is available to us.

14.9 Checkpoints

- *In the test for a difference between means with dependent samples, we analyze whether a measure has an influence on individuals or objects.*
- *In the test for a difference between means with dependent samples, we examine the same test objects twice, once before and once after the measure.*
- *The test assumes that the test variable is metric and normally distributed.*
- *For a small sample and metric but not normally distributed data, and for ordinal data, we resort to the Wilcoxon test for dependent samples.*

14.10 Application

14.1. We are interested in whether a training measure in the field of accounting, according to the organizer of the training measure, leads to less time being spent per month on accounting (data_accounting). Carry out the test by hand at the 10% significance level and interpret the result.

Participant	Pre	Post	Diff	d_i^2
1	17	19	− 2	4
2	9	7	2	4
3	21	15	6	36
4	30	25	5	25
5	29	31	−2	4
6	33	33	0	0
7	5	4	1	1
8	18	10	8	64
9	28	30	−2	4
10	11	8	3	9
11	14	14	0	0
12	6	5	1	1
13	5	7	−2	4
14	12	11	1	1
15	13	15	−2	4

Variable name	Variable description	Values	Missing values	Scale
Participant	Training participant			
Pre	Time for accounting per month before training	In hours	n. d.	Metric
Post	Time for accounting per month after training	In hours	n. d.	Metric
Diff	Difference before and after training	In hours	n. d.	Metric

Source Own survey 2013.

14.2. Repeat the test from application 14.1 again at the 10% significance level using the R Commander and compare the results.

14.3. We are interested in whether the consumption of energy drinks affects the performance of athletes (data_drinks). The study design stipulates that the same athletes run as far as possible twice in one hour each. Once they have the opportunity to consume only water during the run, the other time they are only allowed to consume energy drinks. Carry out the test by hand at the 1% significance level and interpret the result.

Participant	Without	With	Diff	d_i^2
1	17	18	−1	1
2	9	11	−2	4
3	15	15	0	0
4	10	13	− 3	9
5	25	26	−1	1
6	18	16	2	4
7	12	14	−2	4
8	16	17	−1	1
9	22	21	1	1
10	12	11	1	1

Variable name	Variable description	Values	Missing values	Scale
Participant	Participant			
Without	Kilometers run in one hour	In km	n. d.	Metric
With	Kilometers run in one hour	In km	n. d.	Metric
Diff	Difference	In km	n. d.	Metric

Source Own survey 2013

14.4. Repeat the test from application 14.3 at the 10% significance level using the R Commander and compare the results.

15 The Analysis of Variance for Testing Group Differences When There Are More Than Two Groups

15.1 Introduction to the Analysis of Variance

In the test for a difference between means with independent samples, we tested whether two groups differ in terms of a matter, e.g., whether there is a difference in the founding age between men and women (see Chap. 13).

But what do we do when we want to compare more than two groups at the same time? The question could be, for example: Do company founders differ in their founding age according to their nationality? We then certainly have more than two groups to compare at the same time.

The test for a difference between means with independent samples is no longer sufficient in such a case. However, we have the analysis of variance at our disposal, which allows us to compare more than two groups at the same time. The simples form is, when we have one test variable (founding age) and one grouping variable (nationality). The grouping variable is referred to as a factor, and when we only have one factor we speak of a one-way analysis of variance. But we can also have more than one factor, e.g., nationality and gender, then we would have two factors and would accordingly speak of a two-way analysis of variance, and so on.

In this chapter, we deal with the one-way analysis of variance. This provides a good introduction to the method of analysis of variance. In addition, we learn about the concept of how to divide variance observed in an explained and an unexplained part, a very important concept in statistics.

Before we start, a note should be given here. For didactic reasons, we introduce a new data set with data_workload. The reason for this is didactic in nature. In the first edition of the book, the analysis of variance was not yet discussed, and the use of the original data set would only lead to non-significant results, which is not conducive from a didactic point of view.

F. Kronthaler, *Statistics Applied with the R Commander*,
https://doi.org/10.1007/978-3-662-69107-6_15

15.2 The Example: Do Company Founders Differ in Their Workload According to the Founding Motive?

Let's assume we are interested in how much time company founders work after founding their company. For this purpose, we collect the time that founders spend in the company per week. We also suspect that the weekly working time depends on the motive for founding the company. The collected data could look like this and are simulated as before (Table 15.1).

The legend for the data set is shown in Table 15.2.

If we sort the weekly working hours by the founding motive, we can present the data as in Table 15.3. In this, we see the weekly working hours of the founders who have founded out of unemployment (founding motive 1), want to implement an idea with their foundation (founding motive 2), or want to achieve a higher income (founding motive 3).

Based on these data, we can now calculate four arithmetic means. We can calculate the average working time per founding motive. In addition, we can calculate the average working time across all founders.

$$\bar{y} = \frac{\sum y_{gk}}{n} = \frac{936}{18} = 52$$

Table 15.1 Data weekly workload and founding motive

Firm	Workload	Motive	Sex
1	37	1	0
2	40	1	0
3	60	2	1
4	62	2	1
5	43	1	1
6	54	3	1
7	58	3	1
8	60	3	0
9	57	2	0
10	55	2	0
11	55	3	0
12	45	1	0
13	50	2	1
14	64	3	0
15	52	2	1
16	48	1	1
17	57	3	1
18	39	1	0

Table 15.2 Legend for the data weekly workload and founding motive

Variable Name	Variable Description	Values	Missing Values	Scale
Firm	Running index of the surveyed firms			
Workload	Weekly working hours	In hours	n. d.	Metric
Motive	Founder's founding motive	1=Unemployment 2=Implement idea 3=Higher income	n. d.	Nominal
Sex	Sex of the firm founder	0=Man; 1=Woman	n. d.	Nominal

Source Own survey 2021

Table 15.3 Weekly workload in hours, sorted by founding motive

Founding Motive 1	Founding Motive 2	Founding Motive 3
37	60	54
40	62	58
43	57	60
45	55	55
48	50	64
39	52	57

$$\bar{y}_1 = \frac{\sum y_{1k}}{n_1} = \frac{262}{6} = 42$$

$$\bar{y}_2 = \frac{\sum y_{2k}}{n_2} = \frac{336}{6} = 56$$

$$\bar{y}_3 = \frac{\sum y_{3k}}{n_3} = \frac{348}{6} = 58$$

with

g are the groups (these are also referred to as factor levels),
k are the observations per group or per factor level,
$\bar{y}$ is the global mean across all groups or factor levels,
$\bar{y}_g$ are the means for the individual groups or factor levels.

Here we have used for the arithmetic mean $\bar{y}$ and not $\bar{x}$ as we learned it in Chap. 3. This is because in the one-way analysis of variance we denote the factor with X and the test variable with Y. When we calculate the mean for the test variable, it is accordingly denoted with $\bar{y}$. Ultimately, it doesn't matter whether we name a variable with X, Y or Z.

We see, all observed company founders work an average of 52 hours per week. The company founders who started their company out of unemployment work 42 hours per week, those who want to implement an idea work 56 hours per week, and those who want to earn a higher income work 58 hours per week.

In the collected data, or the sample, we have found differences regarding the average weekly working hours. But do these differences also apply to the population? We analyze this question with the one-way analysis of variance.

15.3 The Research Question and the Hypotheses in the Analysis of Variance

The research question for our above example is: Do founders differ in the amount of time they work based on the motive for starting the company? We can write the null hypothesis and the alternative hypothesis as follows:

H_0: *The weekly working hours of the company founders do not differ between the various groups.*

H_A: *The weekly working hours of the company founders differ in at least two groups.*

or

$$H_0\colon\ \mu_1 = \mu_2 = \mu_3$$

$$H_A\colon\ \mu_i = \mu_j \text{ for at least one pair of } ij$$

We see from the alternative hypothesis that we reject the null hypothesis if we already find a group difference between two groups. As always, we set the significance level in advance, which should be 5% here:

$$\alpha = 5\%\ or\ \alpha = 0.05$$

Before we turn to the test statistic, we want to understand the basic principle of the analysis of variance. We will see that we can divide the variation of the values around their means into a deviation explained by the factor and a deviation not explained by the factor.

15.4 The Basic Idea of the Analysis of Variance

We have seen above that we can calculate a global average value $\bar{y}$ for all observations and for each group a group mean $\bar{y}_g$. To explain how to divide the variation in an explained and an unexplained part, it is best to graphically present both the individual observations y_{gk} and the means (see Fig. 15.1).

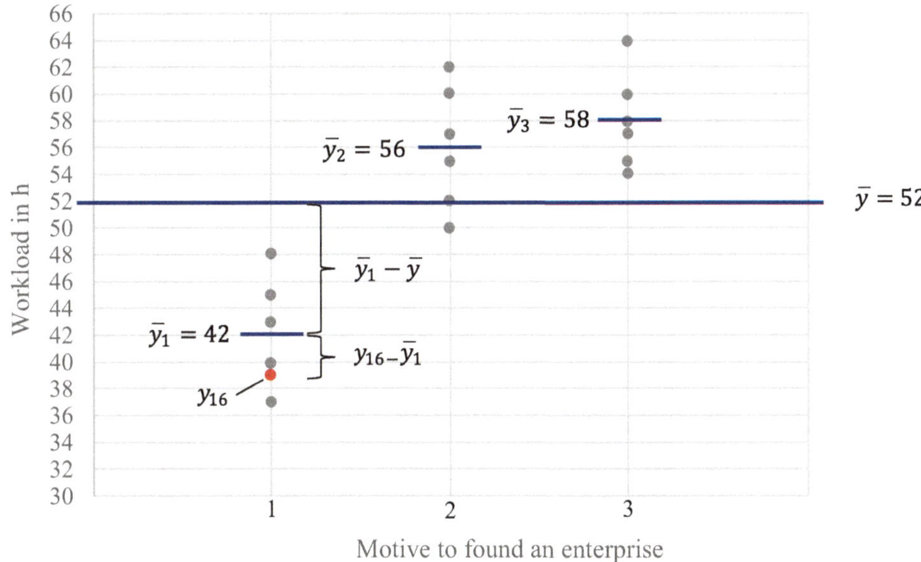

Fig. 15.1 The weekly working hours of company founders according to the motive for starting the company

We see the individual working hours of the company founders, the mean for all company founders, and the group means. Let's focus on the red dot in group 1, the observation y_{16}. The 1 stands for group 1 or founding motive 1 and the 6 for the sixth observation within the group. We see, this company founder works significantly less than the average of the company founders. The deviation is$y_{16} - \bar{y} = 39 - 52 = -13$ hours. This is the total deviation of this company founder from the global mean. If we consider the point further and include the group mean, then we can divide the deviation into $\bar{y}_1 - \bar{y} = 42 - 52 = -10$ hours and $y_{16} - \bar{y}_1 = 39 - 42 = -3$ hours. We refer to the deviation of the group mean from the global mean as the deviation explained by the factor level (founding motive 1), because this is the average group deviation from the mean of all company founders. The deviation of the observation from the group mean, on the other hand, is the unexplained deviation, we cannot explain this part of the deviation by the factor level. What we have just done for this one point, we can do for all points. If we do this and also square the deviations so that negative and positive deviations do not cancel each other out, then we have divided the total squared deviations (it is called sum of squares) into an explained and an unexplained part (see Table 15.4).

Table 15.4 Calculation of the total, explained and unexplained variance

				Total sum of squares	Explained sum of squares	Unexplained sum of squares
Company	Workload	Motive	Averages	$\left(y_{gk}-\bar{y}\right)^2$	$\left(y_g-\bar{y}\right)^2$	$\left(y_{gk}-\bar{y}_g\right)^2$
1	37	1		225	100	25
2	40	1		144	100	4
5	43	1		81	100	1
12	45	1		49	100	9
16	48	1		16	100	36
18	39	1	$\bar{y}_1 = 42$	169	100	9
3	60	2		64	16	16
4	62	2		100	16	36
9	57	2		25	16	1
10	55	2		9	16	1
13	50	2		4	16	36
15	52	2	$\bar{y}_2 = 56$	0	16	16
6	54	3		4	36	16
7	58	3		36	36	0
8	60	3		64	36	4
11	55	3		9	36	9
14	64	3		144	36	36
17	57	3	$\bar{y}_3 = 58$	25	36	1
	$\bar{y} = 52$			1168	912	256

15.5 The Test Statistic

The test statistic of the analysis of variance is derived from the explained and the unexplained sum of squares. The principle is simple, the higher the part of the deviation explained by the groups, the lower the unexplained part and the more likely the grouping variable contributes to the explanation of the found differences. Specifically, the test statistic looks like this:

$$F = \frac{SS_{explained}/(G-1)}{SS_{not\ explained}/(G \times (K-1))}$$

with

F is the value of the test statistic,
$SS_{explained}$ is the explained squared sum of squares,

$SS_{not\ explained}$	is the unexplained squared sum of squares,
$G - 1$	are the degrees of freedom of the numerator df_1,
$G \times (K - 1)$	are the degrees of freedom of the denominator df_2.

We recognize, the test statistic follows the F-distribution, i.e., we can determine the critical value using the F-distribution table.

15.6 The Critical F-value

From Chap. 10 we know that the critical value of the F-distribution depends on two degrees of freedom, in this case that of the numerator df_1 and the denominator df_2 and that the values for the significance levels of 10%, 5% and 1% are tabulated. So we take the specified significance level of 5 %, calculate the two degrees of freedom with $df_1 = G - 1 = 3 - 1 = 2$ and $df_2 = G \times (K - 1) = 3 \times (6 - 1) = 15$ and look for the critical value in the F-distribution table (Appendix 5) using these values. This is 3.68. If our test statistic exceeds this value, we reject the null hypothesis and proceed with the alternative hypothesis.

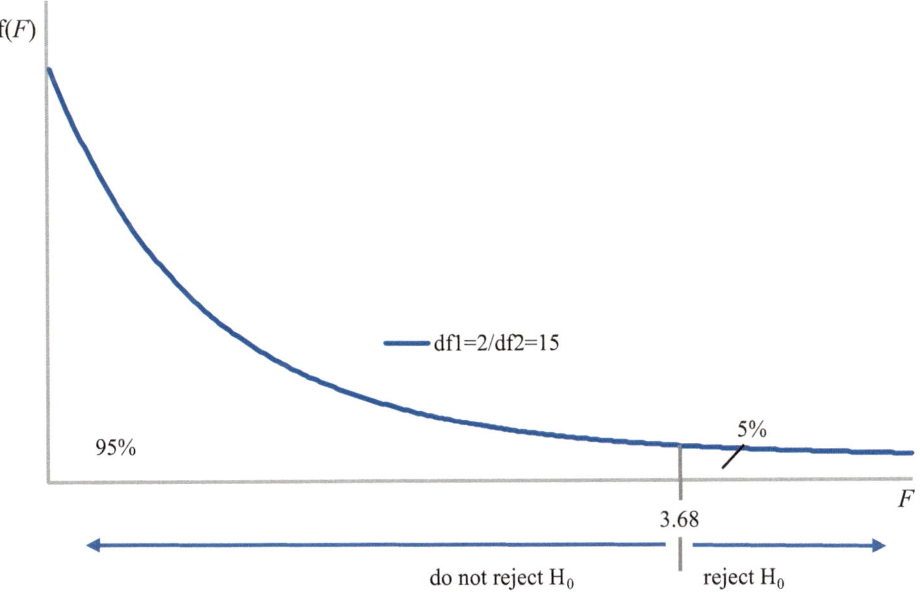

15.7 The F-value and the Decision

To calculate the test statistic, we now only need to insert the values of the sum of squares and the degrees of freedom that we have already calculated above into the formula. By the way, if we divide the sum of squares by the respective degrees of freedom, we get to the mean sum of squares. We will see this when we look at the calculation with the R Commander.

$$F = \frac{SS_{explained}/(G-1)}{SS_{not\ explained}/(G \times (K-1))} = \frac{912/(3-1)}{256/(3 \times (6-1))} = \frac{456}{17.07} = 26.71$$

The F-value is much higher than the critical value, i.e. we reject the null hypothesis and proceed with the alternative hypothesis. We now know that there is a difference in weekly working hours between at least two groups in the population. However, we do not yet know between which groups there is a difference.

15.8 The Analysis of Variance, an Omnibus Test, and the Post-hoc Tests

As we have just shown, we use the analysis of variance to test whether there is a group difference between at least two of the groups involved in the test procedure. We test all groups at once and do not differentiate between which groups there is a difference. The analysis of variance is therefore one of the so-called omnibus test procedures. In statistics, this term refers to test procedures that check for group differences, but do not check between which groups there are differences.

To check between which groups there are differences, so-called post-hoc tests are used following the analysis of variance. To explain what this is all about, let's take a step back.

We already know that a significant result in the analysis of variance indicates group differences in the population. However, we do not know between which groups a difference exists. To determine this, we need to compare the groups pairwise. We can do this, for example, with the help of the independent samples t-test (Chap. 13). However, there is a small problem. If we compare all groups pairwise, we have to perform several tests, the number of tests depends on the number of groups. In our example, we have three groups, i.e., we can compare group 1 with 2, group 1 with 3, and group 2 with 3. So we have to perform three independent samples t-test.

With each of these tests performed, we allow an error to the extent of the specified significance level, leading to the so-called error inflation. Imagine we make a 5% error on the first test, another 5% error on the second, and a 5% error on the third test. Then we have made a total error of 15%. However, we do not necessarily want to increase the original significance level. This is taken into account by the so-called post-hoc tests as they try not to increase the error.

In principle, there are various correction procedures, e.g., the Bonferroni correction, the LSD method, the Tukey method, etc. All methods aim to keep the error at the originally defined level, as described above. Which method is applied depends, among other things, on how much the error inflation is corrected. Widely used methods are, for example, the Bonferroni correction and the Tukey method. The latter is automatically available in the R Commander. How the Tukey method is performed and interpreted will be seen in Sect. 15.10.

15.9 The Effect Size in the Analysis of Variance

Before we finally interpret the results, we want to see how the effect size in the analysis of variance is calculated and determine this for our example.

As with the test for a difference between group means with independent and dependent samples, the correlation coefficient of Bravais-Pearson can again be used. The squared correlation coefficient is calculated as follows:

$$r^2 = \frac{SS_{explained}}{SS_{total}}$$

This value is often referred to as Eta-squared η^2 in the analysis of variance. If we now take the square root, we obtain the correlation coefficient as a measure of the effect size:

$$r = \sqrt{r^2} = \sqrt{\frac{SS_{explained}}{SS_{total}}}$$

We know the interpretation. A

- r less than < 0.3 indicates a small effect,
- a r in the range between 0.3 to < 0.5 indicates a medium effect and
- a r in the range of 0.5 to 1.0 indicates a large effect.

For our example, we get an r of 0.88, i.e. we have a large effect.

$$r = \sqrt{\frac{SS_{explained}}{SS_{total}}} = \sqrt{\frac{912}{1168}} = 0.88$$

The result of our analysis can thus be presented as follows. The analysis of variance shows a statistically significant result. We reject the null hypothesis at the significance level of 5% and assume that there are group differences. The analysis of the effect size shows that we have a large effect or that the motive for founding a company has a strong effect on the weekly working hours. If we compare the individual groups pairwise, we see (we do this in Sect. 15.10), company founders with the founding motive of unemployment work significantly less than company founders with the founding motives of implementing an idea or earning a higher income. The former work an average of 42 hours per week, the latter 56 or 58 hours per week. On average, the difference is thus 14 or 16 hours per week (see also Sect. 15.10).

Freak Knowledge

The analysis of variance requires a metric and normally distributed test variable. In addition, the groups must be independent. If the group sizes vary much, equal variances are required too. So what do we do if these assumptions are violated. If the test variable is metric but not normally distributed or ordinal, the Kruskal–Wallis test is available. If the variances are not equal, there is a solution for this with the Welch F-test. Whether the groups influence each other, i.e. are not independent of each other, must be clarified theoretically. There is no statistical procedure for this.

15.10 Calculating the Analysis of Variance With the R Commander

To calculate the analysis of variance, we have the commands One-way anova and Multi-way anova available in the Statistics tab with the submenu Means. For these to be active and for us to be able to execute them, we must first tell R that the variable motive is a factor (for the multi-way analysis of variance, of course, we need more than one factor). We accomplish this via the Data menu, the submenu Manage variables in active data set, and the Convert numeric variables to factors command. We select the motive variable, possibly rename it, and click OK. We have thus created the grouping variable and the commands are active.

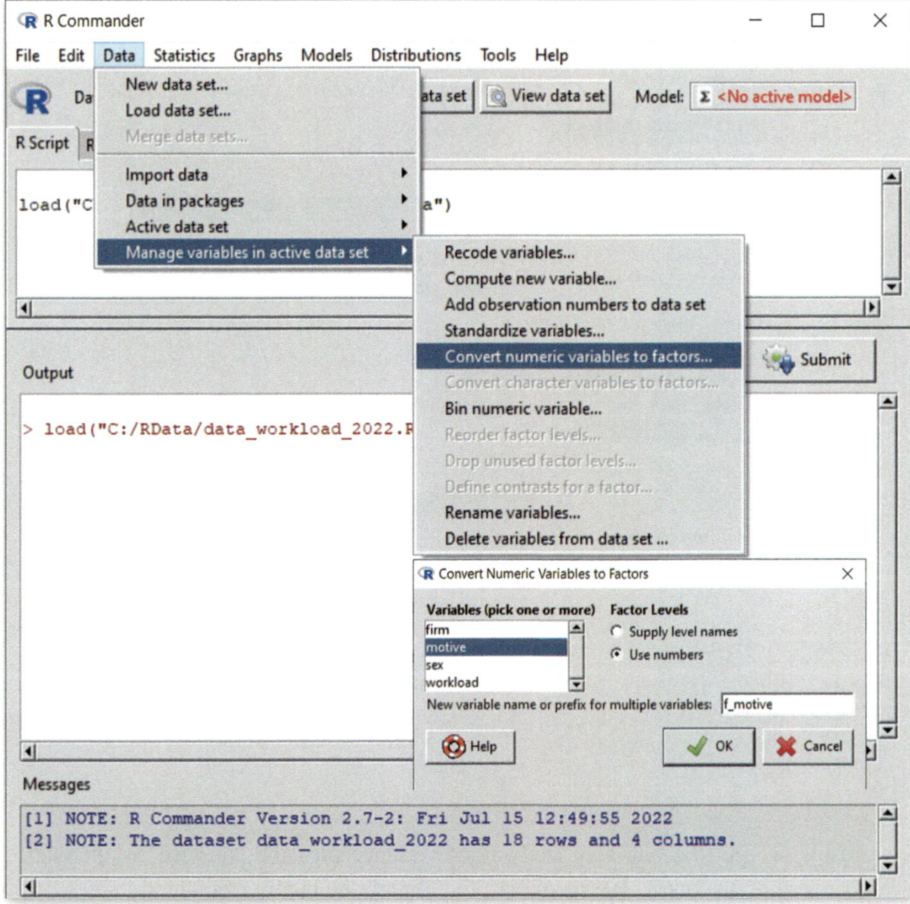

Now we select the Means option and the One-way ANOBA command in the Statistics menu. In the window that opens, we select the grouping variable and the dependent variable, also check the Pairwise Comparison box, and confirm with OK.

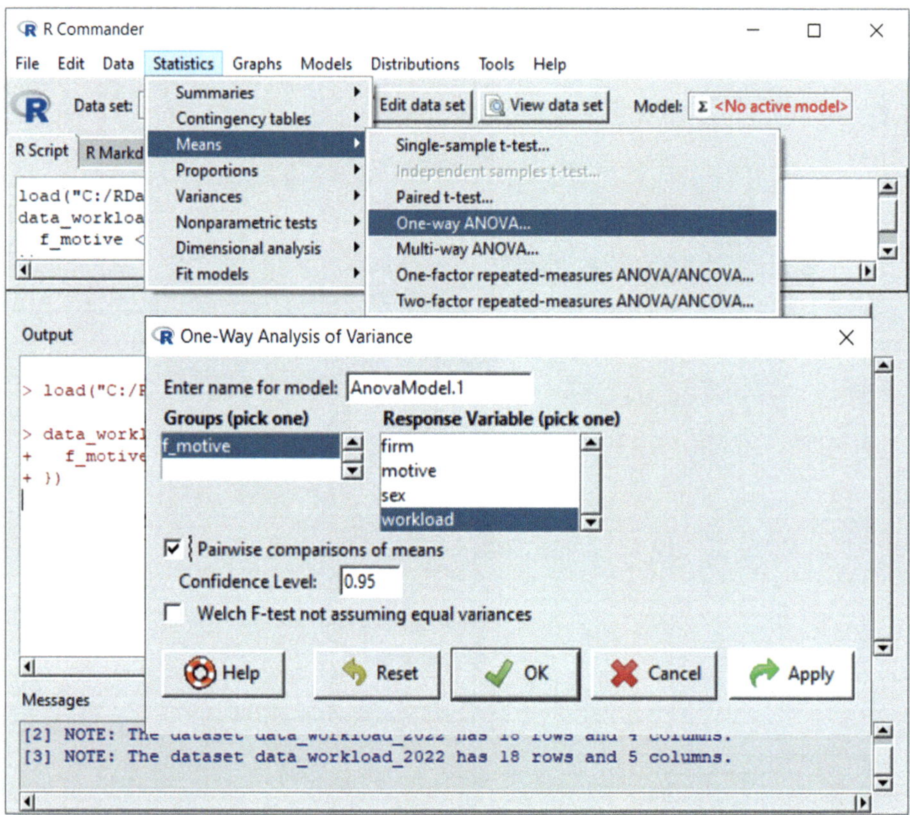

As a result, we get the one-way analysis of variance plus the pairwise comparison of means according to Tukey. We recognize the already manually calculated values of the explained sum of squares amounting to 912, the unexplained sum of squares amounting to 256, the degrees of freedom, and the F-value amounting to 26.72. Next to it, we see the probability (Pr(>F) with which the null hypothesis applies. We recognize, this is very small and we reject it, i.e., we assume the alternative hypothesis and thus group differences. The post-hoc test according to Tukey shows that we have significant group differences between groups 2 and 1 and 3 and 1. The differences is 14 or 16 working hours per week. Here only an excerpt of the presented results is shown. If we look at the output more closely, we also see the descriptive statistics and the confidence intervals.

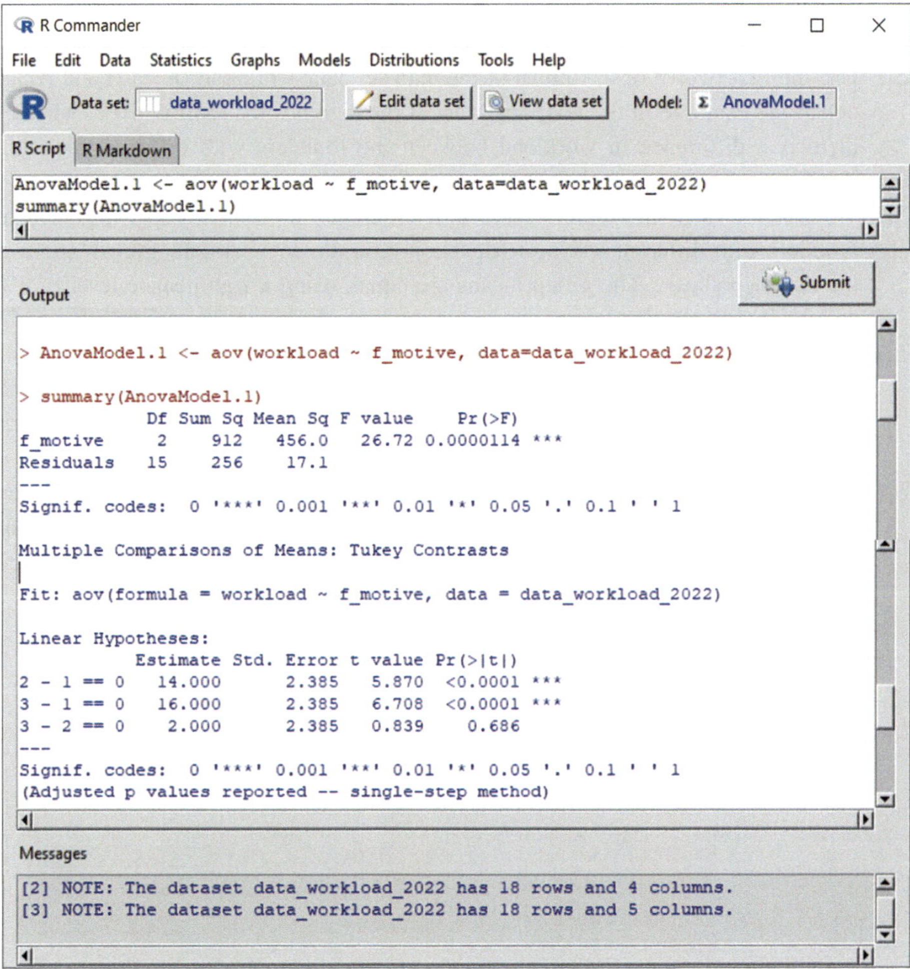

15.11 Checkpoints

- *We use the analysis of variance when we compare more than two groups at the same time.*
- *The analysis of variance is an omnibus test, we analyze whether there are group differences between all compared groups, not between which ones.*
- *Following the analysis of variance, a post-hoc test must be performed, i.e., the groups must be compared pairwise.*
- *In the pairwise comparison of groups, there is the so-called error inflation, therefore error correction is necessary.*
- *The Kruskal-Wallis test is used for metric, but not normally distributed or for an ordinal test variable.*

15.12 Application

15.1. Use the first twelve observations of the data set data_workload (n = 12) and recalculate the example in the textbook. Test at the significance level of 10%, whether there is a difference in workload between the founders with different founding motives.

15.2. We are interested in whether there is a difference in company growth between founders with different levels of education. Use the data set data_growth and test the null hypothesis: The group means are equal, using a significance level of 5% and the R Commander.

15.3. We are investigating whether founders with different educational backgrounds spend different amounts of money on marketing. Use the data set data_growth and test the null hypothesis: The group means are equal, using a significance level of 10% and the R Commander.

15.4. Use the data set data_workload and analyze whether there are differences between female and male founders in terms of workload. Test at the significance level of 5% both with the help of the analysis of variance and with the test for a difference between means with independent samples. Compare the test results.

16 The Test for Correlation with Metric, Ordinal, and Nominal Data

In Chap. 6 we explored the question of the relationship between two variables, i.e., how and whether two variables move together. We examined this for metric, ordinal, and nominal variables. For example, we analyzed a relationship between the variables marketing and innovation or between the variables expectation and self-assessment. We calculated the correlation coefficients based on the sample. With the correlation test, we now check whether the relationship discovered in the sample also applies to the population. Only then is our sample result relevant for recommendations. At the same time, the magnitude of the correlation coefficient also provides information about how strong the effect is. So when we report on the correlation coefficient, we do not need to calculate a measure for the effect size, the correlation coefficient is this measure itself. We indicate the correlation coefficient and whether it is significant. We then interpret the result accordingly.

The following discusses the correlation tests for metric, ordinal, and nominal data. The procedure is the same as for the hypothesis tests we have already learned. We first formulate the null hypothesis and the alternative hypothesis and specify the significance level. Then we determine the test distribution and the test statistic. Next, we determine the rejection area and the critical values. The last steps are to calculate the test value and make the decision.

16.1 The Correlation Test with Metric Data

16.1.1 The Test Situations for a Correlation with Metric Data

When we perform the correlation test with metric data, we are dealing with three different situations. First, we can assume that there is a correlation between two variables

F. Kronthaler, *Statistics Applied with the R Commander*,
https://doi.org/10.1007/978-3-662-69107-6_16

without specifying the direction. Or we assume that we have a positive or a negative correlation.

In the first case, we test non-directional. The null hypothesis and the alternative hypothesis are then as follows:

H_0: *There is no relationship between variable X and variable Y.*
H_A: *There is a relationship between variable X and variable Y.*

No relationship means that the correlation coefficient is equal to zero. A relationship accordingly means that the correlation coefficient is not equal to zero. Therefore, we can also briefly write as follows:

$$H_0: \quad r = 0$$
$$H_A: \quad r \neq 0$$

In the second case, we assume that the correlation is of a positive. The null and alternative hypotheses are then as follows:

H_0: *There is a negative or no relationship between variable X and variable Y.*
H_A: *There is a positive relationship between variable X and variable Y.*

We can also write more briefly:

$$H_0: \quad r \leq 0$$
$$H_A: \quad r > 0$$

If we reject the null hypothesis, we assume that there is a positive relationship between the two variables, the correlation coefficient is greater than zero.

In the third case, due to the assumption of a negative correlation, the null and alternative hypotheses are as follows:

H_0: *There is a positive or no relationship between variable X and variable Y.*
H_A: *There is a negative relationship between variable X and variable Y.*

We can also write more briefly:

$$H_0: \quad r \geq 0$$
$$H_A: \quad r < 0$$

If we reject the null hypothesis, we assume that the correlation coefficient is less than zero and there is a negative relationship.

16.1.2 The Test Statistic and the Test Distribution in the Test

The correlation test for metric data is performed using the Bravais-Pearson correlation coefficient (compare Chap. 6).

$$r = \frac{\sum (x_i - \bar{x})(y_i - \bar{y})}{\sqrt{\sum (x_i - \bar{x})^2 \sum (y_i - \bar{y})^2}}$$

We calculate the Bravais-Pearson correlation coefficient from the sample and then insert the value of r into the following test statistic:

$$t = \frac{r \times \sqrt{n-2}}{\sqrt{1-r^2}}$$

with

r is the correlation coefficient of Bravais-Pearson and
n is the number of observations.

As we can see, the test value is the t-value and our test distribution is the t-distribution with $df = n - 2$ degrees of freedom.

We have now already generated all the knowledge and can explain the test using an example.

16.1.3 Example: Is There a Relationship Between the Effort for Marketing and Innovation in Young Companies?

We want to analyze whether there is a relationship between expenditure on marketing and expenditure on innovation in young enterprises. We do this by considering theoretical aspects and reading the literature. We find that entreprises have a budget constraint and that money they spend in one area must be saved in another. We therefore formulate our null hypothesis and our alternative hypothesis as follows:

H_0: *There is a positive or no relationship between the expenditure for marketing and the expenditure for innovation.*

H_A: *There is a negative relationship between the expenditure for marketing and the expenditure for innovation.*

$$H_0: \quad r \geq 0$$
$$H_A: \quad r < 0$$

So we assume a negative relationship, i.e., if an enterprise spends a lot on marketing, it spends little on innovation and vice versa.

We want to test our hypothesis with a significance level of 5%.

$$\alpha = 5\,\% \quad \text{or} \quad \alpha = 0.05$$

In Chap. 6 we have already calculated the correlation coefficient for both variables for the first six companies. This is $r = -0.91$. At this point, it is certainly useful to recalculate the example (see Chap. 6).

To test whether there is a relationship in the population, we insert the value for the correlation coefficient into our test statistic and calculate the t value. We get

$$t = \frac{r \times \sqrt{n-2}}{\sqrt{1-r^2}} = \frac{-0.91 \times \sqrt{6-2}}{\sqrt{1-(-0.91)^2}} = -4.39$$

Now we only need the critical t value and we can perform our test decision. The test distribution is, as mentioned, the t distribution with $df = n - 2$ degrees of freedom, i.e., we have 4 degrees of freedom with 6 observations. We will reject if our correlation coefficient from the sample falls below the critical value, i.e., -2.132 (compare Appendix 3).

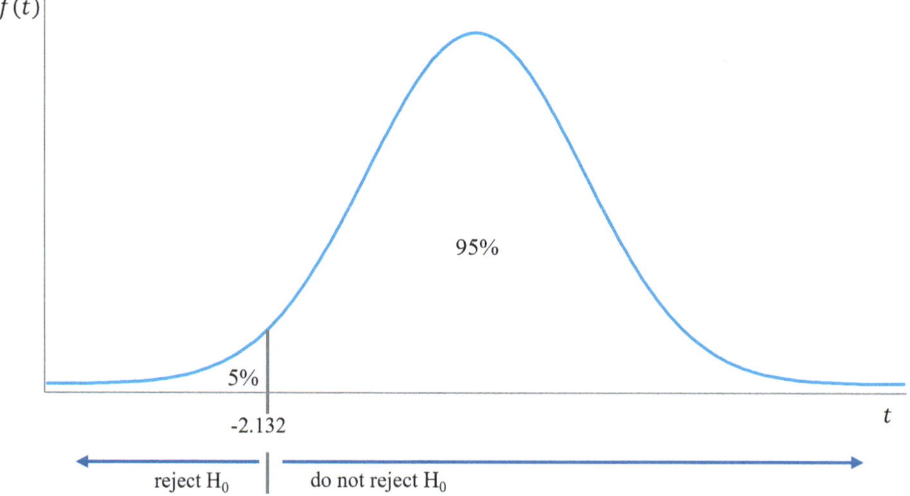

Our calculated test value falls below the critical value, i.e., we reject the null hypothesis in this case and continue with the alternative hypothesis. So, in the population, there is a negative relationship between the expenditure for innovation and expenditure for marketing. The magnitude of the correlation coefficient of -0.91 also indicates that it is a strong relationship or effect. Entrepreneurs who spend a lot on marketing, therefore, spend less on innovation and vice versa. If we want, we have discovered a specialization strategy of companies. Some try to be successful through marketing, others try to position themselves through innovation.

16.2 The Test for a Correlation with Ordinal Data

16.2.1 The Test Situations for a Correlation with Ordinal Data

The test situations in the test for correlation with ordinal data are the same as with metric data. The only difference is that we do not use the Bravais-Pearson correlation coefficient, but the Spearman correlation coefficient (compare Chap. 6):

$$r_{\mathrm{Sp}} = \frac{\sum \left(R_{x_i} - \bar{R}_x\right)\left(R_{y_i} - \bar{R}_y\right)}{\sqrt{\sum \left(R_{x_i} - \bar{R}_x\right)^2 \sum \left(R_{y_i} - \bar{R}_y\right)^2}}$$

Like the correlation coefficient of Bravais-Pearson, the correlation coefficient of Spearman can also be positive or negative. Thus, we can test for both non-directional, as well as for a positive or a negative correlation. We want to write down the null and alternative hypotheses for these three situations in short form once again.

Situation 1:

$$H_0: \quad r_{\mathrm{Sp}} = 0$$
$$H_{\mathrm{A}}: \quad r_{\mathrm{Sp}} \neq 0$$

Situation 2:

$$H_0: \quad r_{\mathrm{Sp}} \leq 0$$
$$H_{\mathrm{A}}: \quad r_{\mathrm{Sp}} > 0$$

Situation 3:

$$H_0: \quad r_{\mathrm{Sp}} \geq 0$$
$$H_{\mathrm{A}}: \quad r_{\mathrm{Sp}} < 0$$

16.2.2 The Test Statistic and the Test Distribution in the Test

The test statistic, with which we perform the test, is t-distributed with $df = n - 2$ degrees of freedom and is as follows:

$$t = \frac{r_{\mathrm{Sp}}}{\sqrt{\frac{1 - r_{\mathrm{Sp}}^2}{n-2}}}$$

with

r_{Sp} is the Spearman correlation coefficient and
n is the number of observations.

With this, we can already perform the test and dedicate ourselves to an example.

16.2.3 Example: Is There a Correlation Between Self-Assessment and Expectation Regarding a Company's Economic Development?

In Chap. 6 we analyzed whether there is a relationship between the variable expectation and the variable self-assessment. The assumption is that there is a relationship between self-assessment regarding industry professional experience and expectation regarding the future development of a company. In Chap. 6 we calculated a positive correlation of $r_{\mathrm{Sp}} = 0.58$ for the first six companies. Now we want to test whether this relationship also applies to the population. The null hypothesis and the alternative hypothesis are as follows:

$$H_0: \quad r_{\mathrm{Sp}} \leq 0$$
$$H_{\mathrm{A}}: \quad r_{\mathrm{Sp}} > 0$$

We want to test these at a significance level of 10%.

$$\alpha = 10\% \quad \text{or} \quad \alpha = 0.1$$

To find out whether the relationship applies based on our sample findings, we insert the value for the correlation coefficient into our test statistic and calculate the t value. We get:

$$t = \frac{r_{\mathrm{Sp}}}{\sqrt{\frac{1-r_{\mathrm{Sp}}^2}{n-2}}} = \frac{0.58}{\sqrt{\frac{1-0.58^2}{6-2}}} = 1.424$$

We now need to compare this value with our critical value, which we read from the t table, with 4 degrees of freedom. We reject the null hypothesis if we exceed a value of 1.533 (compare Appendix 3).

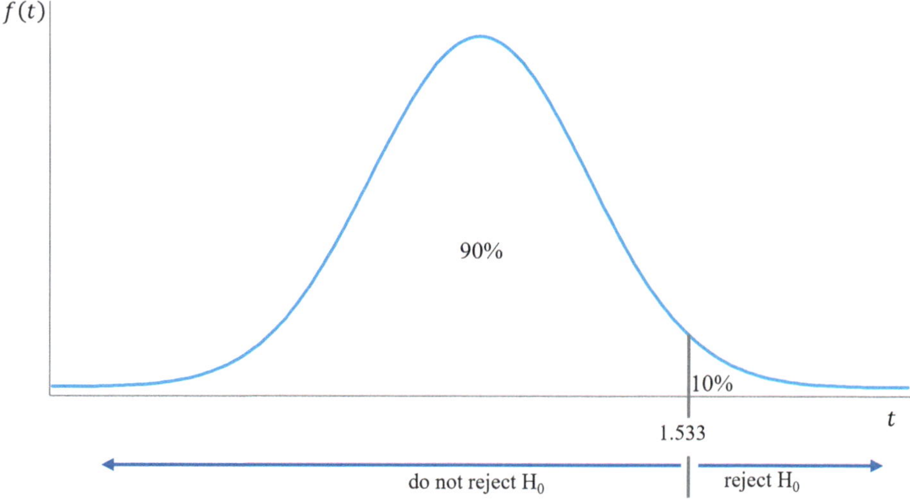

Comparing our calculated test value of 1.424 with the critical value, we see that we do not reject the null hypothesis. This means that we cannot establish a positive relationship for the population based on our sample result. So there is no positive relationship between expectation about the future development of the company and self-assessment regarding industry professional experience. Here is a small note on effect size. The calculated correlation coefficient of 0.58 indicates a large effect size (see Sect. 11.3). However, the sample size of $n = 6$ is relatively small, and is not sufficient to reject the null hypothesis at this level of the correlation coefficient. However, it would probably make sense to look at the matter again with a larger sample.

16.3 The Test for a Correlation with Nominal Data

16.3.1 The Test Situations for a Correlation with Nominal Data

When we test for correlation between two nominal variables, we have two situations. Either we have nominal variables each with two characteristics, or we have nominal variables with more than two characteristics. In both cases, the correlation coefficient does not tell us anything about the direction of the relationship, i.e., we cannot test for positive or negative correlation. We can only test whether we have a relationship or not. For this reason, both tests are often called tests for independence. If the variables are not correlated, then they are independent of each other.

When we have two nominal variables each with two characteristics, the null and alternative hypothesis for the test for independence are as follows:

H_0: *There is no relationship between variable X and variable Y.*
H_A: *There is a relationship between variable X and variable Y.*

Let's remember, the relevant correlation coefficient is the phi coefficient r_Φ (see Chap. 6). r_Φ can take positive and negative values, but the sign does not indicate the direction of the correlation. Therefore, we can briefly write as follows:

$$H_0\colon \quad r_\Phi = 0$$
$$H_\mathrm{A}\colon \quad r_\Phi \neq 0$$

For nominal variables with more than two characteristics, the contingency coefficient C comes into play. The null and alternative hypotheses are the same.

H_0: *There is no relationship between variable X and variable Y.*
H_A: *There is a relationship between variable X and variable Y.*

Since the contingency coefficient only takes values greater than or equal to zero, we briefly write as follows:

$$H_0: \quad C = 0$$
$$H_A: \quad C > 0$$

In both cases, if we reject the null hypothesis, we assume that there is a relationship between the variables.

16.3.2 The Test for Independence with Nominal Variables Each with Two Characteristics

For nominal variables each with two characteristics, we calculated the phi coefficient:

$$r_{\Phi} = \frac{a \times d - b \times c}{\sqrt{S_1 \times S_2 \times S_3 \times S_4}}$$

where

a, b, c, d are the fields of a 2×2 matrix and
S_1, S_2, S_3, S_4 are the row sums and column sums.

		Variable *Y*		
		0	1	
Variable *X*	0	*a*	*b*	S_1
	1	*c*	*d*	S_2
		S_3	S_4	

The value r_{Φ} is simultaneously our test statistic. The squared value of r_{Φ} multiplied by the number of observations is $\chi 2$-distributed with $df = 1$ degrees of freedom:

$$\chi^2 = n \times r_{\Phi}^2$$

With this, we know our test distribution, it is the $\chi 2$-distribution, and we know our test value. We can perform the test for independence.

In Chap. 6 we pursued the question of whether there is a relationship between the sex of the founders and the industry in which they found. We observed the first ten foundations and calculated a r_{Φ} of 0.50 using the following 2×2 *Matrix*.

		Sex		
		0=Man	1=Woman	
Sector	0=Industry	2	0	2
	1=Service	3	5	8
		5	5	

So, we have discovered a moderate relationship between sex and the sector in which they found. Looking at the matrix, we see that women often found service companies, men often found industrial companies. But does the relationship also apply to our population?

For this, we perform the test, with the null and alternative hypotheses as follows:

H_0: *There is no relationship between the sex of the founder and the sector in which they found.*

H_A: *There is a relationship between the sex of the founder and the sectro in which they found.*

$$H_0: \quad r_\Phi = 0$$

$$H_A: \quad r_\Phi \neq 0$$

We test this at a significance level of 5%:

$$\alpha = 5\,\% \quad \text{or} \quad \alpha = 0.05$$

From $r_\Phi = 0.50$ we calculate our test value. This is the $\chi 2$ value:

$$\chi^2 = n \times r_\Phi^2 = 10 \times (0.50)^2 = 2.50 \ \textit{with} \ df = 1.$$

We now compare this value with the critical value of the $\chi 2$ distribution table (see Appendix 4). From this, we can see that we reject the null hypothesis if we exceed a value of 3.841.

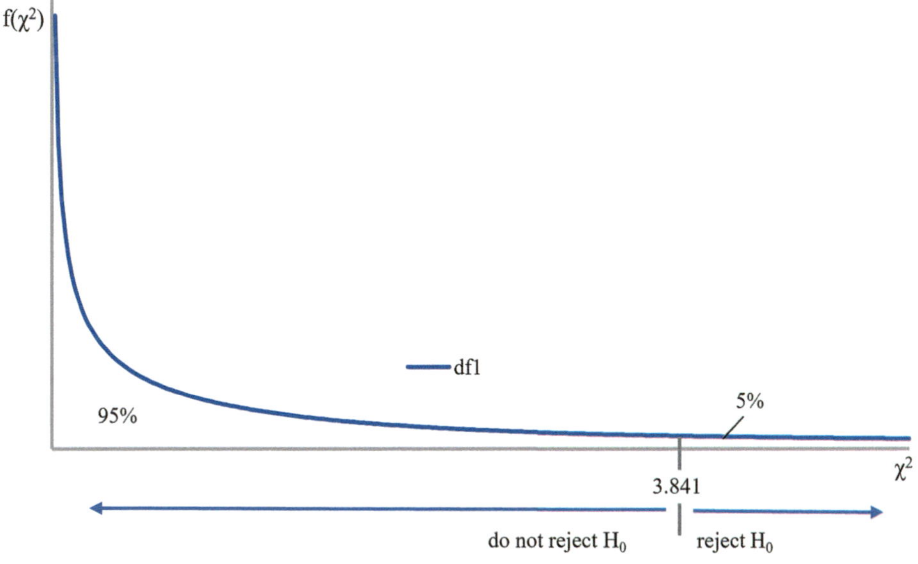

When we compare our calculated test value with the critical value, it becomes apparent that we cannot reject the null hypothesis, i.e., based on our sample findings, we have no relationship between the two variables in the population. Even here, despite a relatively large effect of $r_\Phi = 0.50$, we cannot assume a relationship in the population. The reason, as above in the test with ordinal data, is again the small sample size. In addition, this test only provides reliable results if each cell is occupied by at least 5 test objects. For didactic reasons, we worked with a small number of observations. Therefore, all cells except one are occupied less than 5 times. For us, this means that we need to increase the number of observations in order to be able to carry out the test meaningfully.

16.3.3 The Test for Independence with Nominal Variables with More Than Two Characteristics

For nominal variables with more than two characteristics, the contingency coefficient is calculated (see Chap. 6):

$$C = \sqrt{\frac{U}{U + n}}$$

where

U is the sum of deviations between the observed and theoretically expected values:

$$U = \sum \sum \frac{\left(f_{jk} - e_{jk}\right)^2}{e_{jk}}$$

with.

f_{jk} are the observed and

e_{jk} the theoretically expected row and column values.

For example, if we have a nominal variable with 2 characteristics and one with three characteristics, then the following matrix with the observed frequencies f_{11} to f_{23} results.

Rows	Columns		
	1	2	3
1	f_{11}	f_{12}	f_{13}
2	f_{21}	f_{22}	f_{23}

We calculated the theoretically expected frequencies as follows:

Rows	Columns			
	1	2	3	Row sum
1	$e_{11} = \frac{f_1 \times f_1}{n}$	$e_{12} = \frac{f_1 \times f_2}{n}$	$e_{13} = \frac{f_1 \times f_3}{n}$	f_1
2	$e_{21} = \frac{f_2 \times f_1}{n}$	$e_{22} = \frac{f_2 \times f_2}{n}$	$e_{23} = \frac{f_2 \times f_3}{n}$	f_2
Column sum	f_1	f_2	f_2	

From the observed and the theoretically frequencies, we subsequently calculated the sum of deviations U. The sum of deviations U is our test statistic. This is $\chi 2$-distributed with $df = (j - 1) \times (k - 1)$ degrees of freedom, where j is the number of rows and k is the number of columns.

Thus, we know our test distribution, it is the $\chi 2$-distribution and we know our test value U. With this, we can already perform the independence test.

In our example from Chap. 6 we asked ourselves whether there is a relationship between the founding motive and the sector, in which the enterprise is founded. We observed the following cell frequencies for our 100 companies:

Sector	Motive			
	Unemployment	Implement idea	Higher income	Row sum
Industry	8	15	11	34
Service	9	39	18	66
Column sum	17	54	29	

and calculated the following theoretical values:

Sector	Motive			
	Unemployment	Implement idea	Higher income	Row sum
Industry	$e_{11} = \frac{34 \times 17}{100} = 5.78$	$e_{12} = \frac{34 \times 54}{100} = 18.36$	$e_{13} = \frac{34 \times 29}{100} = 9.86$	34
Service	$e_{21} = \frac{66 \times 17}{100} = 11.22$	$e_{22} = \frac{66 \times 54}{100} = 35.64$	$e_{23} = \frac{66 \times 29}{100} = 19.14$	66
Column sum	17	54	29	

From this, we had our sum of deviations U amounting to 2.42 and a contingency coefficient C amounting to 0.15. Due to the low level of the contingency coefficient, we claimed that there is no relationship between the two variables. We can now check whether the variables are actually independent in the population.

The null and alternative hypotheses are as follows:

H_0: *There is no relationship between the founding motive and the sector in which the enterprise is founded.*

H_A: *There is a relationship between the founding motive and the sector in which the enterprise is founded.*

$$H_0: \quad C = 0$$

$$H_A: \quad C > 0$$

We want to test this at a significance level of 5%:

$$\alpha = 5\% \quad \text{or} \quad \alpha = 0.05$$

Our test value is the sum of deviations U amounting to 2.42. This is $\chi 2$-distributed with $df = (j - 1) \times (k - 1) = (2 - 1) \times (3 - 1) = 2$ degrees of freedom. We can take our critical value from the $\chi 2$-distribution table. It has a value of 5.991 at 2 degrees of freedom (compare Appendix 4).

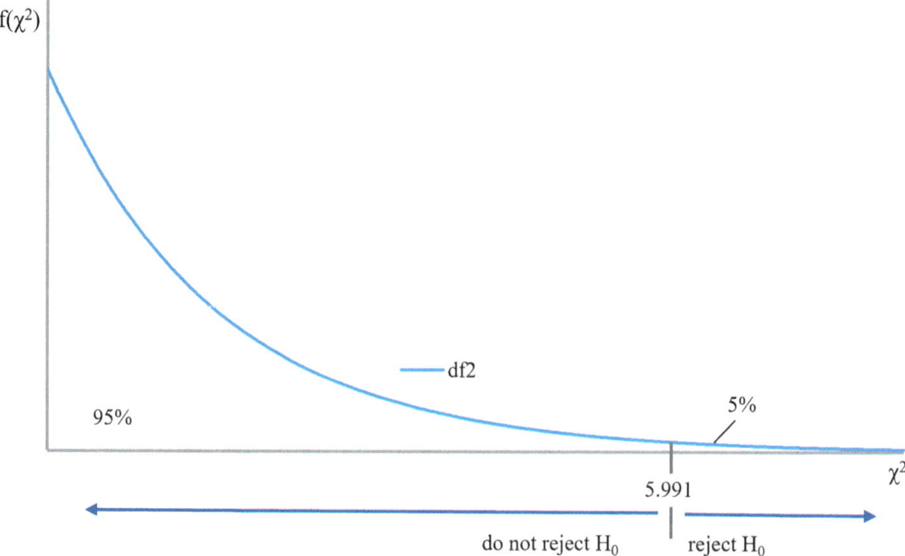

If we compare our test value of 2.42 with the critical value, we find that we do not reject the null hypothesis. Based on the sample findings, we can state that there is no relationship between the founding motive and the sector in which the enterprise is founded. This result was to be expected, as the contingency coefficient, which we can also take as a measure of effect size, is very small in the sample.

Here too, it is important to ensure that each cell is occupied by at least 5 observations. Otherwise, the test may lead to incorrect results.

16.4 Calculation of the Correlation Tests with the R Commander

To calculate the tests for correlation with metric variables, we first draw the scatterplot between the variables and check whether there is a linear relationship between them (see Sect. 6.9). We then perform the test. In the R Commander, we have the command Correlation test available for this. We can use this command for both the test for correlation with metric data and with ordinal data. We find the command in the Statistics menu under Summaries. We select the variables, define the required correlation coefficient, and specify the alternative hypothesis.

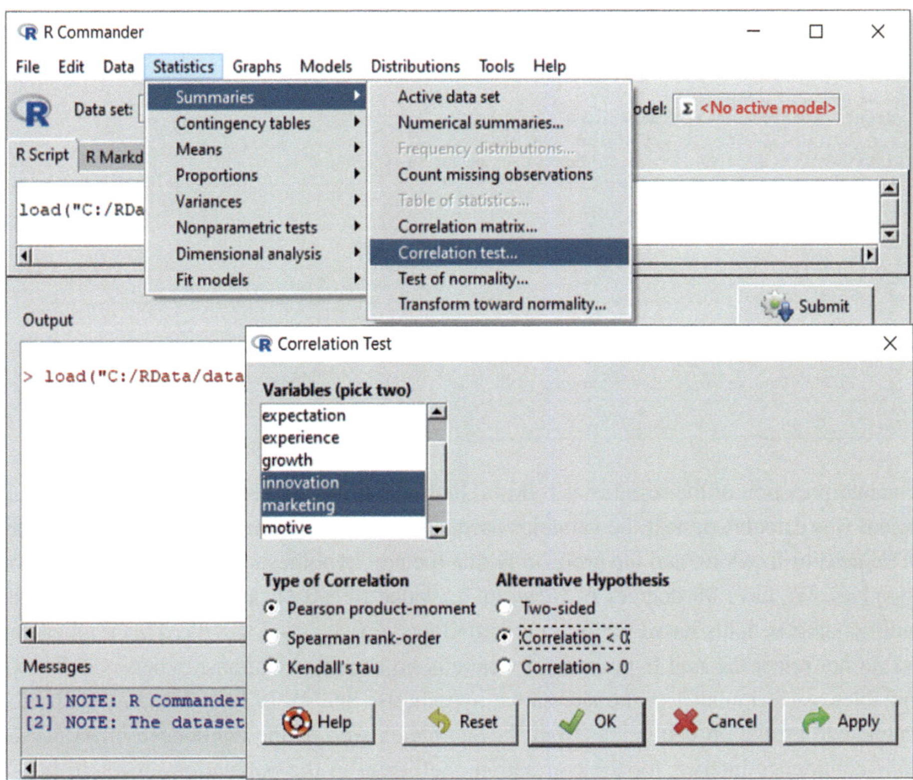

We then confirm with OK and we get the test result.

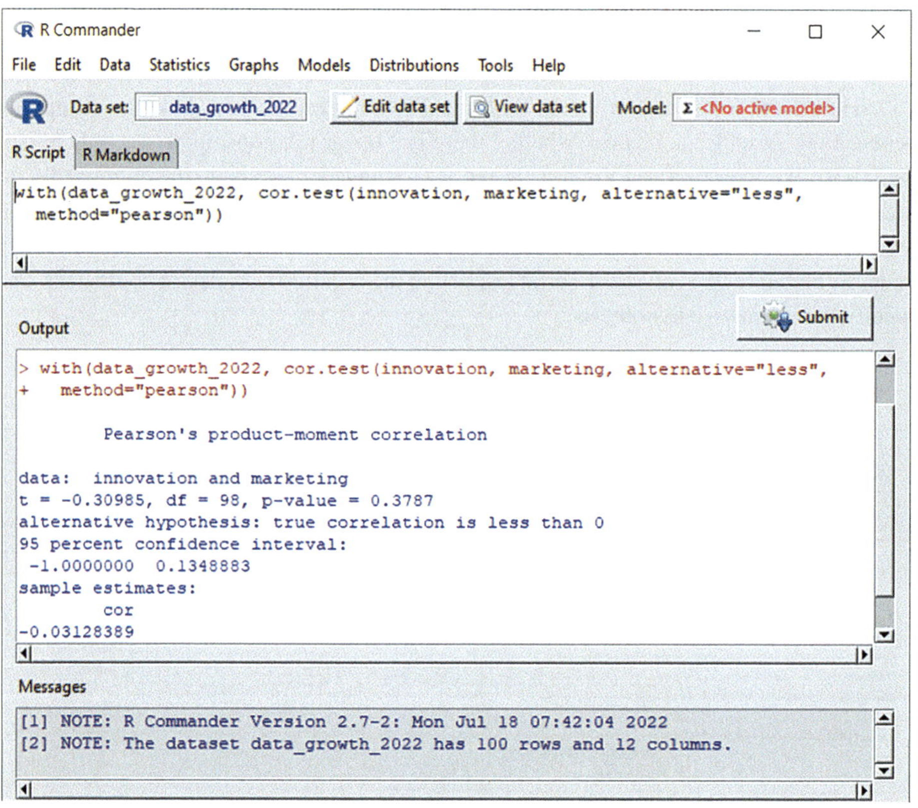

The interpretation of the result is as follows. In the first line, we see which correlation coefficient was calculated, next the variables used, and in the third line our t-value, the number of degrees of freedom, and the probability that the null hypothesis applies to the population (p-value). We have 98 degrees of freedom, a t-value of −0.31, and a probability that the null hypothesis holds based on the sample findings of 0.3787 or 37.87%. This implies that we do not reject the null hypothesis. So there is no negative relationship between the two variables, as formulated in the alternative hypothesis (line 4). In lines 5 to 9, we find the estimated correlation coefficient from the sample as well as the confidence interval. The 95% confidence interval for the correlation coefficient in the population ranges from −1 to 0.13. The calculated correlation coefficient is −0.03. A note: the result deviates from the example in this chapter. The reason is that in the example only the first six observations were calculated ($n = 6$), while here we refer to the complete data set ($n = 100$).

If we want to enter the command directly into the command line, we use the command.

cor.test(data_growth_2022$marketing, data_growth_2022$innovation, alternative = “less”, method = “pearson”).

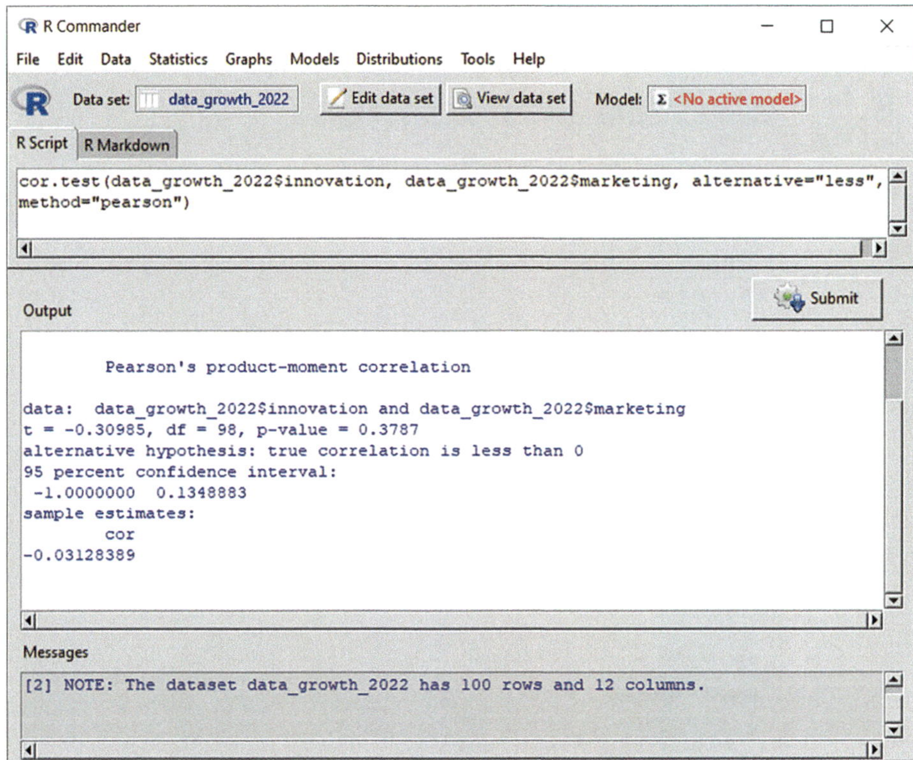

The procedure and interpretation when testing for a correlation with ordinal data are largely identical. The only difference is that we use and interpret the Spearman rank correlation coefficient. Therefore, we will not go into a detailed presentation here. The command changes in that we define spearman as the method:

cor.test(data_growth_2022$expectation, data_growth_2022$self-assessment, alternative = "greater", method = "spearman").

Now we want to look at the test for independence with nominal variables. This is identical for nominal variables with two characteristics as well as for nominal variables with more than two characteristics. To perform the test, we first have to tell R again that we have nominal variables. We do this with the command Convert numeric variables to factors in the Data menu and the submenu Manage variables in active data set (see, for example, Sect. 6.9). After we have told R which are the nominal variables for which we want to perform the test, we find the corresponding command in the Contingency tables submenu in the Statistics menu. The command is called by clicking on Two-way table. In the window that opens, we mark the row and column variable and in the Statistics tab the corresponding options. We are particularly interested in the Chi-square test of independence and the expected frequencies.

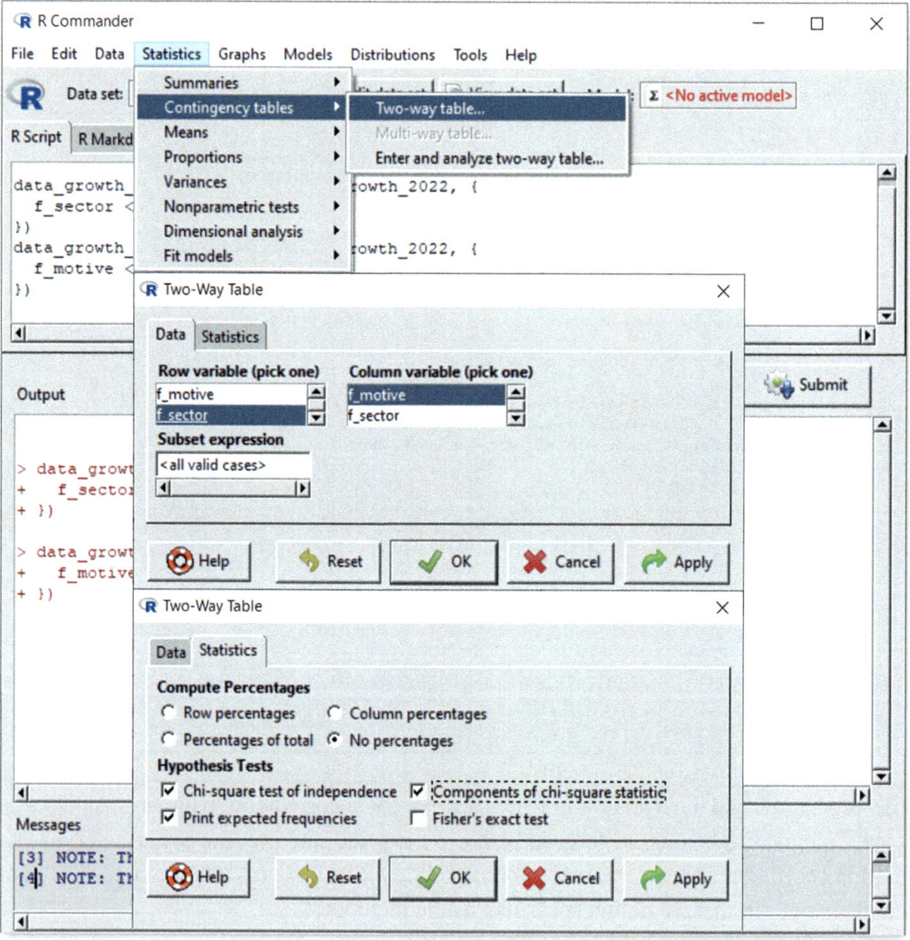

After entering, we confirm with OK and receive the result of the test.

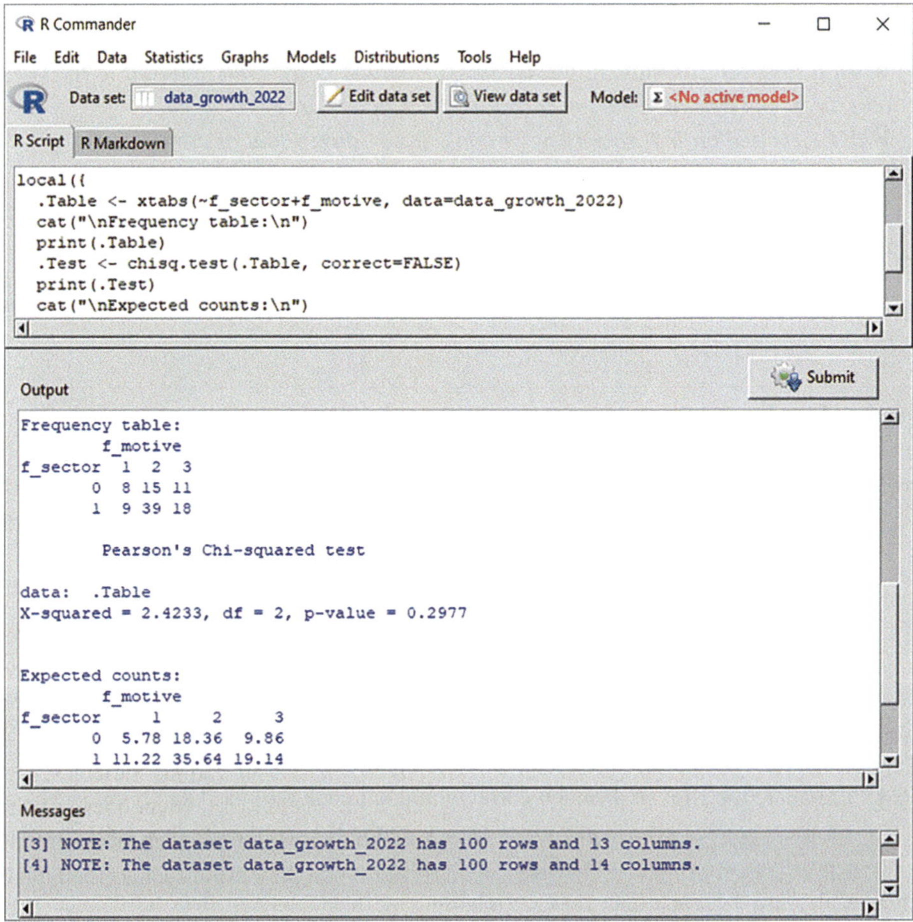

We are almost familiar with all the numbers again, we have seen them in the book several times. The first output area contains the observed frequencies for the founding motive in the respective sector (Frequency table). In the lower output area, we receive the expected frequencies (Expected counts). In between, we find our test for independence. This begins with its name (Pearson's Chi-squared test). The data are specified below. These are the frequencies in the table. Following this, we receive our sum of deviations U or the χ2-square value of 2.42, the number of degrees of freedom, and the probability that the variables are independent of each other (p-value). This is 0.298 or 29.8% here. This is the probability of obtaining a test value of 2.42 (see example above), assuming that the null hypothesis represents the population. As above, we therefore do not reject the null hypothesis based on the findings. We would only reject if this value would fall below our specified significance level.

16.5 Checkpoints

- *When testing for correlation, we use the correlation coefficients relevant to the variables, e.g., the Bravais-Pearson correlation coefficient for metric data.*
- *With metric and ordinal variables, we can test non-directional or directional.*
- *With metric and ordinal variables, the test variable is t-distributed.*
- *With nominal variables, we test whether a relationship exists. These test procedures are often called tests for independence.*
- *With nominal variables, the test variable is $\chi 2$-distributed, the test procedures are often also referred to as $\chi 2$-independence tests.*

16.6 Application

16.1. Calculate the Bravais-Pearson correlation coefficient for our data set data_growth for the variables growth and marketing for the first eight observations (n = 8) by hand and test the result non-directional at the significance level of 5%.

16.2. Calculate the Bravais-Pearson correlation coefficient for our data set data_growth for the variables growth and marketing with the R Commander (n = 100) and test the result non-directional at the significance level of 5%.

16.3. Calculate the Bravais-Pearson correlation coefficient for our data set data_growth for the variables growth and experience for the first eight observations (n = 8) by hand. Test for a positive correlation at the significance level of 1%.

16.4. Calculate the Bravais-Pearson correlation coefficient for our data set data_growth for the variables growth and experience with the R Commander (n = 100). Test for a positive correlation at the significance level of 1%.

16.5. Calculate the Spearman rank correlation coefficient for our data set data_growth for the variables growth and self-assessment for the first eight observations (n = 8) by hand and test for a positive correlation at the significance level of 10%.

16.6. Calculate the Spearman rank correlation coefficient for our data set data_growth for the variables growth and self-assessment with the R Commander (n = 100) and test for a positive correlation at the significance level of 10%.

16.7. Calculate with the R Commander whether there is a relationship between the variables sex and sector in the data set data_growth (n = 100). Test the result at the significance level of 5%.

16.8. Calculate the contingency coefficient for our data set data_growth for the variables sex and motive with the R Commander (n = 100) and test the result at the significance level of 10%.

16.9. We are interested in whether there is a correlation between three stock values and we obtain historical values for VW, Daimler and SAP (see Chap. 6).

Date	VW	Daimler	SAP
01.04.2020	99.27	25.74	97.40
31.03.2020	105.48	27.14	101.48
30.03.2020	105.50	27.21	101.70
27.03.2020	106.50	27.32	100.78
26.03.2020	112.90	29.21	103.88
25.03.2020	115.50	30.25	100.00

Test at the significance level of 5% between which stocks a relationship exists.

More Tests for Nominal Variables 17

By now, we have become familiar with a whole range of test procedures. What we still lack is an important strand, the test procedures for nominal data. We want to take a look at these now. The test procedures for nominal data are often briefly called $\chi 2$-tests, as the test distribution is usually the $\chi 2$-distribution. With the $\chi 2$-test for independence, we have already learned one of the procedures for nominal data. We will now also look at how the tests with nominal data and one sample, with two independent samples, and with two dependent samples.

17.1 The χ^2-Test with One Sample: Does the Proportion of Female Founders Correspond to the Gender Ratio in Society?

In the χ^2-test with one sample, we examine whether the observed frequencies in a sample match the frequencies that we have based on our prior knowledge. Looking at our data set, we can, for example, ask ourselves whether the proportion of female founders corresponds to the gender ratio in society or not, i.e., whether significantly fewer or significantly more women start a company. But we could also ask whether the proportion of female founders in the sample corresponds to the proportion of female founders in the population. In the latter case, we would check our sample for representativeness.

Let's go through the procedure using the first question. We ask ourselves whether more or fewer women start a company than men. The null and alternative hypotheses are then as follows:

F. Kronthaler, *Statistics Applied with the R Commander*,
https://doi.org/10.1007/978-3-662-69107-6_17

H_0: The frequency of companies founded by men and women corresponds to the gender ratio in society.
H_A: The frequency of companies founded by men and women does not correspond to the gender ratio in society.

We know that there are approximately equal numbers of women and men in society, so we can also write more briefly as follows:

$$H_0: \quad \pi_1 = \pi_2 = 0.5$$
$$H_A: \quad \pi_1 \neq \pi_2 \neq 0.5$$

with
π_1 and π_2 as the proportions of men and women in the population.

We now want to test this at a significance level of 10%:

$$\alpha = 10\% \quad \text{or} \quad \alpha = 0.1$$

The test statistic is the sum of deviations, i.e., we examine how the observed values deviate from the theoretically expected values:

$$U = \sum \frac{(f_i - e_i)^2}{e_i}$$

where
f_i are the observed values and
e_i are the theoretically expected values.

The test statistic is χ^2-distributed with $df = c - 1$ degrees of freedom. c is the number of categories that can occur in the variable.

Now we need the observed values and the theoretically expected values. In our sample of 100 companies, we observed a total of 35 female founders and 65 male founders. Based on our gender ratio in society, we would expect $100 \times 0.5 = 50$ female founders and 50 male founders. We can insert these numbers into our formula:

$$U = \sum \frac{(f_i - e_i)^2}{e_i} = \frac{(65 - 50)^2}{50} + \frac{(35 - 50)^2}{50} = 9.0$$

where we have $df = c - 1 = 2 - 1$ degrees of freedom. There are two possible outcomes, male and female, i.e., we have two possible response categories.

We now only need to determine our critical χ^2 value and can then carry out the test decision. The critical χ^2 value for one degree of freedom and a significance level of 10% is 2.706 according to the χ^2 distribution table (see Appendix 4).

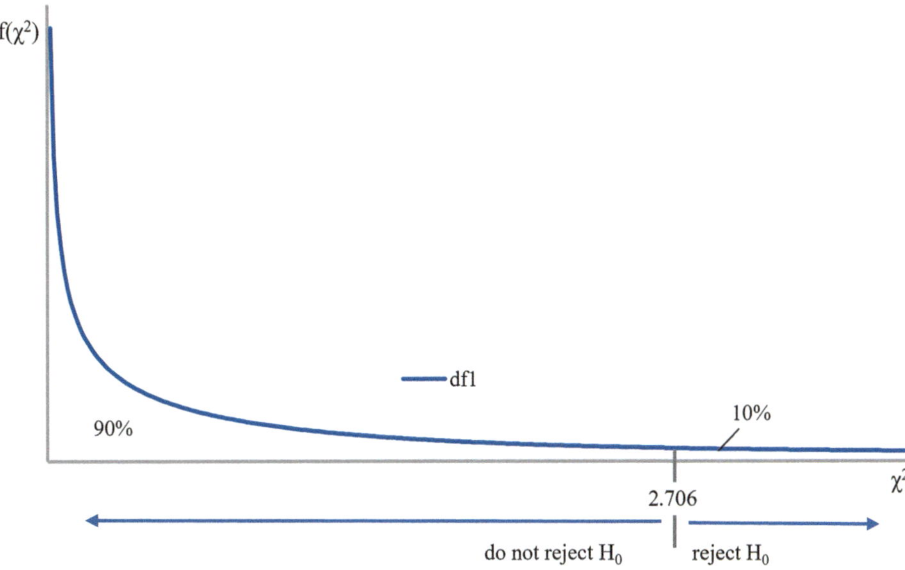

If we compare our calculated value for the sum of deviations U of 9.0 with our critical value of 2.706, we reject the null hypothesis. This implies that the frequency of companies founded by men and women does not correspond to the gender ratio in the total population. Men found companies more often than women. The size of the difference can be clearly seen from the numbers. In absolute terms, we have 35 female founders and 65 male founders in the sample, i.e., 35% of the foundations are by women and 65% by men.

Freak Knowledge

Measures for calculating the effect size in χ^2 -tests are, for example, the Phi coefficient or Cramer's V.

17.2 The χ^2 -Test with Two Independent Samples: Are the Founding Motives the Same for Sservice and Industrial Companies?

With the χ^2test with two independent samples, we examine whether two groups differ in terms of the frequency proportions of a nominal variable. We compare whether the respective frequencies in the respective groups deviate from the theoretically expected frequencies.

We explain this best using the example used in Chap. 16 for the independence test. Here we had examined whether there is a relationship between the founding motive and the sector. We observed the following cell frequencies for our 100 companies:

Sector	Motive			
	Unemployment	Implement idea	Higher income	Row sum
Industry	8	15	11	34
Service	9	39	18	66
Column sum	17	54	29	

The cell frequencies can be read as follows: In the industry sector, 8 out of 34 companies (23.5%) are founded out of unemployment, 15 out of 34 (44.1%) of the founders want to implement an idea and 11 out of 34 (32.4%) want to earn a higher income. For service sector startups, the ratios are 9 to 66 (13.6%), 39 to 66 (59.1%) and 18 to 66 (27.3%).

The test is supposed to help us determine whether the two groups differ in terms of the proportions of founding motives. We test the following null and alternative hypothesis:

H_0: The proportions of the motives to found an enterprise are the same for industrial and service companies.
H_A: The proportions of the motives to found an enterprise are not the same for industrial and service companies.

The procedure of the test corresponds to the procedure of the independence test. Here too, we compare the observed frequencies with the theoretically expected frequencies. The test statistic is the same, it is again the sum of deviations U:

$$U = \sum \sum \frac{\left(f_{jk} - e_{jk}\right)^2}{e_{jk}}$$

where U is again χ^2-distributed with $df = (j - 1) \times (k - 1)$ degrees of freedom.

Thus the test is the same a discussed in Chap. 16 . The only difference lies in the formulation of the hypotheses, once we test for independence, the other time for a difference between two groups. We take the calculation of the deviation sum from Chap. 16, it has a value of 2.42 with 2 degrees of freedom. Thus, the test decision is the same as above. We do not reject our null hypothesis.

17.3 The χ^2Test with Two Dependent Samples: Is my Advertising Campaign Effective?

Sometimes we wonder whether a measure changes the behavior of individuals or objects, for example, whether an advertising measure carried out by our company has an influence on whether people buy our product or not. The null and alternative hypotheses would be as follows:

H_0: The advertising campaign has no influence on whether people buy our product.
H_A: The advertising campaign has an influence on whether people buy our product.

We want to test these hypotheses at a significance level of 1%:

$$\alpha = 1\,\% \quad \text{or} \quad \alpha = 0.01$$

To carry out the test, we draw a random sample of $n = 300$ people and ask them before our advertising campaign whether they are buyers or non-buyers of our product. After the campaign, we ask the same people again. Subsequently, we can build the following 2×2 matrix:

		Survey after campaign	
		Buyer	Non-buyer
Survey before campaign	Buyer	$f_a = 80$	$f_b = 50$
	Non-buyer	$f_c = 100$	$f_d = 70$

where the rows represent the situation before our campaign and the columns represent the situation after our campaign. f_a to f_d are the observed frequencies in the cells a to d.

We can read the matrix as follows: Before the campaign, we had $80 + 50 = 130$ buyers and $100 + 70 = 170$ non-buyers in the sample. After the campaign, we have $80 + 100 = 180$ buyers and $50 + 70 = 120$ non-buyers. Looking at the matrix more closely, we find that of our original 130 buyers, 50 have switched to non-buyers and of our original 170 non-buyers, 100 have switched to buyers. In total, we have 150 switchers in the sample.

The test procedure, which now compares whether this change occurs randomly, or whether our advertising measure has an influence on it, is the McNemar χ^2-test. It only considers the switchers, i.e., the cells at the top right (b) and bottom left (c) in the matrix and assumes that half of the switchers will switch from buyers to non-buyers and the other half from non-buyers to buyers. To generalize this a bit, we can also say that one half of the switchers switch from 0 to 1 and the other half from 1 to 0.

With this knowledge, we can calculate the theoretically expected values for the cells b and c:

$$e_b = e_c = \frac{f_b + f_c}{2}.$$

The theoretically expected frequencies for the cells *b* and *c* is therefore:

$$e_b = e_c = \frac{f_b + f_c}{2} = \frac{50 + 100}{2} = 75.$$

With the observed frequencies and the theoretically expected frequencies, we can again calculate our sum of deviations, which simultaneously serves as test statistic. It is χ^2-distributed and has df = 1 degrees of freedom:

$$U = \sum \frac{(f_i - e_i)^2}{e_i} = \frac{(f_b - e_b)^2}{e_b} + \frac{(f_c - e_c)^2}{e_c} = \frac{(50 - 75)^2}{75} + \frac{(100 - 75)^2}{75} = 16.7.$$

Now we just need to determine our critical χ^2-value. We take this from the χ^2-distribution table at a level of 6.635 (compare appendix 4).

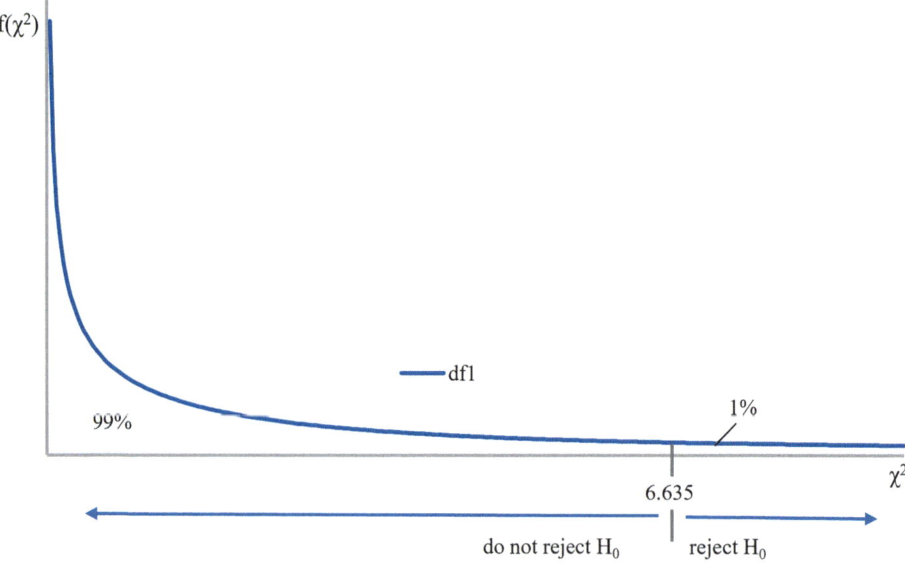

If we compare the calculated value with the critical value, we see that we reject the null hypothesis. We can therefore assume that our advertising campaign works. We have significantly more buyers after the campaign, with the switch from non-buyers to buyers being significantly higher than the switch from buyers to non-buyers.

Finally, we want to make an important marginal note. For the χ^2-test procedures to be effective, each cell occupancy should include at least 5 observations. If this is not the case, the test results are not reliable.

Freak-Knowledge

In the χ^2-test procedures discussed here, we have each tested non-directional. For χ^2-tests, this is the norm. If a 2×2 matrix is present, we can also test directional. In such a case, we follow the logic of the non-directional test procedures and double the significance level.

17.4 Calculation with the R Commander

We can find the command for calculating the χ2-tests in one sample in the Statistics menu under the Proportions submenu. To activate the command, we first need to convert the variable required for the test procedure into a factor using the Data menu and the Manage variables in active data set submenu. After we have called up the command Single-sample proportion test, a window opens in which we select the variable. In the Options tab, we also enter the information for the test procedure. We specify the alternative hypothesis and the proportion value assumed in the alternative hypothesis, possibly the confidence interval and the type of test. As a rule, we take the version that is preset as standard.

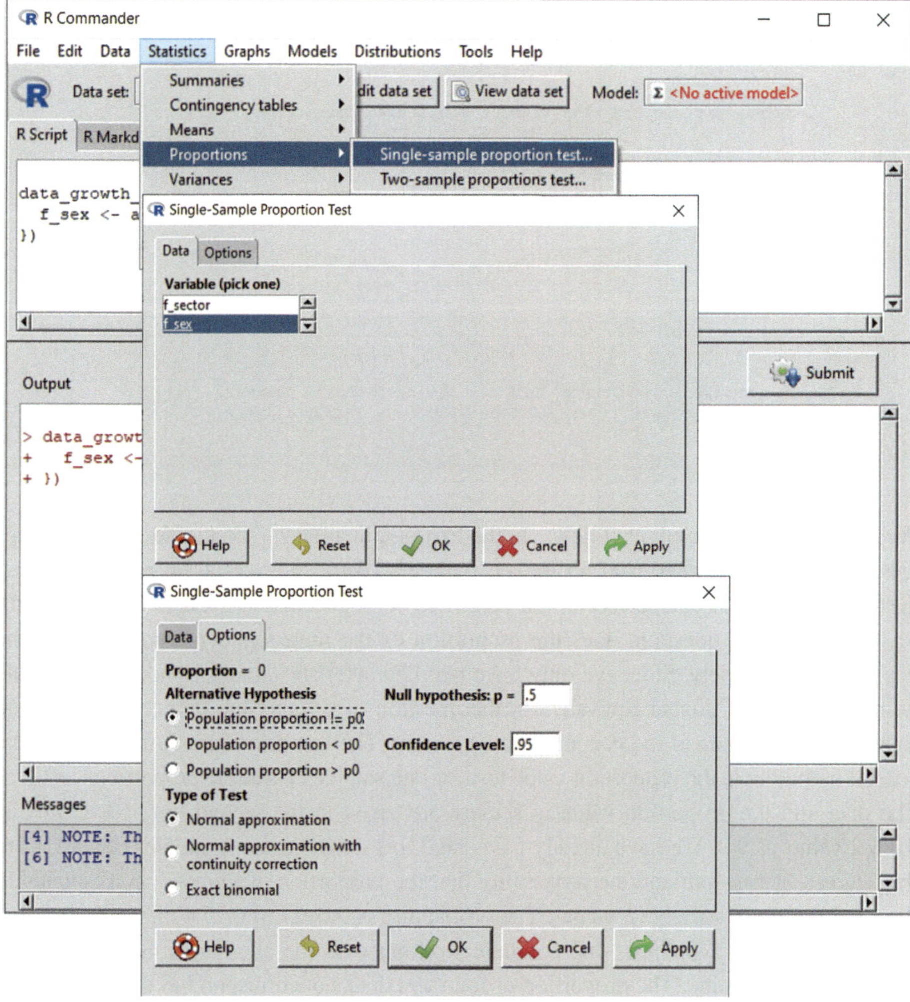

We then confirm with OK and we receive the test result.

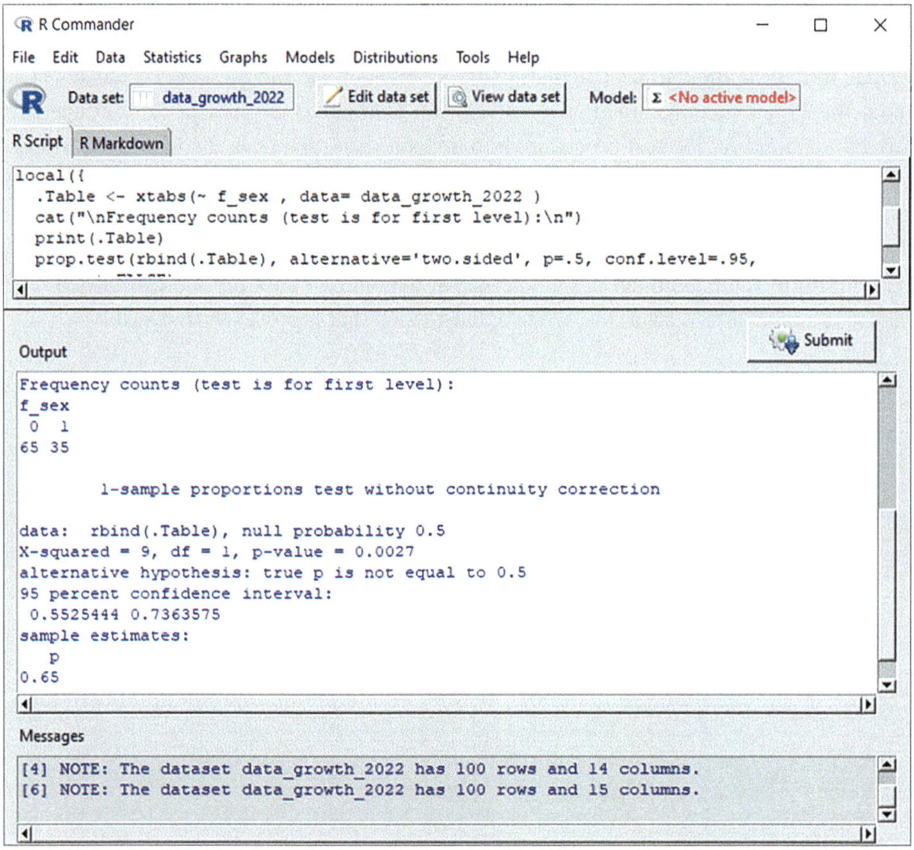

R Commander

File Edit Data Statistics Graphs Models Distributions Tools Help

Data set: data_growth_2022 Edit data set View data set Model: Σ <No active model>

R Script R Markdown

```
local({
  .Table <- xtabs(~ f_sex , data= data_growth_2022 )
  cat("\nFrequency counts (test is for first level):\n")
  print(.Table)
  prop.test(rbind(.Table), alternative='two.sided', p=.5, conf.level=.95,
```

Submit

Output

```
Frequency counts (test is for first level):
f_sex
 0  1
65 35

        1-sample proportions test without continuity correction

data:  rbind(.Table), null probability 0.5
X-squared = 9, df = 1, p-value = 0.0027
alternative hypothesis: true p is not equal to 0.5
95 percent confidence interval:
 0.5525444 0.7363575
sample estimates:
   p
0.65
```

Messages

```
[4] NOTE: The dataset data_growth_2022 has 100 rows and 14 columns.
[6] NOTE: The dataset data_growth_2022 has 100 rows and 15 columns.
```

We can interpret the result as follows. In the first output area, we see the observed frequencies, we have 65 founders and 35 female founders. We also see that we have conducted the test for our first characteristic, the zero, i.e., for the founders (test is for first level). So we answer the question, does the proportion of the male founders correspond to the gender ratio in society. Since we only have two characteristics, male founders and female founders, this is of course equivalent to the question of whether the proportion of female founders corresponds to the gender ratio in society. The test procedure is indicated in the second output area, the proportion value test for one sample (1-sample proportions test …). The data and the proportion value in society are given in the second line. Next, we see the $\chi 2$ value of 9.0. We have already calculated this above in our example. Behind it are the degrees of freedom and the probability that the proportion of founders corresponds to the gender ratio in society (p-value). The probability is 0.0027 or 0.27%. This implies that we reject the null hypothesis and continue working with the alternative hypothesis indicated in the fourth line. The proportion of founders does not correspond to the gender ratio in society. In lines 5 to 9 we find the 95% confidence interval and the proportion value estimated for the population from the sample. The latter is 65%, the confidence interval ranges from 55.2% to 73.6%. The $\chi 2$ test for two independent samples can be carried out in the same way as the test for independence, as described in Sect. 16.5. For the $\chi 2$ test for two dependent samples, we do not have a command via the menu navigation. In this case, we need to know the command. At this point, I would like to encourage you to try to search for the command structure on the internet, either with the help function of the R Commander or with a search engine. The command we find probably has the following structure:

mcnemar.test(x, y, correct = FALSE).

Applied to our example, the command is accordingly

mcnemar.test(data_advertising_2022$f_pre, data_advertistin_2022$f_post, correct = FALSE).

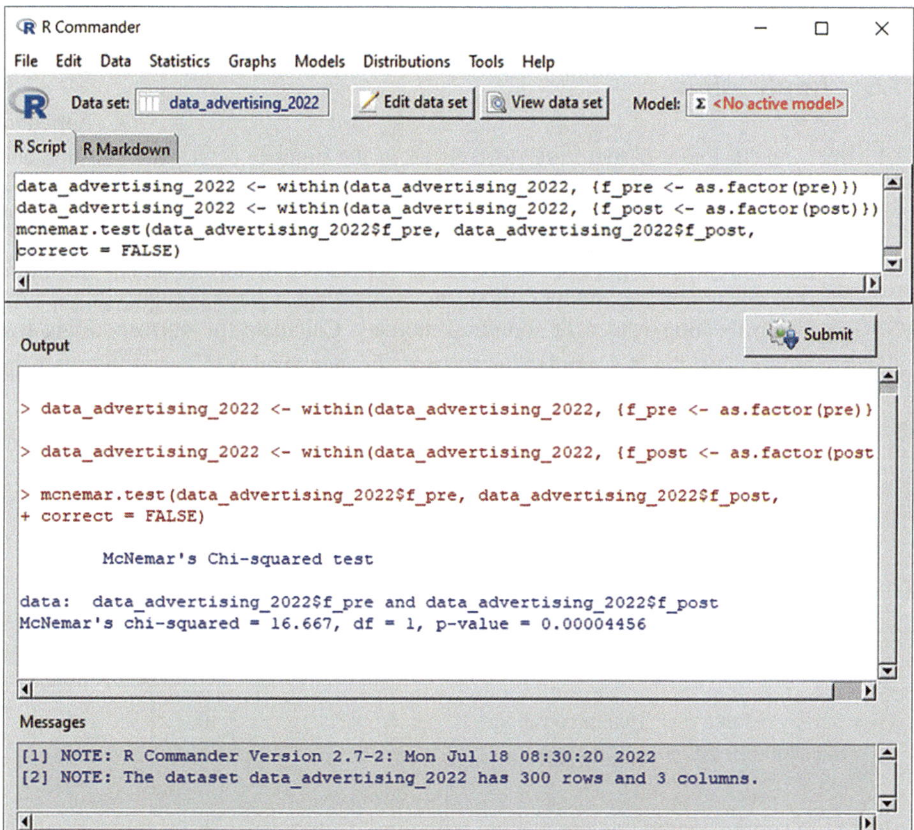

In the result, we see that we have conducted the McNemar $\chi 2$ test. The $\chi 2$ value of 16.7, which corresponds to our deviation sum calculated by hand above. We have 1 degree of freedom and the probability that our advertising campaign has no influence on whether people buy our product is very small, 0.000045 or 0.0045%. This means that we reject our null hypothesis, the advertising campaign is effective. Just a small note, it could be that our advertising campaign works, but in the wrong direction, i.e., it deters people from buying our product. It is therefore always useful to look closely at the frequency table given above in the example.

17.5 Checkpoints

- *The test procedures for nominal data are based on the analysis of frequencies.*
- *The test procedures for nominal data are often simply referred to as χ^2-tests.*
- *The cell frequencies should comprise at least five observations per cell in χ^2-test procedures.*

17.6 Application

17.1. We want to know if there are differences in the number of service startups and industrial startups. We consider the following null hypothesis: Service startups and industrial startups occur equally often. Calculate the corresponding test procedure with the R Commander at a significance level of 5% and interpret the result.

17.2. We are interested in whether there are differences between men and women regarding the importance of founding motives. Calculate the corresponding test procedure with the R Commander at a significance level of 5% and interpret the result.

17.3. We are conducting an educational campaign in our company regarding nutrition in the workplace. Our goal is to promote health-conscious nutrition in the workplace. We randomly observe 200 employees and ask them about their nutrition in the workplace before and after the campaign (data_awareness_campaign). With the help of these results, we can create the following 2×2 matrix:

		Survey after education	
		Health conscious	Non-health conscious
Survey before education	Health conscious	60	40
	Non-health conscious	70	30

Calculate the corresponding test procedure with the R Commander at a significance level of 10% and interpret the result.

18 Summary Part IV—Overview of the Test Procedures

At the end of the fourth part of the book, it makes sense to summarize the discussed test procedures in a table. This table can serve as a tool to choose a test procedure, alongside Fig. 11.5. It must be noted as a limitation that the table does not, of course, present all test procedures (Tab. 18.1).

F. Kronthaler, *Statistics Applied with the R Commander*,
https://doi.org/10.1007/978-3-662-69107-6_18

Tab. 18.1 Test procedures and examples

Test	When is the test used?	Examples
Examination of one group		
One-sample *t*-test	Examination of a mean for a metric and normally distributed variable	What is the average age of company founders?
Wilcoxon test	Examination of a mean for a metric non-normally distributed variable or the median for an ordinal variable	How much professional experience do founders have on average?
χ^2-test	Comparison of observed frequencies with theoretically expected frequencies for a nominal variable	Does the proportion of female founders reflect the gender share in society?
Examination of two independent groups		
t-test with independent samples	Testing for a difference between groups for a metric and normally distributed variable	Is there an age difference between men and women when founding companies?
Mann-Whitney test	Testing for a difference between groups for a metric non-normally distributed variable or an ordinal variable	Is there a difference in the assessment of industry experience between male and female founders?
χ^2-test	Testing whether two groups differ in terms of the frequency proportions for a nominal variable	Do founders of service companies and those of industrial companies differ in their founding motives?
Examination of more than two independent groups		
Analysis of variance	Testing for a difference between more than two groups for a metric and normally distributed variable	Do company founders differ in workload according to different founding motives?
Examination of two dependent groups		
t-test for dependent samples	Testing whether there are differences in a group before and after a measure, for a metric and normally distributed variable	Does an advertising measure have an influence on the image of my product (measured metrically)?
Wilcoxon test for dependent samples	Testing whether there are differences in a group before and after a measure, with a metric non-normally distributed variable or with an ordinal variable	Does an advertising measure have an influence on the image of my product (measured ordinally)?

(continued)

Tab. 18.1 (continued)

Test	When is the test used?	Examples
McNemar χ^2- test	Investigation of whether there are differences in a group before and after a measure, with a nominal variable	Does the advertising measure have an influence on whether people buy my product or not?
Testing for a correlation between two variables or for independence		
Bravais-Pearson	Testing whether there is a relationship between two metric variables	Is there a relationship between the expenditure on marketing and the expenditure on innovation?
Spearman	Testing whether there is a relationship between two ordinal variables	Is there a relationship between healty nutrition and education?
χ^2 -independence test	Testing whether there is a relationship between two nominal variables	Is there a relationship between the sector in which the company is founded and the motivation for founding?

Part V
Regression Analysis

Regression Analysis: The Ability to Predict What Will Happen

Regression analysis is an extremely popular tool that is used in many different situations in research. For instance, we could ask whether marketing expenditures have an impact on company growth and what happens if we increase or decrease these expenditures. We could also attempt to determine which factors influence the performance of football teams or which European football club is likely to win the Champions League: Barcelona, Bayern, Real Madrid or perhaps Dortmund? The examples demonstrate that the application possibilities are diverse. Generally, we use regression analysis to explain and predict. To illustrate this a little bit, here are a few questions:

- What factors influence the performance of a football team? Is the total value of the football team or the presence of a superstar central?
- What influence do my marketing expenditures have on my sales figures? What happens if I increase marketing expenditures?
- What influence do research and development expenditures have on a country's rate of innovation? What happens if a country increases its expenditures for basic research?
- What effect does development aid have on the growth of developing countries? What happens if the payments are increased?
- What influence does violence on television have on the propensity for violence among young people? What happens if television programs become more violent?

The possibilities for application are manifold. Unfortunately, answering such questions is not always easy, even with sophisticated statistical methods, and as we will see, requires thorough theoretical considerations in addition to statistical methods.

19 The Simple Linear Regression

19.1 Objectives of Regression Analysis

The objective of regression analysis is to explain the influence of a variable X on a variable Y and to forecast what happens to the variable Y when the variable X changes or is changed. Y is accordingly the dependent variable and X the independent variable:

$$Y \longleftarrow X$$

dependent variable (metric) *independent variable (metric)*

We can read the above relationship as follows. The variable Y, the dependent variable, is influenced by the variable X, the independent variable. Both variables are metric.

Freak-Knowledge

In addition to the terms dependent and independent variable, numerous other terms are used, depending on the field of study:

dependent variable	*independent variable*
endogenous variable	*exogenous variable*
response variable	*explanatory variable*
criterion variable	*predictor variable*
regressand	*regressor*

F. Kronthaler, *Statistics Applied with the R Commander*,
https://doi.org/10.1007/978-3-662-69107-6_19

The distinction into dependent and independent variables is not a statistical, data-based classification. It is of a theoretical nature. This means that the distinction is based on the theoretical justification that X has an influence on Y. This theoretical foundation is central to regression analysis. If we do not have this, the result of the regression analysis is worthless. We cannot then say what happens to Y when we change X, as we do not know whether X has an influence on Y or whether the relationship is the other way around.

Let's discuss this again using an example. The question we ask is as follows: Does the industry professional experience of the founder have an influence on turnover growth? To answer the question, we first describe the relationship using existing theories and literature. We may then come to the research hypothesis that industry professional experience of a founder has an influence on turnover growth. Perhaps founders with more industry professional experience know the market better and can therefore offer better products, etc. The industry professional experience is thus the independent variable and turnover growth the dependent variable.

$Y \longleftarrow X$

turnover growth (metric) *professional experience (metric)*

If we look more closely at the question of causality, we notice that the direction here is clear. The industry professional experience at the time of founding can have an influence on turnover growth. The reverse is not possible, as turnover growth follows the time of founding and thus the industry professional experience of the founder at the time of founding. The question of causality is unfortunately not always so simple.

19.2 The Linear Regression Line and the Method of Ordinary Least Squares

The objective of linear regression is to determine the linear relationship between the dependent and independent variable. We explain the procedure best using our example. As discussed above, we define the growth rate of companies as the dependent variable Y and the experience of the company founders as the independent variable X.

$Y \longleftarrow X$

growth rate (metric) *experience (metric)*

We start with the scatterplot between the two variables. The scatterplot displays the relationship between the dependent and the independent variable and provides information about whether the relationship is linear (Fig. 19.1).

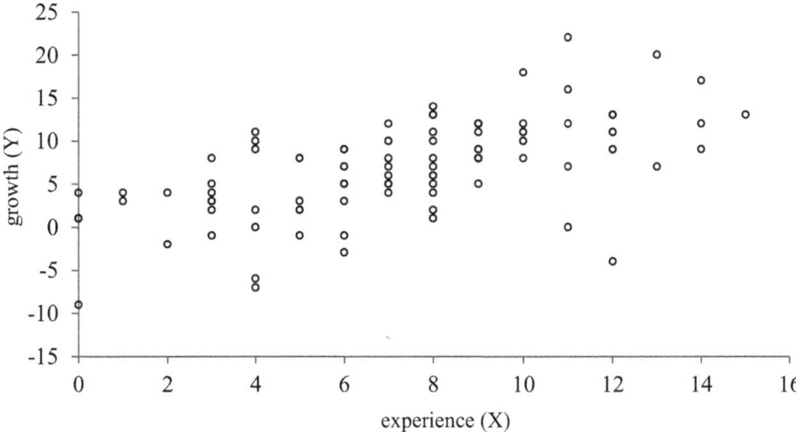

Fig. 19.1 Scatterplot for the variables growth and experience

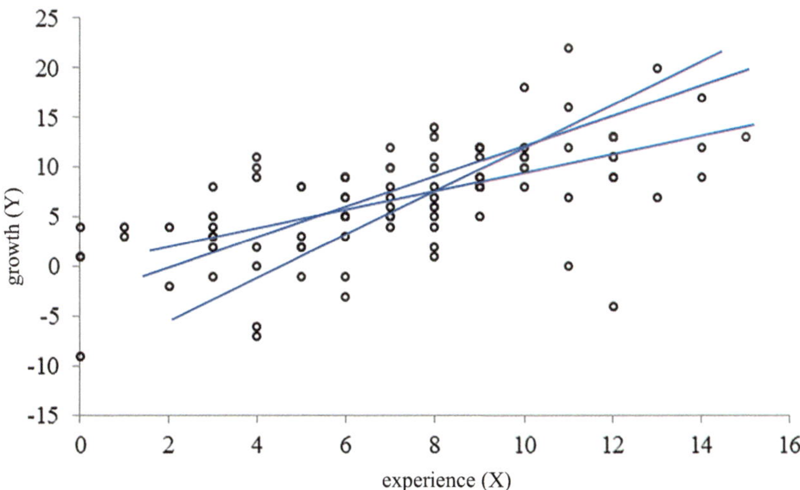

Fig. 19.2 Optimal line for the scatterplot for the variables growth rate and experience?

The objective of simple linear regression is to find an optimal straight line which describes the cloud of points. As with the Bravais-Pearson correlation coefficient, this assumes that the relationship is linear and a straight line fits to the data points

The question is now how we can lay a straight line in the cloud of points so that it optimally describes the relationship. One possibility is to draw the line freehand to the best of our knowledge and belief. In Fig. 19.2, three possibilities are drawn how the line could look like.

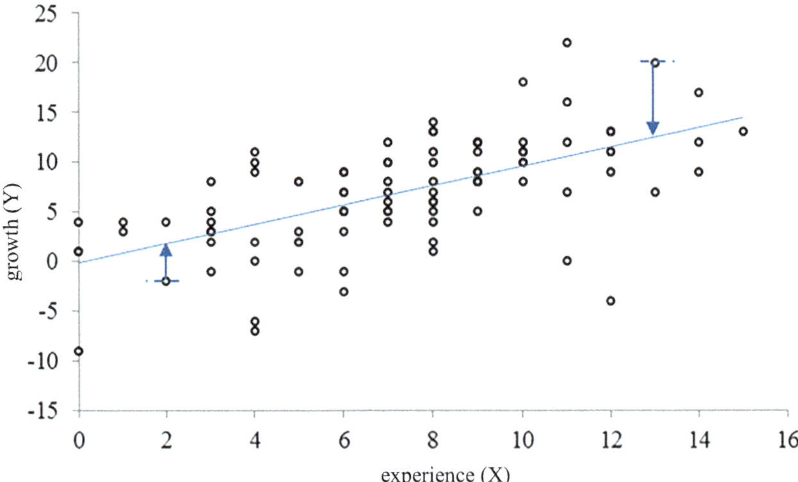

Fig. 19.3 Optimal line for the scatterplot of the variables growth rate and experience: ordinary least squares method

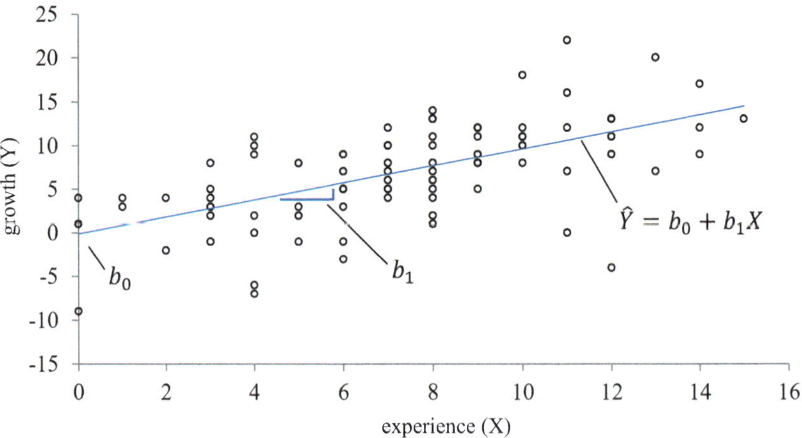

Fig. 19.4 Intercept and slope of the least squares line

But which is the correct straight line? Visually, it's hard to decide. We need a better method. One of the best we know is the method of least squares.

With the help of the method of least squares, the distance between our observations and the straight line we are looking for is minimized (Fig. 19.3). When this distance is minimal, we have found the optimal straight line.

Mathematically, a straight line can be written as follows:

$$Y = b_0 + b_1X$$

where

b_0 is the intercept, i.e., the point where the line intersects the Y-axis, and
b_1 is the slope of the line.

The slope describes how much the variable Y changes when X increases by one (Fig. 19.4).

Since we estimate the straight line using the method of least squares, we write for the estimated straight line (we also call it the regression line) as follows:

$$\hat{Y} = b_0 + b_1 X$$

where

$\hat{Y}$ (Y-hat) are the estimated Y-values of X with the help of the regression line,
b_0 is the intercept, and
b_1 is the slope.

Now we can turn to the minimization rule. As already said, the goal of the least squares method is to minimize the distance between observed y_i values and estimated $\hat{y}_i$ values. We can also write the distance as follows:

$$e_i = y_i - \hat{y}_i$$

where
e_i is the distance between observed and estimated values.

When we sum up all the distances, we arrive at the following formula:

$$\sum e_i = \sum (y_i - \hat{y}_i)$$

In addition, we can replace $\hat{y}_i$ with the regression line equation $\hat{y}_i = b_0 + b_1 x_i$ and we get:

$$\sum e_i = \sum (y_i - (b_0 + b_1 x_i))$$

Now comes one last thought step and we will understand why the method is called the least squares method. The line that we lay into the cloud of points using the minimizing method is the average line that best describes the cloud of points. Since it is the average line, the sum of the deviations is always zero, we have positive and negative deviations. Therefore, we help ourselves by not minimizing the deviations, but the squared deviations. The rule of the least squares method is therefore to minimize the sum of the squared deviations:

$$\sum e_i^2 = \sum (y_i - (b_0 + b_1 x_i))^2 \rightarrow \text{min!}$$

We find this minimum when we form the partial derivatives with respect to b_0 and b_1 . When we do this, we get the following formulas for the intercept and the slope, with which we can calculate the regression line from our observations:

$$b_0 = \frac{\sum x_i^2 \sum y_i - \sum x_i \sum x_i y_i}{n \sum x_i^2 - \left(\sum x_i\right)^2}$$

$$b_1 = \frac{n \sum x_i y_i - \sum x_i \sum y_i}{n \sum x_i^2 - \left(\sum x_i\right)^2}$$

Alternatively, the slope and the intercept can be calculated using the Bravais-Pearson correlation coefficient, the standard deviations, and the means. The formulas for this are as follows:

$$b_0 = \bar{y} - b_1 \times \bar{x}$$

$$b_1 = r \times \frac{sd_y}{sd_x}$$

with

$\bar{y}$ and $\bar{x}$ are the means of the dependent and independent variable,
r is the correlation coefficient of Bravais-Pearson between X and Y,
sd_y and sd_x are the standard deviations of X and Y.

The method applied does not matter, the result is the same.

We now want to perform the calculation for the first eight companies in the data set, using the method discussed first. It is best to create a table for this and calculate the required values.

Company	y_i	x_i	x_i^2	$x_i y_i$
1	5	7	49	35
2	8	8	64	64
3	18	10	100	180
4	10	7	49	70
5	7	6	36	42
6	12	10	100	120
7	16	11	121	176
8	2	4	16	8
Σ	78	63	535	695
		$\left(\sum x_i\right)^2 = 63^2$		

We insert the values according to the formulas and obtain the intercept and the slope of the regression line:

$$b_0 = \frac{\sum x_i^2 \sum y_i - \sum x_i \sum x_i y_i}{n \sum x_i^2 - \left(\sum x_i\right)^2} = \frac{535 \times 78 - 63 \times 695}{8 \times 535 - 63^2} = -6.61$$

$$b_1 = \frac{n \sum x_i y_i - \sum x_i \sum y_i}{n \sum x_i^2 - \left(\sum x_i\right)^2} = \frac{8 \times 695 - 63 \times 78}{8 \times 535 - 63^2} = 2.08$$

With this, we can write down the regression line and calculate estimated Y values for certain X values.

$$\hat{Y} = b_0 + b_1 X = -6.61 + 2.08X$$

For example, we would like to calculate the estimated growth rate for a company whose founder has 11 years of industry professional experience. We simply insert the value 11 for our X and obtain:

$$\hat{Y} = -6.61 + 2.08 \times 11 = 16.27$$

The growth rate estimated using the regression line is 16.27%. Remember, we said that the slope is the change that occurs when we increase the X value by one. Let's check this and calculate the estimated growth rate for 12 years.

$$\hat{Y} = -6.61 + 2.08 \times 12 = 18.35$$

The change is $18.35 - 16.27 = 2.08$, which exactly corresponds to the slope. We see that we can relatively easily determine the regression line from our observed values and use it to explain the situation and make forecasts. But it's not quite that simple. First, we have to ask how well our calculated regression line explains the relationship.

19.3 How Well and How Much Can We Explain, the R^2

The problem with the regression line is that we can calculate an optimal line for any cloud of points, no matter how much and how well the line explains the relationship. This is illustrated in Fig. 19.5. On the left side, we have a line that explains the relationship between X and Y relatively well, the points scatter closely around the line. On the right side is a line that explains the relationship poorly, the points are far away from the line. Since the points show a large variation, we can only say with great uncertainty what happens to Y when we change X.

The measure we use to determine the explanatory power of the regression line is the coefficient of determination R^2. It indicates what percentage of the change of the depend-

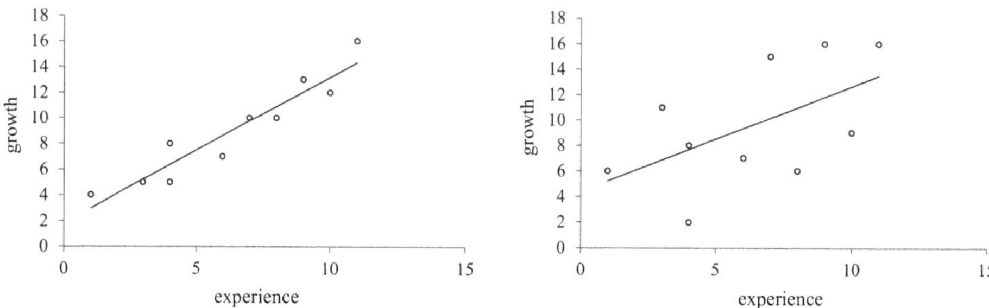

Fig. 19.5 Fit of the regression line

ent variable Y is explained by the independent variable X, technically speaking, what percentage of the variance of Y is explained by X:

$$R^2 = 1 - \frac{\text{unexplained variance}}{\text{total variance}} = 1 - \frac{\sum (y_i - \hat{y}_i)^2}{\sum (y_i - \bar{y})^2}$$

The coefficient of determination is defined between 0 and 1. 0 means that the regression line explains 0% of the movement (variance) of Y. 1 means that it explains 100% of the movement (variance) of Y. Accordingly, for example, an R^2 of 0.3 means that the regression line explains 30% of the movement of Y, the remaining 70% cannot be explained by the regression line.

Freak-Knowledge
The calculation of the R^2 is based on the analysis of variance. The variation of the Y values around their average value can be broken down into a part explained by the regression line and a part not explained by the regression line. If the explained part is high, the unexplained part is small, and vice versa.

To calculate the R^2, we first need to calculate the mean value $\bar{y}$ and the estimated $\hat{y}_i$ values. Then we determine the sum of the deviations of the squared deviations of the y_i from the mean and the estimated y values.. The result shows us what percentage of the variance of Y is explained by the regression line. We want to continue the example from above. Here we had estimated the following regression line:

$$\hat{Y} = b_0 + b_1 X = -6.61 + 2.08X$$

If we insert our x_i values into this, we get the estimated $\hat{y}_i$ values (see table) and we can determine the required sums once we have also calculated the average value.

Company	y_i	x_i	$\hat{y}_i$	$y_i - \hat{y}_i$	$(y_i - \hat{y}_i)^2$	$y_i - \bar{y}$	$(y_i - \bar{y})^2$
1	5	7	7.95	−2.95	8.7025	−4.75	22.5625
2	8	8	10.03	−2.03	4.1209	−1.75	3.0625
3	18	10	14.19	3.81	14.5161	8.25	68.0625
4	10	7	7.95	2.05	4.2025	0.25	0.0625
5	7	6	5.87	1.13	1.2769	−2.75	7.5625
6	12	10	14.19	−2.19	4.7961	2.25	5.0625
7	16	11	16.27	−0.27	0.0729	6.25	39.0625
8	2	4	1.71	0.29	0.0841	−7.75	60.0625
Σ	78				37.772		205.5
	$\bar{y} = 9.75$						

If we fill in the calculated sums into our formula, we get an R^2 of 0.82:

$$R^2 = 1 - \frac{\sum (y_i - \hat{y}_i)^2}{\sum (y_i - \bar{y})^2} = 1 - \frac{37.772}{205.5} = 0.82$$

Therefore, we can explain 82% of the movement of Y through the regression line.

As another example, let's calculate the R^2 for the left and right side of Fig. 19.5. For this, we first need to read the values from the figure. We get a value of 0.9 for the left side and a value of 0.32 for the right side. This means that the regression line in the left half of the figure explains about 90% of the variation of Y and the regression line in the right half explains about 32%.

In summary, we can say that the higher the R^2 is, the better the regression line fits the data and the more accurately we can predict what happens to Y when we change X. A high R^2 is therefore an indication of the quality of the regression analysis. In social sciences, however, we are often already satisfied with a seemingly small R^2, e.g. of 0.2 or 0.3. The reason for this is that we often work with concepts that are difficult to measure and prone to errors in data collection. The data is therefore often not precise and we have a so-called noise in the data. The R^2 is then inevitably often relatively small. As we will see in Chap. 20, not only the R^2 is decisive for the quality of the regression line, but we can also test whether the regression line as a whole has explanatory power.

19.4 Calculation with the R Commander

The calculation of simple linear regression is incredibly straightforward with the R Commander. First, we check with the help of the scatterplot whether there is a linear relationship between the dependent and independent variable. We look for the command Scatterplot in the Graphs menu . In the window that opens, we select the variables. For the X-variable, we choose the independent variable and for the Y-variable, the dependent variable. In the Options tab, we remove all check marks. The visual impression should not be disturbed by additional information and lines.

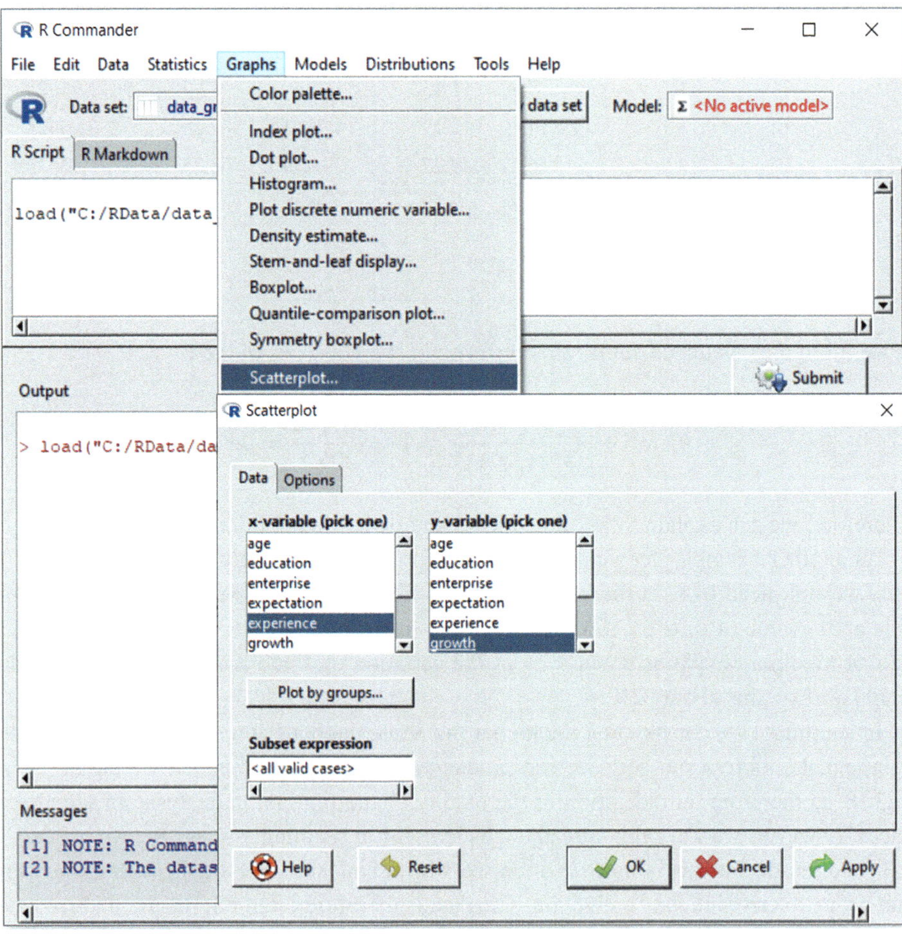

We confirm everything with OK and receive the scatterplot. We then evaluate this for a linear relationship between the dependent and independent variable.

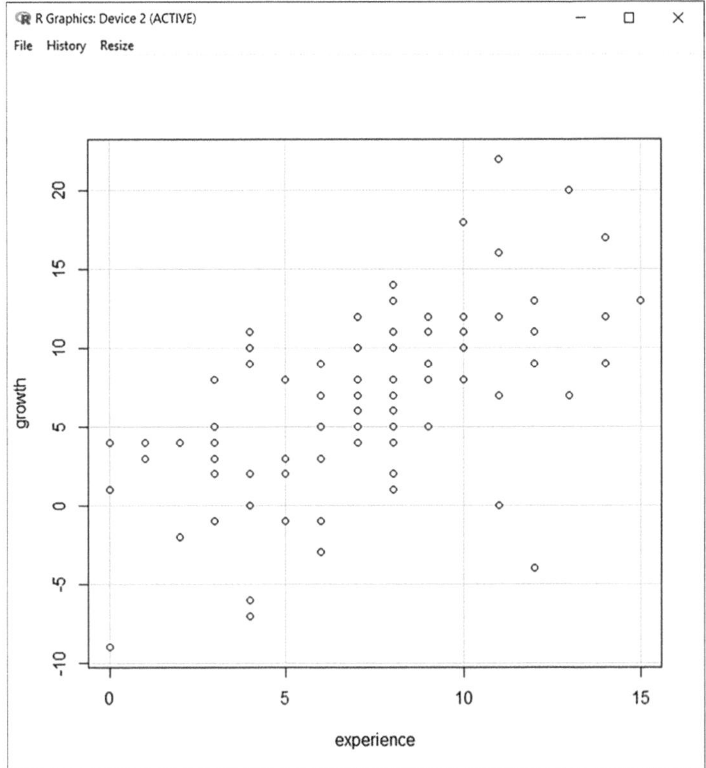

In this case, it seems possible without any problems to draw a straight line into the cloud of points, so nothing stands in the way of calculating the simple linear regression. To calculate this, we use Fit models in the Statistics menu's and in the submenu the Linear regression command. In the window that opens, we select the dependent and independent variable and confirm with OK.

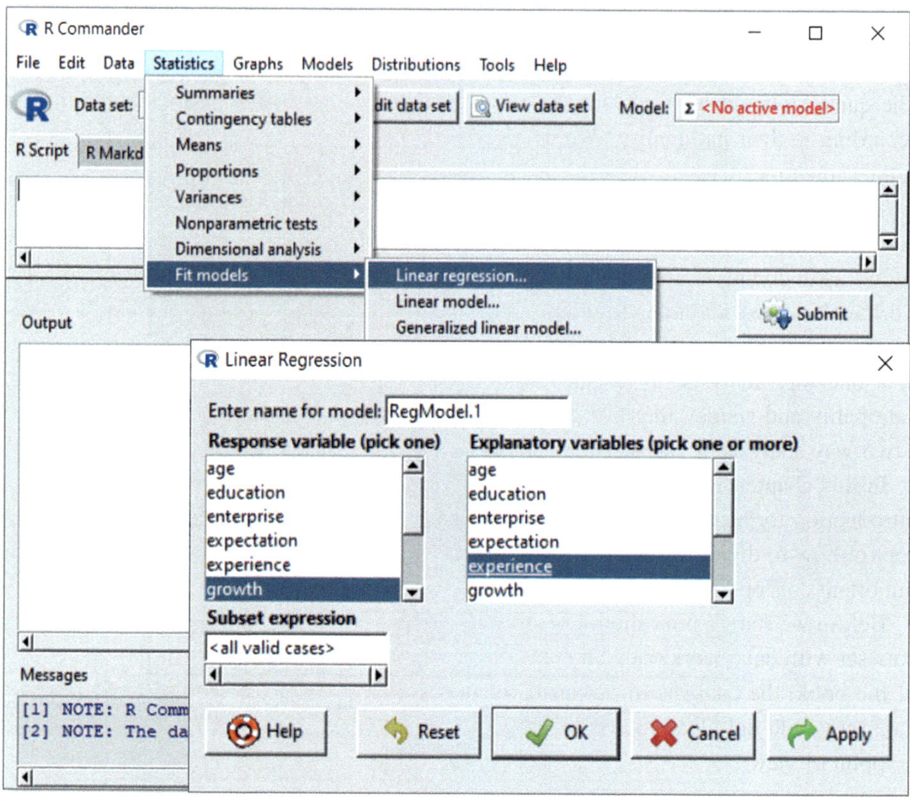

We thus obtain the results for the regression analysis.

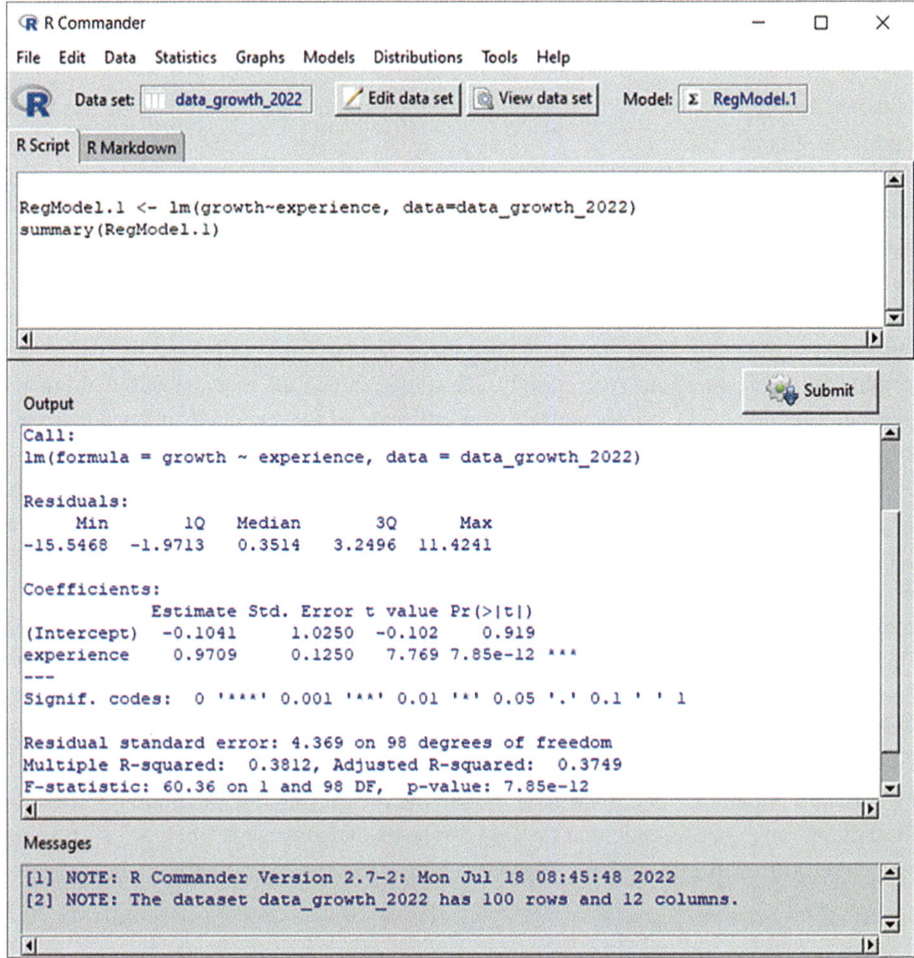

When interpreting the results, we focus on the values we already know, the slope, the intercept, and the coefficient of determination R^2. The slope is located behind the independent variable experience with the heading Estimate. Estimate stands for estimated value, as it is a value estimated from the sample. We find the intercept behind the expression Intercept. Here too, of course, it is an estimated value. The coefficient of determination is behind the expression Multiple R-squared. Our regression line is accordingly $\hat{Y} = -0.1041 + 0.9709X$ with an explanatory power of about 38%.

19.5 Is One Independent Variable Enough, Out-of-Sample Predictions, and More Warnings

At the end of this section, a few warnings and notes on the applicability of simple linear regression should be given.

One independent variable to explain our dependent variable will usually not be sufficient. We can easily imagine that company growth does not only depend on industry experience, but perhaps also on marketing efforts or efforts for innovation. Similarly, the sector could be relevant, and it may make a difference whether the founder is a woman or a man. The founder's age or the motive for founding could also be significant. This means that we usually have several independent variables that influence the dependent variable for each question. Accordingly, we must include all relevant variables in the regression model. If we do not do this, the influence of the included independent variables may be overestimated, and our forecasts are inaccurate.

Another problem with forecasts is that we should normally only make them in the range in which our observations are located. If we make forecasts outside our observation range, we speak of out-of-sample predictions or predictions outside our sample. The problem with out-of-sample predictions is primarily that we must assume that the relationship we have found in our data also applies beyond the observed range of values. This may or may not be the case. If this is not the case, then we make a false prediction. Compare this to the corresponding figure (Fig. 19.6).

In the figure, the observed range of values for the independent variable extends from one to eleven years. For this range of values, we have a close linear relationship between the variables experience and growth. If we use the relationship for prediction outside the observation range, we assume that the line can be extended (orange line). So at 16 years, we predict a growth rate of 20%. However, if the relationship actually changes (green

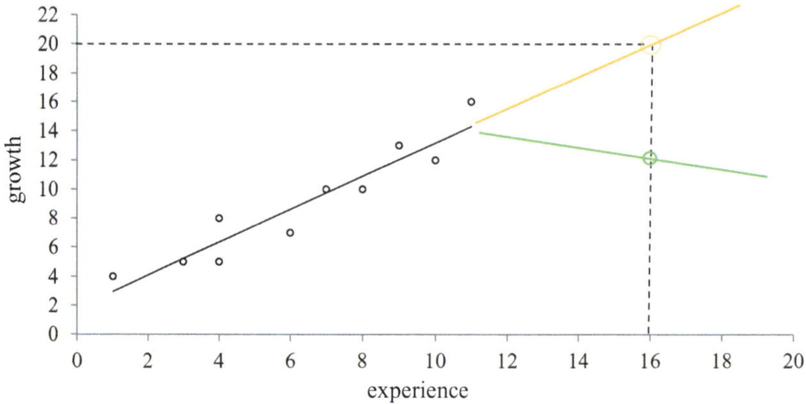

Fig. 19.6 Out-of-Sample prediction

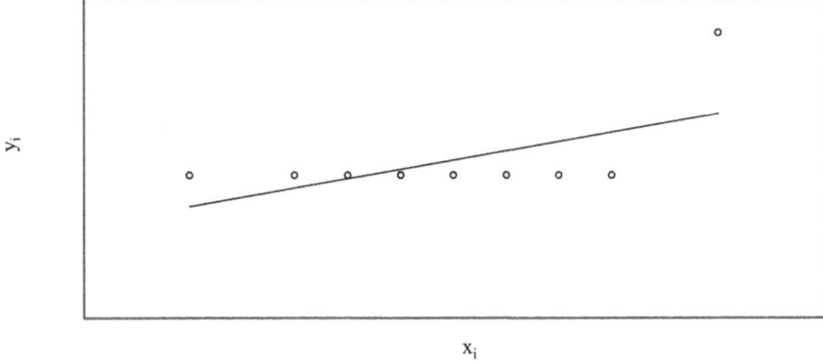

Fig. 19.7 No relationship between dependent and independent variable - #1

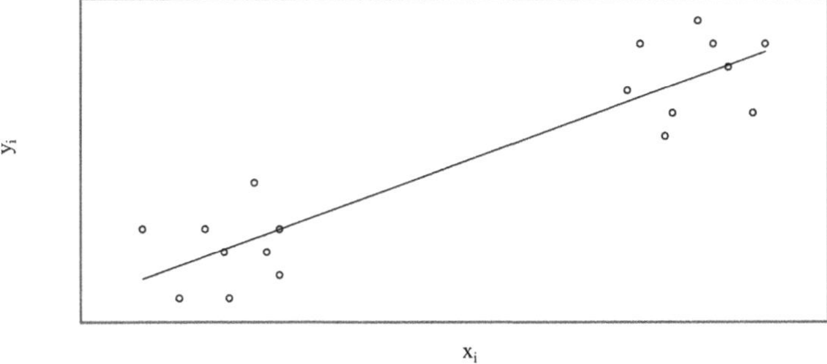

Fig. 19.8 No relationship between dependent and independent variable - #2

line), then the true value would not be 20%, but about 13%. Accordingly, for out-of-sample predictions, we must have a good theoretical justification for why the observed trend can be continued, and we must be cautious in interpreting the prediction.

One last remark. Linear regression calculates a straight line. However, this also means that a linear relationship must exist in the data. If this is not the case, linear regression is unsuitable for representing the relationship.

In Fig. 19.7 there is no relationship between *X* and *Y*. *Y* does not change when *X* changes. However, we have an outlier in the top right. The regression line takes this point into account and accordingly calculates a linearly ascending line.

In Fig. 19.8 two point clouds are shown. In the bottom left, *Y* changes unsystematically with a change in *X*, the same is true in the top right. So we have no linear relationship between *X* and *Y*, but we do have a break. If *X* exceeds a certain value, then *Y* jumps. The simple regression line tries to calculate a line between the point clouds and is therefore not suitable for representing the situation.

In Fig. 19.9 we see a non-linear relationship between X and Y. Linear regression tries to represent the relationship with a straight line, which is obviously not correct in this case.

Freak Knowledge
If there is a non-linear relationship between variables, e.g. in the form of a u-shaped or exponential relationship, there are mathematical transformation possibilities that convert such a relationship into a linear relationship. Linear regression can then be used again.

19.6 Checkpoints

- *With the help of simple linear regression, a linear relationship between two metric variables is estimated.*
- *Applying regression analysis requires a theory-guided approach. We need a solid theoretical foundation that shows us that the independent variable has an influence on the dependent variable.*
- *The calculation of the regression line is done using the method of least squares.*
- *The coefficient of determination* R^2 *is a measure of the strength of the relationship. It explains how much of the change of the dependent variable is explained by the independent variable.*
- *Usually, one independent variable is not sufficient to adequately explain the dependent variable, i.e. typically more than one independent variable is necessary.*
- *Out-of-sample predictions are based on the assumption that the discovered relationship also applies beyond the observation range, and should therefore be used with caution.*

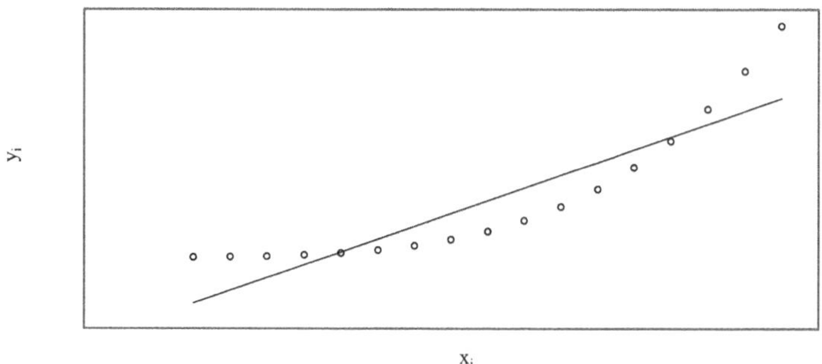

Fig. 19.9 Non-linear relationship between dependent and independent variable

19.7 Application

19.1. Why are theoretical considerations necessary before a regression analysis should be conducted?

19.2. Draw and calculate by hand the scatterplot, the regression line, and the coefficient of determination for the variable growth and marketing for the first eight companies of our dataset data_growth. Use the variable growth as the dependent variable and the variable marketing as the independent variable. Interpret the result.

19.3. Use the result from task 2 and predict the growth rate for marketing expenses of 20%. Is there a problem with the forecast?

19.4. Draw and calculate by hand the scatterplot, the regression line, and the coefficient of determination for the variable growth and innovation for the first eight companies of our dataset data_gGrowth. Use the variable growth as the dependent variable and the variable innovation as the independent variable. Interpret the result.

19.5. Use the result from task 4 and predict the growth rate for expenses on innovation of 20%. Is there a problem with the forecast?

19.6. Draw and calculate for our data set data_growth using the R Commander the scatterplot, the regression line, and the coefficient of determination for the following pairs of variables: growth and marketing, growth and innovation, and growth and age. Use the variable growth as the dependent variable. Interpret the results.

19.7. Describe with regard to task 6, why one independent variable is not sufficient to explain the growth rate.

20 The Multiple Regression Analysis

20.1 The Multiple Regression Analysis - More Than One Independent Variable

In the previous chapter, we saw that usually one independent variable is not enough to adequately describe the dependent variable. Normally, several factors have an influence. Hence, typically we need the multiple regression analysis, also called multivariate regression analysis, to describe a n issue. With this, we simultaneously analyze the influence of several independent variables on a dependent variable.

$Y \leftarrow X_1, X_2, \ldots X_J$

dependent variable — *independent variables*

(metric) — *(metric and dummy-variables)*

The dependent variable is metric, the independent variables are also metric or so-called dummy variables. Dummy variable is the technical term for nominal variables with the characteristics 0 and 1, e.g. man and woman, industrial company and service company, Catholic and non-Catholic.

Freak-Knowledge
Nominal variables with more than two characteristics or ordinal variables can be included in the multiple regression analysis if they are recoded into dummy variables.

F. Kronthaler, *Statistics Applied with the R Commander*,
https://doi.org/10.1007/978-3-662-69107-6_20

The objective of multiple regression analysis is to describe the influence of a set of independent variables X_j on a dependent variable Y and to determine what happens to Y when a variable X_j changes or is changed. The difference to linear simple regression is that we are not only examing only one independent variable, but several independent variables simultaneously.

Although we are simultaneously examing several independent variables, we are usually only interested in one of these variables. We include the other independent variables in the regression analysis because they also have an influence on the dependent variable. To determine the real influence of the variable of interest, we must control for the influence of the other variables. Technically, we therefore also speak of control variables. We determine the influence of an independent variable on Y under control of all other variables relevant to the issue.

We want to illustrate this again with an example. We are interested in the influence of industry professional experience on company growth. The theory indicates that industry professional experience has an influence. At the same time, however, we also take from the theory that not only industry professional experience is important for company growth, but also marketing, research and development, industry, etc. could play a role. To determine the real influence of industry professional experience, we therefore need to control for these factors.

The procedure for estimating the regression function is identical to the procedure for simple linear regression. We minimize the squared deviations of the regression function from the observed y_i values using the method of ordinary least squares:

$$\sum e_i^2 = \sum \left(y_i - \left(b_0 + b_1 x_{1i} + b_2 x_{2i} + \ldots + b_j x_{ji}\right)\right)^2 \rightarrow \text{min!}$$

With the help of the minimization rule, we obtain the regression function with the estimated coefficients:

$$\hat{Y} = b_0 + b_1 X_1 + b_2 X_2 + \ldots + b_J X_J$$

where

b_0 is the constant term,
b_1 to b_J are the estimated coefficients for the independent variables X_1 *to* X_j,
J is the number of independent variables,
$\hat{Y}$ are the values estimated by the regression function.

We want to illustrate this with our example. The theory has shown that in addition to industry professional experience, the effort for marketing and innovation, the sector, the age of the company founders, and the sex of the founders also have an influence. Thus, we have six independent variables that we need to include in our regression function. Two of them, the sectpr and the sex variables, are dummy variables. We can therefore specify our model as follows:

$$\hat{Y} = b_0 + b_1X_1 + b_2X_2 + b_3X_3 + b_4X_4 + b_5X_5 + b_6X_6$$

or

$$\hat{g} = b_0 + b_1\text{mark} + b_2\text{inno} + b_3\text{sec} + b_4\text{age} + b_5\text{exp} + b_6\text{sex}$$

where

$\hat{g}$ are the growth rates estimated using the regression function,
mark is the effort for marketing,
inno is the effort for innovation,
sec is the sector in which the enterprise operates,
age is the age of the company founders,
exp is the industry professional experience of the founders, and
sex is the sex of the founders.

Before we now estimate this model using the R Commander (the calculation by hand is slowly becoming cumbersome), we need to briefly discuss the various measures used to determine the goodness of the fit of the regression model.

20.2 F-Test, t-Test and Adjusted-R²

First of all, it should be clear that we estimate the regression function from a sample. Hence, we receive the regression model for the sample. But since we want to make a statement about the population, we need to test the regression model. We can test both the entire regression model and the individual regression coefficients for the independent variables. The null and alternative hypotheses for the regression model are as follows:

H_0: *The regression model does not explain the dependent variable.*

H_A: *The regression model explains the dependent variable.*

The significance level is usually $\alpha = 1\,\%$ or $\alpha = 0.01$.

If we reject the null hypothesis, we assume that the regression function explains the dependent variable in the population. The test statistic we use is defined as follows:

$$F = \frac{R^2/(J)}{(1 - R^2)/(n - J - 1)}$$

where

R^2 is the coefficient of determination,
J is the number of independent variables and
n is the number of observations.

The test value is F-distributed with $df_1 = J$ and $df_2 = n - J - 1$ degrees of freedom.

The test, briefly called F-test, thus uses the explained variance in relation to the unexplained variance. If the explanatory power is large, F becomes large and we reject the null hypothesis. The F-test is the first thing we look at after calculating the regression model.

For the independent variables, we each set up our own null and alternative hypothesis:

H_0: *The independent variable X_j does not contribute to the explanation of the dependent variable.*

H_A: *The independent variable X_j contributes to the explanation of the dependent variable.*

We specify the significance level in advance of the investigation with $\alpha = 10\,\%, 5\,\%, 1\,\%$ or $\alpha = 0.1, 0.05, 0.01$.

If we reject the null hypothesis, we assume that the independent variable X_j has explanatory power, using the following test statistic:

$$t = \frac{b_j}{se_{b_j}}$$

b_j is the estimated coefficient for the independent variable X_j,
se_{b_j} is the standard error for the estimated coefficient.

Freak Knowledge
The coefficients are estimated using one of many possible samples. Therefore, there are many possible coefficients that vary around the true value in the population. The standard error represents this variation.

With the t-tests, we check which of the variables explain the dependent variable. We can dedicate ourselves to these tests before or after we have examined the coefficient of determination.

We are already familiar with the coefficient of determination R^2. It compares the explained variance with the total variance and indicates how much of the change of the Y-variable is explained by the regression model.

$$R^2 = 1 - \frac{\text{unexplained variance}}{\text{total variance}} = 1 - \frac{\sum (y_i - \hat{y}_i)^2}{\sum (y_i - \bar{y})^2}$$

When analyzing the regression model, especially when comparing different regression models, the simple R^2 is often not used, but rather the so-called "adjusted-R^2". It is to be interpreted in the same way as the simple R^2. The background is as follows: It is possible to increase the R^2 by including more independent variables. For reasons we will not discuss here, the addition of another independent variable never leads to a reduction of the R^2. Thus, by adding independent variables, we can theoretically increase the R^2 at will and pretend an increased explanatory power of the model. For this reason, a measure has been suggested that controls for the number of variables included. The measure is calculated as follows:

$$R^2_{\text{adj}} = R^2 - \frac{J \times (1 - R^2)}{n - J - 1}$$

From the formula it follows that the numerator becomes larger and the denominator smaller the more variables we include in the regression model. Accordingly, the R^2_{adj} can become smaller if more variables are included in the model. If we have included all relevant independent variables in the model, typically we use the simple coefficient of determination for interpretation. The adjusted coefficient of determination is mainly used when we are unsure about the number of independent variables we should use. We can then use the R^2_{adj} to compare whether one regression model is better than the other.

20.3 Calculation with the R Commander

We can now estimate and interpret our multiple regression function with the R Commander. The execution is done exactly as in Chap. 19, only that we now specify more than one independent variables. In the Statistics menu, we use the submenu Fit models and then the command Linear regression.

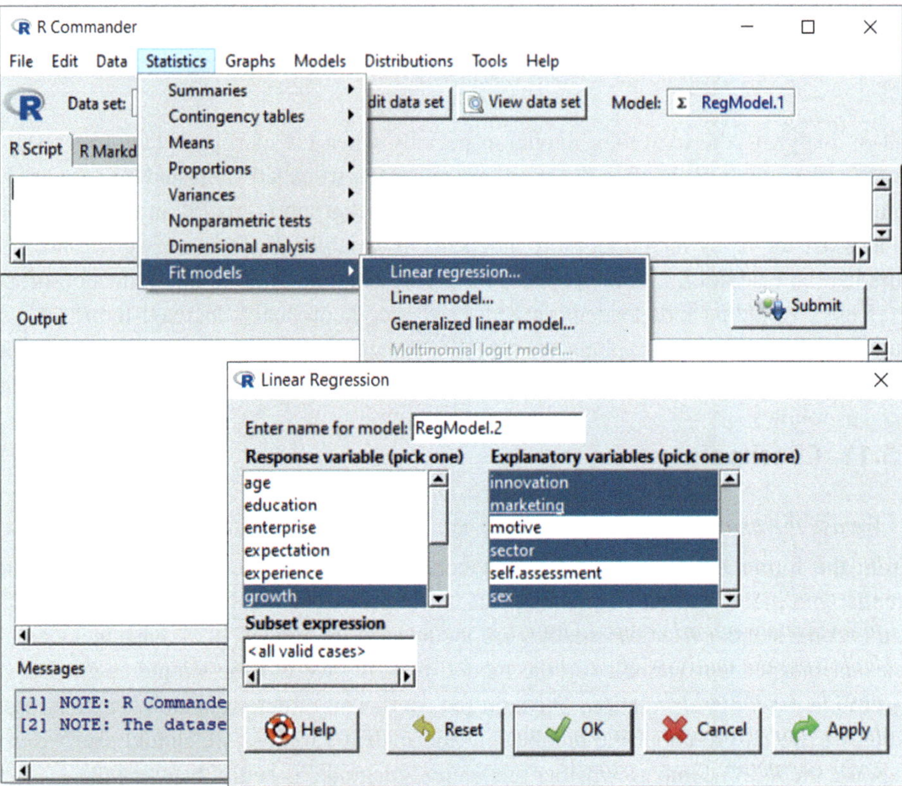

In the window that opens, we mark the dependent variable and, in combination with the mouse and the control key, the relevant independent variables. After confirming with OK we get the result of the regression analysis.

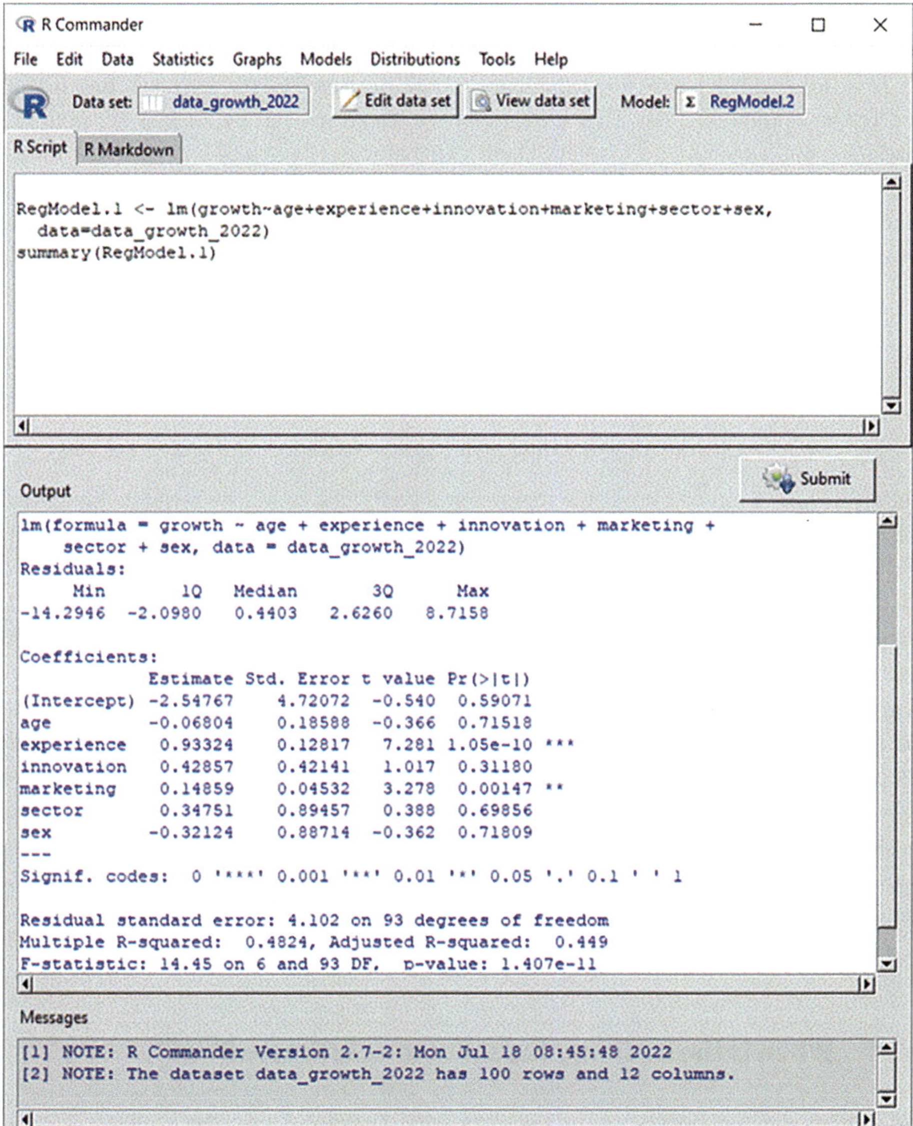

We can now turn to the interpretation of the results. In the last line, we find the F-value and the probability of obtaining such a large F-value if the null hypothesis is correct (p-value). In our case, the probability is much lower than 1%, i.e., we reject the null hypothesis. Accordingly, we assume the alternative hypothesis that the regression function explains the dependent variable. The R^2 in the line above shows us that our regression model explains about 48% of the change of the dependent variable, which is already a relatively high value in social science research. Now we can look at the individual independent variables. We find these in the coefficient block of the output. For each variable, including the intercept, the estimate, the standard error, the t-value, and the probability of whether the null hypothesis for the respective variable is holds (Pr(>|t|)) are given. We first focus on the probability. For the variable age, the probability is 71.52%, for sector 69.86%, for experience 0.00%, for sex 71.81% for marketing 0.15% and for innovation 31.18%. This means that we reject our null hypothesis for both the marketing and experience variables. These variables contribute to the explanation of the growth rate. For the other variables, we cannot reject the null hypothesis. In the first column after the variable name, we also find our estimates for the coefficients. With these, we can set up the regression model. Our regression function is:

$$\hat{g} = b_0 + b_1\text{mark} + b_2\text{inno} + b_3\text{sec} + b_4\text{age} + b_5\text{exp} + b_6\text{sex}$$

We can now supplement this with the estimated coefficients (rounded to two places here):

$$\hat{g} = -2.55 + 0.15\text{mark} + 0.43\text{inno} + 0.35\text{sec} - 0.07\text{age} + 0.93\text{exp} - 0.32\text{sex}$$

Let's now take a closer look at the regression model. First of all, it should be noted that we have set up the regression function for all estimated independent variables, regardless of whether they explain the dependent variable or not. This is always the case, we have estimated the regression model with these variables. b_0 is the value of the dependent variable when all independent variables have the value zero. In our case, the growth rate is −2.55%, when we insert zero for all independent variables, e.g., when zero percent of sales are spent on marketing, etc. The coefficients b_1 to b_6 indicate by how much the growth rate changes when we increase the respective variable by one. The prerequisite is that all other variables remain unchanged. The coefficient can only be applied to those variables that explain the dependent variable. In our case, these are the variables marketing and experience. For example, we can say what happens when we increase the marketing variable by one, the growth rate then increases by 0.15 percentage points. If the experience variable increases by one, the growth rate increases by 0.93 percentage points.

20.4 When is the Ordinary Least Squares Estimate BLUE?

We carried out the estimation of the regression function using the method of ordinary least squares. This method leads to the best possible results when certain assumptions are met. We still need to look at and discuss these assumptions, which are called Gauss-Markov assumptions:

1. The regression function is correctly specified and contains all relevant independent variables. The relationships between the independent variables and the dependent variable are linear.

2. The deviations of the observed Y-values from the estimated $\hat{Y}$-values have an expected value of zero.
3. The deviations are not correlated with the independent variables.
4. The variance of the deviations is constant.
5. Two or more independent variables are not correlated with each other.
6. The deviations are uncorrelated with each other.
7. The deviations are normally distributed.

Under these assumptions, the least squares estimator is BLUE, it is the best, linear, unbiased, and most efficient estimator we know.

Freak-Knowledge

Unbiased means that the expected value of the coefficients corresponds to the true values of the coefficients in the population. Efficient means that the coefficients are estimated with the smallest possible deviation.

We analyze most of these assumptions using the deviation of the observed Y -values from the estimated $\hat{Y}$ -values, also known as residuals. We can easily instruct the R Commander to add the residuals to our data set. To do this, after estimating the regression model, we select the command Add observation statistics to data from the Models menu and select the Studentized residuals) by checking the box, as well as the Fitted values and the Observation indices. We also select the case labeling and the predicted values. Two small notes: First, the studentized residuals are the standardized residuals (cf. to the standardization of a variable Chap. 10), but the value to be standardized does not go into the calculation. Second, we use the studentized residuals in the analysis, as violations of assumptions are easier to detect with the studentized residuals than would be the case with only standardized values.

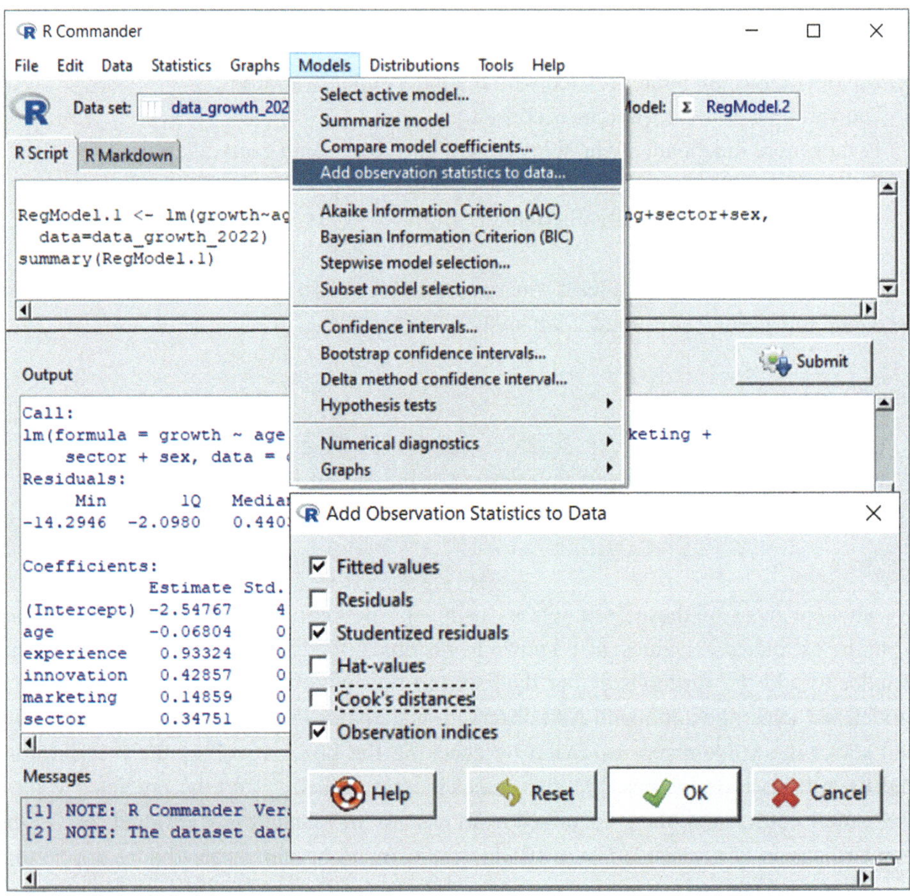

We confirm with OK and R adds the selected values to our data set. We are thus able to discuss the assumptions as follows using the graphical skills we have already learned. We are already familiar with the graphics used, the scatterplot, the histogram, and the correlation matrix, and know how to create them with the R Commander. We therefore no longer go into the R Commander. The reader is challenged to create the graphics with the R Commander and analyze them as described below.

Assumption 1) The regression function must be correctly specified and linear in the relationships between the independent and dependent variables. Correctly specified means that we have included all relevant independent variables in the regression model. This requires that we have worked theoretically well when setting up the regression model. Whether there is a linear relationship between the dependent and the independent variables, we can check with the help of scatterplots between the dependent and the independent variables. For this, we have to draw the scatterplots for all metric independent variables and then visually inspect whether there is a linear relationship between the respective independent variable and the dependent variable (Fig. 20.1).

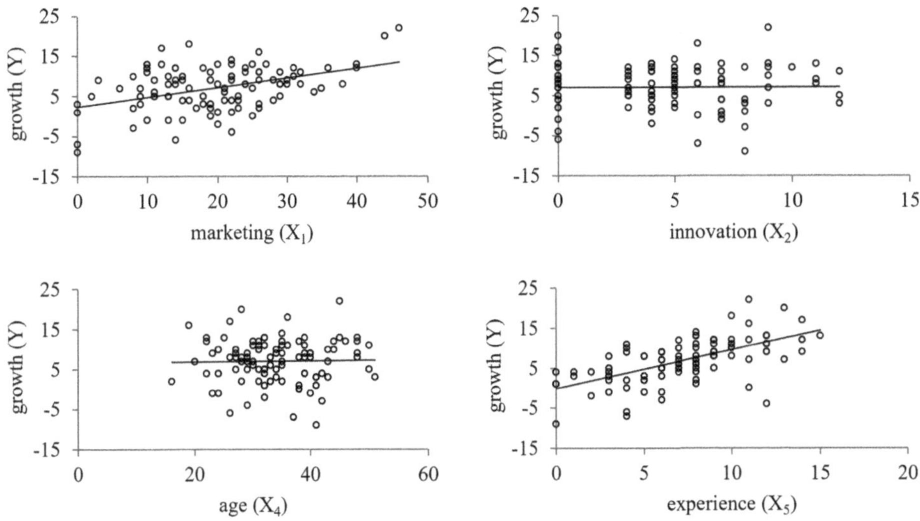

Fig. 20.1 Examing linearity between the dependent and the independent variables

The drawn scatterplots for the metric independent variables show in our case either a linear relationship (marketing and experience) or no relationship (age and innovation). A non-linear relationship, e.g., in the form of an arc, cannot be identified. From this, we can conclude that the assumption of linearity is not violated. Scatterplots for our dummy variables sector and sex are not necessary, as the x values can only take zero and one and thus no linear relationship between these variables and the dependent variable is possible.

2) The second assumption demands that the deviation of the observed Y-values from the estimated $\hat{Y}$-values has an expected value of zero. This is a requirement which, if requirement 1 is met, mainly concerns data quality. The observed Y-values must not be systematically measured too high or too low. If this were the case, then our constant value b_0 would be biased. If the Y-values are systematically measured too high, our b_0 would also be too high and vice versa. However, there is no influence on the coefficients of the independent variables. If we are only interested in these and we have no interest in b_0, we do not have to worry about the assumption.

Assumption 3) This assumption demands that the deviations are not correlated with the independent variables. We can evaluate this again with the scatterplot. We plot the residuals or the standardized residuals against the independent variables. If the assumption is not violated, no relationship between the variables should be recognizable. The residuals should scatter unsystematically. Looking at the scatterplots, this seems to be the case in our example. No trend or pattern can be recognized (Fig. 20.2).

The assumption would be violated if we were to recognize, for example, an ascending or a descending trend in the cloud of points.

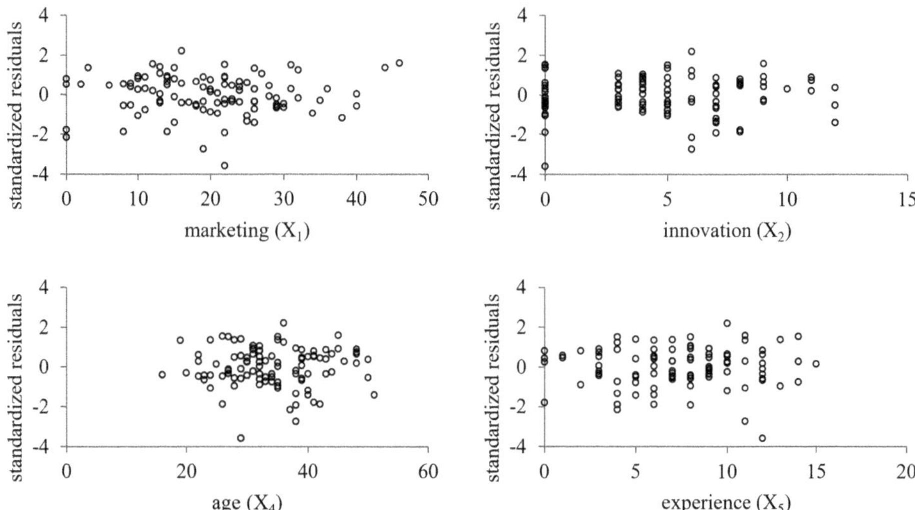

Fig. 20.2 Examing independency between the residuals and the independent variables

Assumption 4) The fourth assumption demands that the variance of the deviations is constant. Technically, we speak of homoscedasticity or heteroscedasticity. Homoscedasticity means that the variance is constant over the range of estimated $\hat{Y}$ -values. Heteroscedasticity exists when the variance is not constant. We can evaluate the fourth assumption again with the help of our residuals. For this, we draw the scatterplot for the residuals or the standardized residuals against the estimated $\hat{Y}$-values. If the residuals scatter unsystematically within the same range, i.e., the deviations do not systematically become smaller or larger, then the requirement is met. In our case, we see in the picture that the residuals scatter approximately within the same range (Fig. 20.3).

If we, for example, notice that the variation of the residuals becomes funnel-shaped smaller or larger, then the assumption is violated (see Fig. 20.4) and the coefficients are no longer efficiently estimated. This means that the actual size of the coefficients is more unknown than necessary.

Assumption 5) The fifth assumption demands that two or more independent variables are not correlated with each other. Technically, it is required that there is no multicollinearity between the independent variables. One way to check for multicollinearity is to calculate the correlation coefficients between each pair of variables. If a correlation coefficient is higher than 0.8, we tend to have a problem with multicollinearity.

	Marketing	Innovation	Age	Experience
Marketing	1			
Innovation	− 0.0312839	1		
Age	− 0.0393392	0.954949346	1	

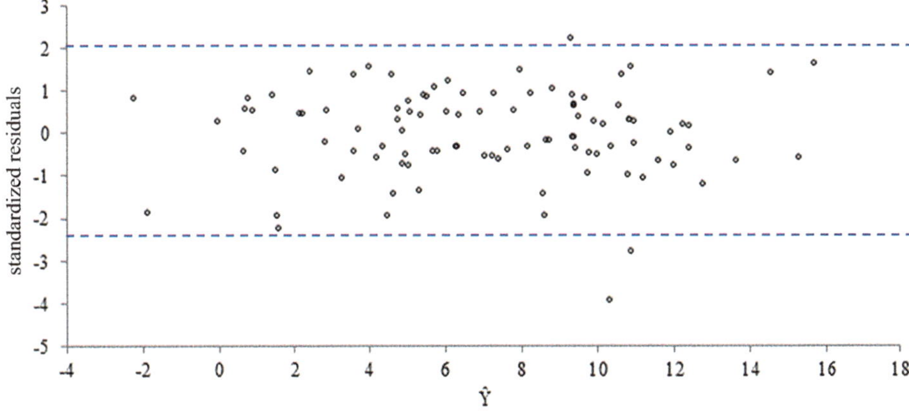

Fig. 20.3 Examing the variance of the residuals (heteroscedasticity)

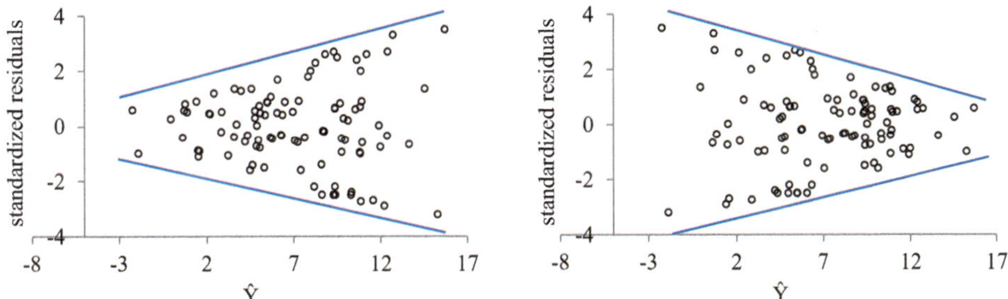

Fig. 20.4 Examples of heteroscedasticity

	Marketing	Innovation	Age	Experience
Experience	0.26496172	−0.266397792	−0.220079	1

In our case, the correlation coefficient between the variable innovation and the variable age is higher than 0.8, i.e., we have a very high correlation between innovation and age. Therefore, our ordinary least squares estimator can no longer correctly assign the effect that the variables age and innovation have on the dependent variable. The coefficients for the respective variables are potentially incorrect. The solution is to re-estimate the regression model and omit one of the highly correlated variables. In our case, we should remove either the variable age or the innovation variable. We need to consider which variable is more important to us. Another way to check for multicollinearity is to calculate the variance inflation factor. Variance inflation factors greater than 4 indicate problems with multicollinearity.

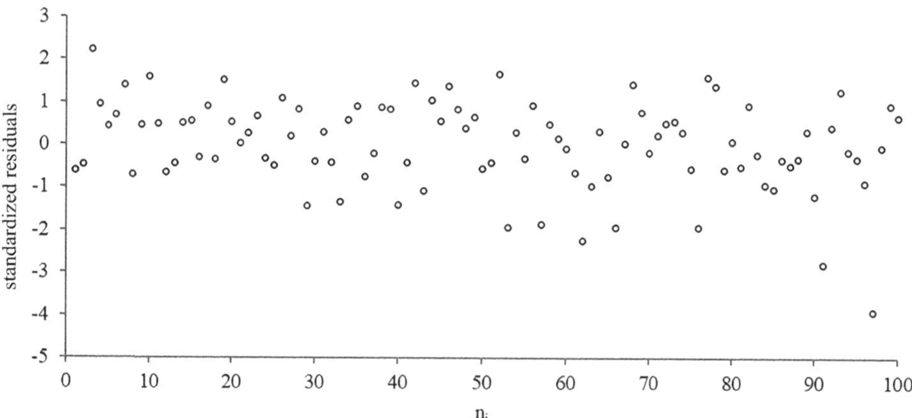

Fig. 20.5 Examination of the independence of the residuals from each other (autocorrelation)

Assumption 6) The sixth assumption demands that the deviations are not correlated with each other, i.e., that successive deviations do not become larger if the preceding deviation is large or smaller if the preceding deviation is large and vice versa. Technically speaking, this is the question of autocorrelation. We can analyze this again with the scatterplot of the residuals. In this, we plot the observations in sequence on the x-axis and the residuals on the y-axis. If the residuals scatter unsystematically and we do not recognize a pattern, the assumption is not violated. The following image is a good example of this (Fig. 20.5).

One more note: the problem of autocorrelation exists primarily when we analyze time series.

7) The last assumption demands that the residuals are normally distributed. This assumption is especially important with a small sample. If we have a small sample and the assumption is violated, then our F-test and our t-tests are no longer valid and we do not know if we can actually reject the null hypotheses. We can evaluate this requirement visually, for example, with the help of the histogram (Fig. 20.6).

The histogram should look approximately bell-shaped. In our case, the distribution is slightly skewed to the left, but does not deviate much from a normal distribution. Based on the visual impression, the assumption does not seem to be seriously violated. This can be further investigated with more advanced methods that go beyond this book.

If one of these assumptions is seriously violated, our regression results are no longer trustworthy. The ordinary least squares method is then no longer BLUE, and we know other methods that yield better results. The discussion of these methods goes far beyond a basic statistics book. At this point, I would like to refer to the advanced statistical literature. At the end of the book, I provide some hints in this regard.

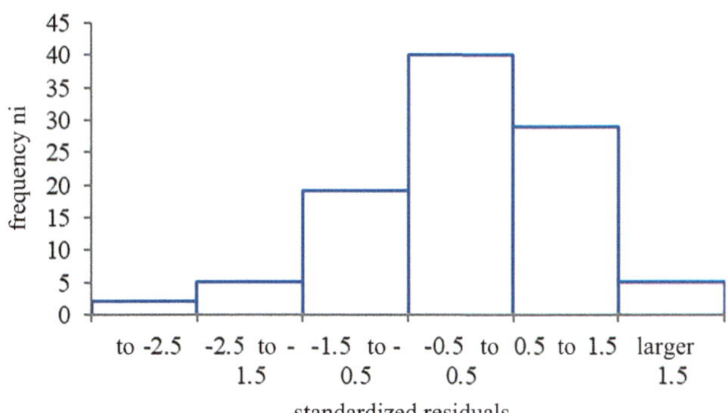

Fig. 20.6 Examination of the normal distribution of residuals

20.5 Checkpoints

- *With the help of multiple regression, we analyze the influence of several independent variables on a dependent variable.*
- *The dependent variable is metric, the independent variables are metric or dummy variables.*
- *With the help of the F-test, we test whether the entire regression model explains the dependent variable.*
- *With the help of the t-test, we test whether the independent variables explain the dependent variable.*
- *The method of ordinary least squares for estimating the regression function is BLUE if certain assumptions are met.*
- *If one or more assumptions are violated, then there are estimation methods that yield better results than the ordinary least squares method.*

20.6 Application

20.1. Calculate the regression model from this chapter without the independent variable age.

20.2. Test the assumptions for the multiple regression model in application 1.

20.3. Have we solved the problem of multicollinearity by omitting the age variable?

Part VI

What Happens Next?

21 How to Present Results

21.1 Contents of an Empirical Paper

After analyzing data, we usually have to present the results in writing, in the form of a study paper, a bachelor's or master's thesis, or a journal article. The following points always need to be discussed:

1. the problem,
2. the question,
3. the literature,
4. the data and the method,
5. the empirical results,
6. the summary and the conclusions.

The problem addresses why the topic is being dealt with. If possible, there is a hook in the form of a current political discussion, etc. The research question is directly derived from the problem. The reader of the report thus recognizes why this question was chosen. Following this, the state of knowledge is presented in a literature section and hypotheses are possibly formulated in relation to the question and the literature. Subsequently, the method and the data chosen to answer the question or test the hypotheses are discussed. The next step involves the presentation of the empirical results. Finally, the results are summarized, related to the literature, and it is discussed what conclusions can be drawn from them. To illustrate this, we want to use two examples of different questions. It should not be forgotten that these are not complete research reports, but only "fictional" sketches/short reports. No real literature analysis was carried out and the results are based on simulated data. That means, both the results and the short reports have nothing, absolutely nothing, to do with reality.

F. Kronthaler, *Statistics Applied with the R Commander*,
https://doi.org/10.1007/978-3-662-69107-6_21

21.2 Example I for a Report: Is There a Difference in Founding Age Between Male and Female Founders (Fictional)

In the current socio-political discussion, it is problematized whether women and men differ in age when founding companies (see, for example, daily newspaper of 12.01.2015). Against this background, the aim of the present analysis is to determine whether this actually corresponds to the facts. Specifically, the question is pursued whether female and male founders differ in their average age at the time of founding a company.

In the literature, there are indications on the one hand that women are older at the time of founding a company than men. It is argued that women take longer to build up the necessary professional experience for a company formation. They interrupt their professional career more often, for example, to take a baby break (see Author X 2014). In contrast, it is argued that women are more decisive, more willing to take entrepreneurial risks, and therefore tend to found a company earlier than men (see Author Y 2012). [...].

[...] Since there are theoretically different opinions, we want to empirically test the following hypothesis at the 10% significance level: Female and male founders differ in age at the time of company formation.

To test the hypothesis, a representative data set on company foundations is used. The data set was collected in June 2013. A random sample of 400 companies founded in 2007 was drawn. The response rate was 25%, so the data set contains 100 companies. The data set informs with regard to the research question about the gender and the age of the founders, among other firm characteristics. Various tests carried out suggest that the sample is representative.

The test variable age is metric and normally distributed, accordingly the test for a difference of means with independent samples is used. The following table contains the test result with equal variances.

Test variable: Age (in years) by gender			
Male founders	$\bar{x} = 34.77$	$sd = 7.55$	$n = 65$
Female founders	$\bar{x} = 33.29$	$sd = 7.76$	$n = 35$
$t = 0.93$	$df = 98$	$t_{crit} = \pm 1.66$	$P(T \leq t) = 0.36$
Effect size $r = 0.09$			

The result clearly shows that we cannot reject the null hypothesis, there is no difference in the founding age between male and female founders, based on our sample findings. The probability that there is no difference in the population is relatively high. Furthermore, the effect size indicates at most a small influence and the difference found in the sample is also relatively small at about 1.5 years. Therefore, for the population, we assume that male and female founders are the same age at the time of founding and that there is no difference between the two groups.

21.3 Example II for a Report: Professional Experience and Company Performance (Fictional)

In the current economic policy discussion, there is an increasing push to limit the support of business start-ups to those founders who bring sufficient professional experience (see, for example, daily newspaper of 22.11.2013). Supporters of this measure hope that only promising companies will be supported and that state funding will become more efficient and effective.

Against this background, the aim of the work is to investigate the relationship between professional experience and growth of newly founded companies. Specifically, the following question is pursued: What impact does professional experience have on the company growth in the first five years of existence of an independent company?

The theory and literature on professional experience and company growth assume that professional experience has a positive influence on the growth of young companies. A founder with professional experience knows the market better and can thus better recognize market opportunities (Author X 2011). Furthermore, the literature shows that founders with professional experience more easily obtain loans from banks and venture capital. Since they have more capital available for investments, they are able to grow faster (Author Y 2009). [...].

[...] In summary, the following hypothesis can be formulated: The professional experience at the time of founding has a positive impact on the company growth of newly founded companies.

To test this hypothesis, a primary data collection was carried out in June 2013. A sample of 400 companies was randomly selected from the companies founded in 2007. These companies were surveyed on various characteristics needed for the study. Of the 400 companies, 100 participated in the survey, which corresponds to a response rate of 25%. Various tests carried out suggest that the sample is representative.

To answer the question, the dependent variable is the average company growth from 2007 to 2012. The independent variable of interest is professional experience of the company founders at the time of founding. In addition, various control variables are implemented, as it can be assumed that company growth is correlated with other variables. Control variables are the expenditures for marketing and for innovation, the age of the founders, the sector, and the gender. Overall, the following regression function is estimated using the method of ordinary least squares OLS:

$$\hat{g} = b_0 + b_1\text{mark} + b_2\text{inno} + b_3\text{sec} + b_4\text{age} + b_5\text{exp} + b_6\text{sex}$$

The tests of the assumptions of the method of ordinary least squares show that there is a tendency for a problem with multicollinearity. The variables age and innovation are highly correlated with a correlation coefficient of 0.95. To solve the problem, the variable age is therefore removed and the following reduced model is estimated:

$$\hat{g} = b_0 + b_1\text{mark} + b_2\text{inno} + b_3\text{sec} + b_4\text{exp} + b_5\text{sex}$$

The results of the estimate show that we can reject the null hypothesis for the model. The model explains 48% of the variation of the dependent variable (see the following table).

	Ordinary least squares estimation; dependent variable: turnover growth
Intercept	−4.18***
Marketing	0.15***
Innovation	0.28**
Sector	0.33
Experience	0.93***
Sex	−0.28
Number of observations	100
R^2	0.48
F-Test	17.47***

*, **, *** indicate statistical significance at the 10, 5, and 1% level

Looking at the independent variable of interest, professional experience, we see that it is statistically significant at the 1% level.

The results of the study are consistent with the theoretical findings and existing literature. The professional experience of the founders at the time of founding has a positive effect on the growth of newly established companies. It should be noted, however, that the study did not determine what the higher growth is specifically due to. It remains open whether the higher growth can be attributed to better market knowledge or whether it is due to better financing opportunities in the initial phase of the foundation. Before final political conclusions can be drawn, this question needs to be investigated further in additional studies.

21.4 Application

Select five journal articles from your field of study that work with data, and check them against the content described above.

22 Advanced Statistical Methods

With multiple regression analysis, we have learned about one of the most important multivariate analysis methods. However, it is not the only one. In addition, there are other multivariate analysis methods that are used for specific questions. We will give a brief overview of the most important ones here. In the next section, we will then turn to the literature that can be used to delve deeper into regression analysis and the other methods of multivariate analysis.

Logistic Regression

With the help of logistic regression, similar to multiple regression analysis, the influence of several independent variables on a dependent variable is examined. The difference is that in the case of logistic regression, the dependent variable only has the values 0 and 1, an event does not occur or an event occurs. The applications are diverse, for example, it can be analyzed which factors were decisive for surviving or not surviving the Titanic disaster. Was the factor gender decisive, the age or the class in which one traveled? The question can also be which factors lead to a loan defaulting or not defaulting. Another example is the investigation of people's purchasing decisions. What characteristics lead to people buying a certain product or not.

Discriminant Analysis

With the help of discriminant analysis, group differences are analyzed. The results often serve to make a prediction about which group newly added individuals or objects are likely to belong to. For example, banks try to classify borrowers in advance of lending in terms of their risk into bad risks (potential loan default) and good risks (loan is unlikely to default). For this, the differences between good risks and bad risks are determined and based on these results, the new customers are classified.

F. Kronthaler, *Statistics Applied with the R Commander*,
https://doi.org/10.1007/978-3-662-69107-6_22

Factor Analysis

We can use the method of factor analysis to analyze the relationships between a larger number of metric variables. We can use factor analysis to discover dimensions that lie behind the correlation of variables, but we can also use factor analysis to reduce the number of variables. An example of this would be the attempt to measure the quality of restaurants. We can measure the quality of restaurants based on various variables, e.g. taste, temperature of the dish, freshness, waiting time, cleanliness, friendliness of the staff, etc. Afterwards, we could use factor analysis to investigate whether the multitude of variables can be reduced and whether there is a common dimension such as food quality and service quality behind various variables.

Cluster Analysis

The method of cluster analysis aims to group objects or individuals based on certain criteria. The aim is to group objects or individuals in such a way that they are as similar as possible and differ from the other groups. For example, one could identify consumer types with regard to price sensitivity, quality sensitivity, sustainability needs, etc., or one could try to classify regions based on certain criteria such as gross domestic product, unemployment, infrastructure, industrialization, etc.

Conjoint Analysis

Conjoint analysis is a tool that can be used to analyze which features and which combinations of features determine the utility of a product or service. With the help of this method, the user can, for example, analyze which attributes of a product, e.g. color, price, quality, size, etc., lead to a selection decision being made and which preferences prevail among consumers. Producers can use this knowledge to design products that match consumer preferences.

23 Interesting and Advanced Statistical Textbooks

Finally, I would like to give some hints about interesting and advanced statistics books. This section is intended to provide a little help for those who want to delve deeper into statistics. However, it cannot replace your own research for suitable literature. In addition to the mentioned books, there are a number of other good textbooks.

Döring, N.; Bortz, J.; unter Mitarbeit von S. Pöschl (2016), Forschungsmethoden und Evaluation in den Sozial- und Humanwissenschaften, 5. Aufl., Springer: Heidelberg
"Forschungsmethoden und Evaluation in den Sozial- und Humanwissenschaften" is the German-language classic for questions about the research process, quantitative and qualitative data collection, hypothesis formation and hypothesis testing. Almost all problems are discussed in these areas.

Bortz, J.; Schuster, Ch. (2010), Statistik für Human- und Sozialwissenschaftler, 7. Aufl., Springer: Berlin Heidelberg
"Statistik für Human- und Sozialwissenschaftler" is an outstanding presentation of descriptive statistics, inferential statistics, and multivariate analysis methods. Like the book "Research Methods and Evaluation", it is a German classic of quantitative data analysis.

Backhaus, K.; Erichson, B.; Plinke, W.; Weiber, R. (2018), Multivariate Analys emethoden, 15. Aufl., Springer: Berlin Heidelberg
The book "Multivariate Analysemethoden" offers an in-depth introduction to the methods of multivariate data analysis. The reader finds an understandable presentation of

F. Kronthaler, *Statistics Applied with the R Commander*,
https://doi.org/10.1007/978-3-662-69107-6_23

regression analysis, time series analysis, analysis of variance, factor analysis, cluster analysis, etc. All methods are explained using specific data sets with the help of the statistical software SPSS. Again it is a German book.

Greene, W.H: (2018), Econometric Analysis, 8th Edition, Pearson: New York
"Econometric Analysis" is the standard work of economic application of regression analysis. On over 1000 pages, the reader finds almost all problems and solutions that can occur in the procedures of regression analysis in the social science context. Here too, the reader needs a deeper understanding of mathematics.

Gujarati, D. (2015), Econometrics by Example, 2nd Edition, Palgrave Macmillan: Basingstoke
For those who want to delve into regression analysis, the book "Econometrics by Example" is recommended. The book manages to answer the central questions of regression analysis simply using concrete examples and to show solutions for problems without using much mathematics. It is a book that non-technicians have been waiting for a long time.

Hair J.F.; Black W.C.; Babin, B.J.; Aderson, R.E. (2018), Multivariate Data Analysis, 8th Edition, Pearson: Harlow, GB
"Multivariate Data Analysis" is a simple and understandable textbook on methods of multivariate data analysis. The authors succeed in explaining the concepts and problems of the methods and applying them to concrete examples without using mathematics. After reading the book, the reader is able to apply factor analysis, regression analysis, cluster analysis, and other multivariate methods in depth. The chapter on data preparation is particularly noteworthy.

Kronthaler, F; Zöllner, S. (2021), Data Analysis with RStudio – An Easygoing Introduction, Springer: Berlin
Those who deal intensively with data analysis will not be able to avoid dealing more intensively with R and looking beyond the R Commander. Therefore, this simple introduction to RStudio was created together with a colleague. The introduction enables the start into professional work with R.

Noelle-Neumann, E.; Petersen, T. (2005), Alle nicht jeder: Einführung in die Methoden der Demoskopie, 4. Aufl., Springer: Berlin Heidelberg
Finally, a German classic of demographic research should be introduced: "All not everyone" explains simply and understandably the problems of opinion research, interviews, questionnaire construction, sampling, fieldwork, and data evaluation. The book is an

excellent bedtime and holiday reading for those who have always wanted to deal with opinion research or want to know how opinion researchers find out who will be elected next Sunday.

Another Dataset for Practice 24

A central feature of the book is the application of the discussed methods to other data sets. To illustrate this, another data set is depicted below. Students can use this data set to practice. Instructors have the option to use this data set in class instead of the one used in the book if it better suits their field of study. The complete simulated data set is available at www.statistik-kronthaler.ch. Over time, additional data sets will be provided on this platform, which can be used in different fields of study.

Let's imagine we are an intern at the company WebDESIGN. The company programs websites and internet solutions for other companies. WebDESIGN has data from a customer satisfaction survey in which 120 random WebDESIGN customers were surveyed. As an intern, we are tasked with analyzing the data and finding out how satisfied the customers are and what influences customer satisfaction. The data for the first twelve customers and the legend are shown in the following tables (Tabs. 24.1 and 24.2).

From the perspective of WebDESIGN, various questions can be examined using the dataset. The following questions should facilitate an introduction to the analysis of the dataset. In addition, other questions are conceivable:

- What is the number of observations? How many metric, ordinal, and nominal variables are included in the data set?
- Calculate the corresponding average values for all variables in the data set and interpret the results.
- Calculate the corresponding measures of variation for all variables in the data set and interpret the results.
- Draw the corresponding box plots for the metric and ordinal variables and interpret the results.
- Create histograms for the variables design and functionality. Which of the two variables is more symmetric?

F. Kronthaler, *Statistics Applied with the R Commander*,
https://doi.org/10.1007/978-3-662-69107-6_24

Tab. 24.1 WebDESIGN data set

Customer	Customer type	Sector	Employ-ees	Region	Design	Function-ality	Technical problems	Price	Satisfac-tion	Recommen-dation	Future orders
1	1	1	13	0	1.6	1.9	2.5	0.8	1	1.7	0
2	1	3	11	0	7.7	4.7	8.3	7.3	3	9.0	1
3	1	1	1	1	7.7	5.3	8.3	6.6	3	9.9	1
4	1	2	22	0	9.8	4.2	9.2	8.9	3	9.8	1
5	1	3	12	0	7.6	6.3	6.7	6.1	3	9.9	1
6	1	1	22	1	8.2	6.1	8.5	6.9	2	7.7	1
7	1	3	20	0	7.2	5.2	5.7	6.0	2	7.0	1
8	1	3	13	0	8.2	2.4	10.0	8.4	1	9.1	1
9	1	1	34	1	9.8	5.9	7.8	6.5	3	9.9	1
10	1	3	5	0	9.9	4.7	8.3	7.5	3	8.2	1
11	0	3	27	0	9.8	4.1	9.9	8.7	3	9.3	0
12	0	3	4	0	8.9	5.1	9.8	8.5	3	10.0	1

Tab. 24.2 Legend to the WebDESIGN data set

Variable name	Variable description	Values	Missing values	Scale
Customer	Index questionnaire			
Customer type	Customer type	0=New customer 1=Regular customer	n.d	Nominal
Sector	Sector in which the customer operates	1=Industry 2=Trade 3=Service	n.d	Nominal
Employees	Number of company employees	In full-time equivalents	n.d	Metric
Region	Region in which the company is located	0=Switzerland 1=Germany	n.d	Nominal
Design	Satisfaction with the design of the website	Rating on a line from 0 (dissatisfied) to 10 (very satisfied)	n.d	Metric
Functionality	Satisfaction with the functionality of the website		n.d	Metric
Technical problems	Satisfaction with the fixing of technical problems		n.d	Metric
Price	Satisfaction with the price-performance ratio		n.d	Metric
Satisfaction	Overall satisfaction with WebDESIGN	1=very dissatisfied 2=dissatisfied 3=satisfied 4=very satisfied	n.d	Ordinal
Recommendation	It is likely that you recommend WebDESIGN to other companies?	Rating on a line from 0 (no) to 10 (yes)	n.d	Metric
Future orders	Will WebDESIGN be considered for future orders?	0=no 1=yes	n.d	Nominal

Source: Own survey (2013)

- Draw the scatterplots and calculate the Bravais-Pearson correlation coefficient for the variable pairs recommendation and employees, recommendation and design, recommendation and functionality, recommendation and technical problems, and recommendation and price. Interpret the results.
- Calculate the Spearman correlation coefficient for the variable pair recommendation and satisfaction and interpret the result.

- Calculate the contingency coefficient for the variables customer type and future orders. Interpret the result.
- Calculate the phi coefficient for the variables sector and future orders and interpret the result.
- On average, how large are our customer's companies and what range does the 90%, 95%, and 99% confidence interval cover?
- WebDESIGN is interested in finding out whether new customers or regular customers would be more likely to recommend the company. Test this considering all relevant steps at the significance level of 10%.
- In addition, WebDESIGN wants to know if there is a difference between Swiss customers and German customers regarding the recommendation. Test this considering all relevant steps at the significance level of 5%.
- Test the already calculated correlation coefficients at the significance level of 5% considering all relevant steps.
- Calculate the regression function with the dependent variable recommendation and the independent variables customer type, employees, region, design, functionality, technical problems, and price. Test whether the assumptions are met so that the method of ordinary least squares yields the best unbiased and efficient result. If necessary, perform the estimation again.

Appendix

Appendix 1: Solutions to the Applications

Chapter 1

1.1 Firstly, statistics generate knowledge and we can generate knowledge ourselves. Secondly, statistics enable us to make informed decisions. Thirdly, with knowledge about statistics, we can better assess analyses, studies, and statements based on data and recognize attempts at manipulation.

1.2 The data set contains the following three main pieces of information: about which persons or objects do we have information; what information do we have about the persons or objects; the information itself.

1.3 Nominal, ordinal, and metric: A nominal scale level allows us to distinguish the objects of study. An ordinal scale level gives us information about the ranking in addition to the distinction. A metric scale level allows a distinction, gives a ranking, and enables precise distance determination.

1.4 The variable "sector" is nominal, the variable "self-assessment" is ordinal, and the variable "turnover" is metric.

1.5 The variable "education" with the characteristics secondary school, high school, etc. is ordinal, the characteristics allow a ranking.

1.6 One possibility is to measure the number of years spent in an educational institution. An objection could be that the number of years someone has spent in educational institutions has little to do with their level of education. The objection is valid, but can also be raised if we measure the level of education with the school leaving certificate. With a variable, we try to represent a fact as well as possible.

1.7 The solution here is individual depending on the simulated data set.

1.8 The scale level of a variable determines, in addition to the question, which statistical procedure we can use.

1.9 A legend is necessary because the data is usually coded. With the help of the legend, we decode the coding and can thus understand the meaning of the data at any time, even at a later date.

F. Kronthaler, *Statistics Applied with the R Commander*,
https://doi.org/10.1007/978-3-662-69107-6

1.10 The data set comprises 100 observations. It contains three nominal, two ordinal, and five metric variables (the running index was not counted).

1.11 This is determined by the seriousness of the source and our knowledge about it. This is not a real data set, but a data set simulated for teaching purposes.

1.12 A solution can be found in Chap. 8.

Chapter 3

3.1 The mode is used with nominal, ordinal, and metric data. The calculation of the median makes sense with ordinal and metric data. The calculation of the arithmetic mean requires metric data. The geometric mean requires metric-ratio-scaled data.

3.2 $\bar{x}_{\text{innovation}} = 6\,\%$; $\bar{x}_{\text{marketing}} = 20\,\%$. The mean is higher for the variable marketing, i.e., companies on average allocate a higher proportion of their revenue for marketing.

3.3 $\text{Median}_{\text{self-assessment}} = 4$; $\text{Median}_{\text{education}} = 1.5$. The median for the variable self-assessment is 4. 50% of the companies have a higher value, 50% have a lower value. The median for the variable education is 1.5. 50% of the companies have a higher value, 50% have a lower value.

3.4 $\text{Mode}_{\text{sex}} = 0\,\&\,1$; $\text{Mode}_{\text{expectation}} = 2$. The most common values for the variable sex are 0 & 1. We have an equal number of male and female founders. The most common value for the variable expectation is 2. Most companies do not expect any change in future development.

3.5 No, the variable is nominally scaled.

3.6 The results are schematically represented below only for the growth rate. It should be noted that for metric variables, the arithmetic mean, the median, and the mode be calculated, for ordinal variables the median and the mode, and for nominal variables only the mode.

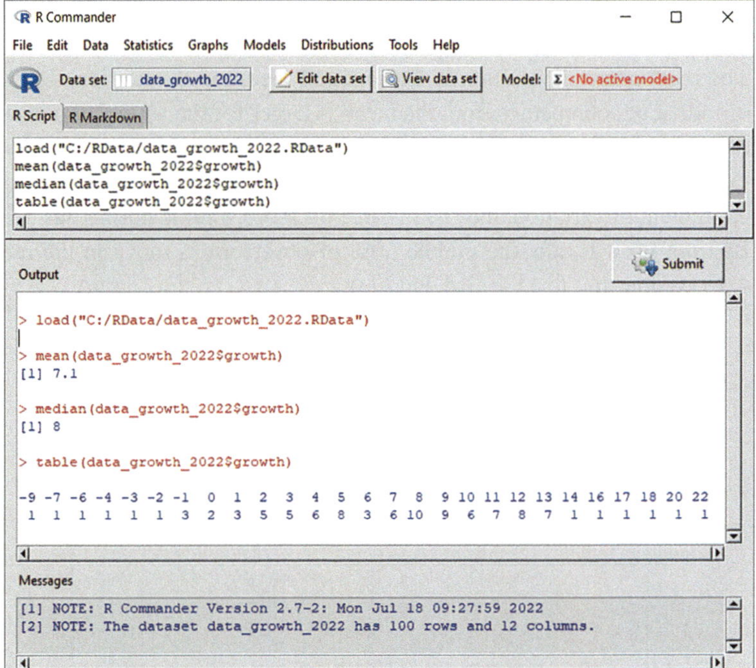

3.7 The average growth factor is 1.18, the average growth rate is 18%.

3.8 The average growth rate is approximately 10%. Based on a rounded growth factor of 1.10, a projected turnover for the year 2016 of 283,232 Fr. is obtained.

3.9 According to the World Bank's figures, India has grown on average by 1.096% during this period, and China by 0.516%. If a forecast is made on this basis, then India will have a higher population than China for the first time in 2025, with 1,459,865,193 inhabitants compared to China's 1,443,820,867 inhabitants.

3.10 The arithmetic mean is sensitive to outliers, as all values are included in the calculation. With the median and the mode, extreme values are not taken into account, therefore these do not react sensitively to outliers.

Chapter 4

4.1 The range and the interquartile range can be used with ordinal and metric data. The standard deviation and the variance require metric data.

4.2 No measure of dispersion is needed for nominal data, as the data does not scatter around a value.

4.3 $ra_{age} = 29.0$; $sd_{age} = 9.62$; $var_{age} = 92.57$; $IQR_{age} = 14.0$; $cv_{age} = 27.49\%$.

4.4 $ra_{marketing} = 21.0$; $sd_{marketing} = 7.67$; $var_{marketing} = 58.86$; $IQR_{marketing} = 12.5$; $cv_{marketing} = 35.68\%$.

4.5. $ra_{innovation} = 9.0$; $sd_{innovation} = 2.83$; $var_{innovation} = 8.00$; $IQR_{innovation} = 3.5$; $cv_{innovation} = 56.67\%$.

4.6 The coefficient of variation is the required measure of variation. The deviation of individual companies from the mean is larger for the variable innovation.

4.7 $ra_{Novartis} = 5.51$; $sd_{Novartis} = 2.08$; $var_{Novartis} = 4.34$; $cv_{Novartis} = 2.85\%$; $ra_{UBS} = 1.27$; $sd_{UBS} = 0.50$; $var_{UBS} = 0.25$; $cv_{UBS} = 6.25\%$. According to the coefficient of variation, the UBS share was more volatile during this period.

4.8–9 Boxplot growth rate, the middle 50% of observations move in the range from 4% (1st quartile) to 11% (3rd quartile).

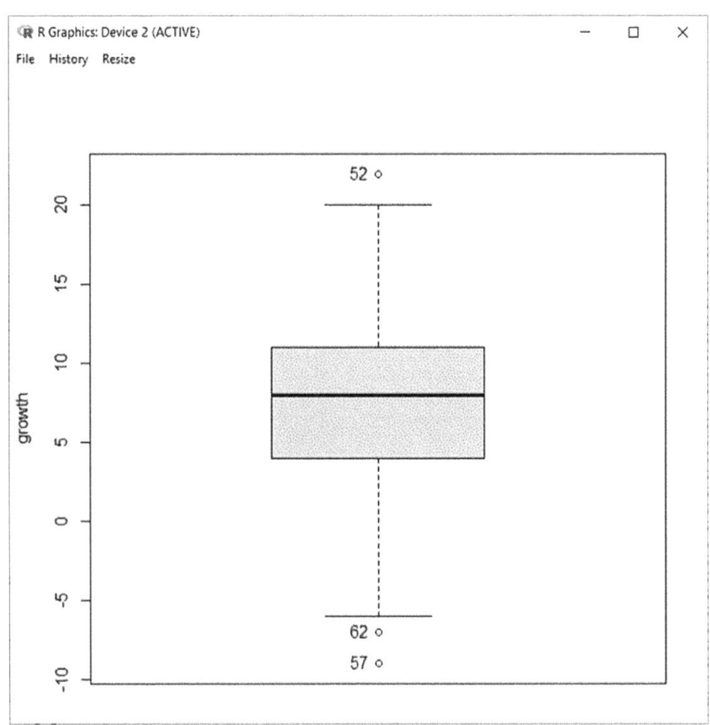

4.10–11 Boxplot growth rate by motive of company foundation

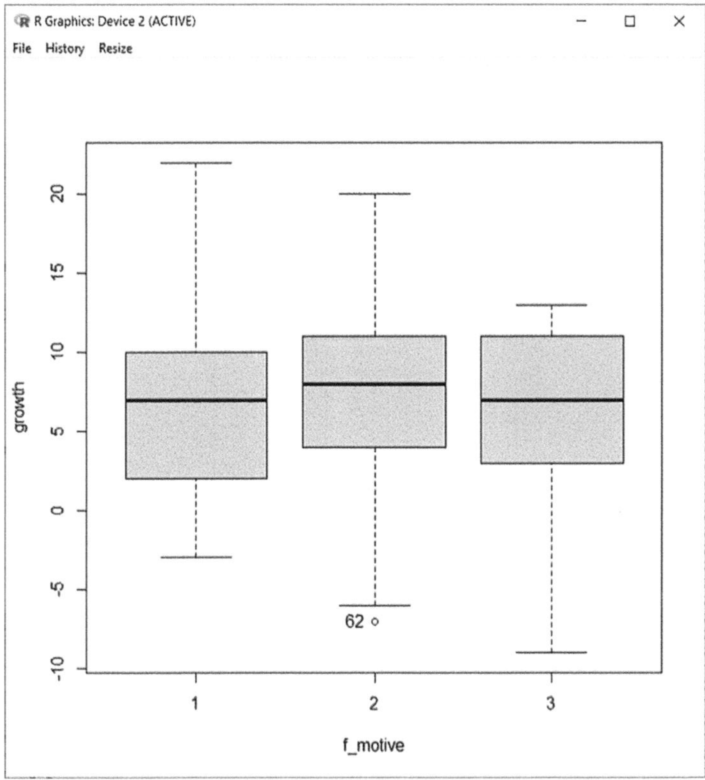

4.12 The results are schematically presented below only for the variable marketing. It should be noted that for metric variables, the smallest value, the largest value, the interquartile range, and the standard deviation can be calculated, for ordinal variables the smallest value, the largest value, and the interquartile range, and for nominal variables nothing.

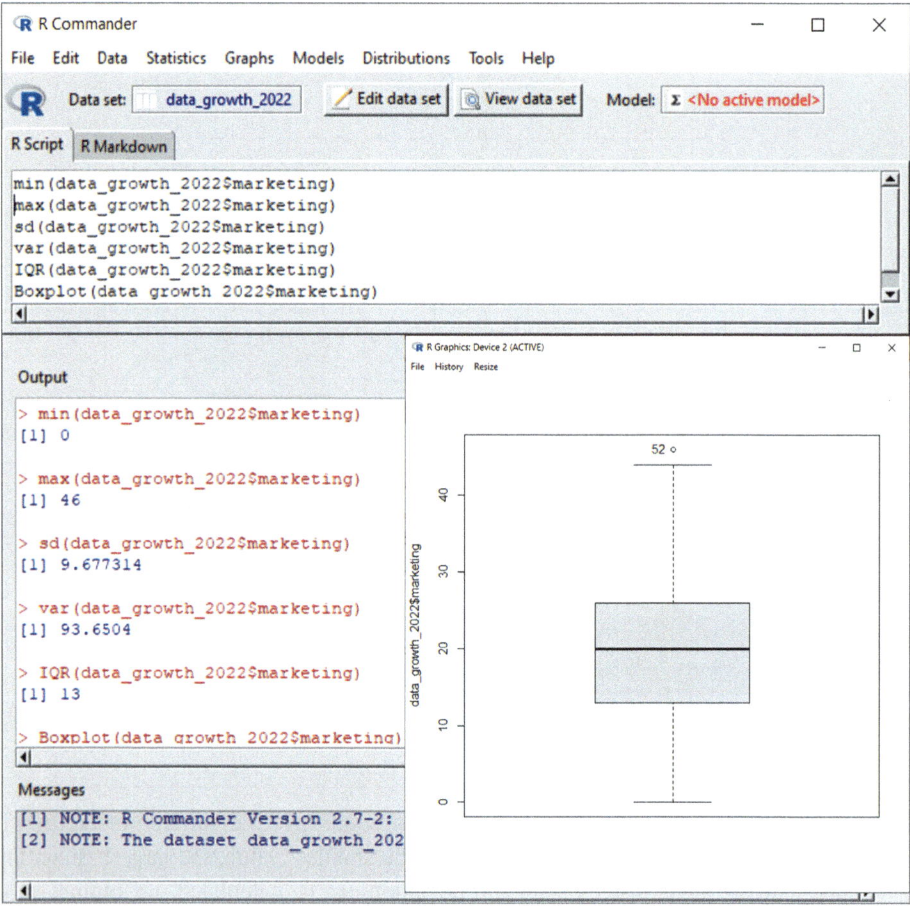

Chapter 5

5.1 Using the geometric mean, company 1 has a calculated average growth rate of approximately 5%, and company 5 has 7%.

5.2 Charts help to convey numbers and create lasting impressions, moreover, graphs often provide clues about the behavior of persons and objects or trends.

5.3 Frequency charts quickly give us an overview of the distribution of data, respectively how often individual values occur.

5.4 When class widths vary, the histogram is preferable. The histogram takes into account the different class widths.

5.5 Frequency table:

Expectation	Number	Relative share	Cumulative relative share
x_i	n_i	f_i	F_i
1	2	0.33	0.33

Expectation	Number	Relative share	Cumulative relative share
2	3	0.50	0.83
3	1	0.17	1.00

Absolute and relative frequency representation:

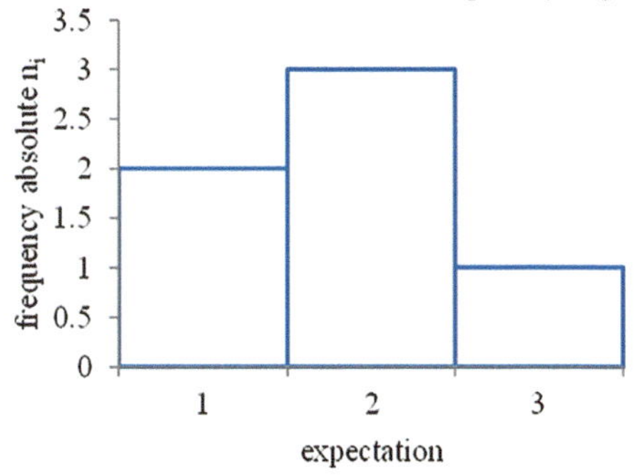

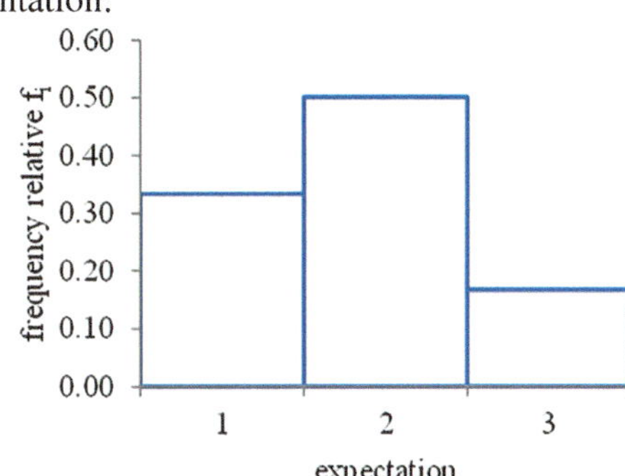

5.6 Relative frequency representations:

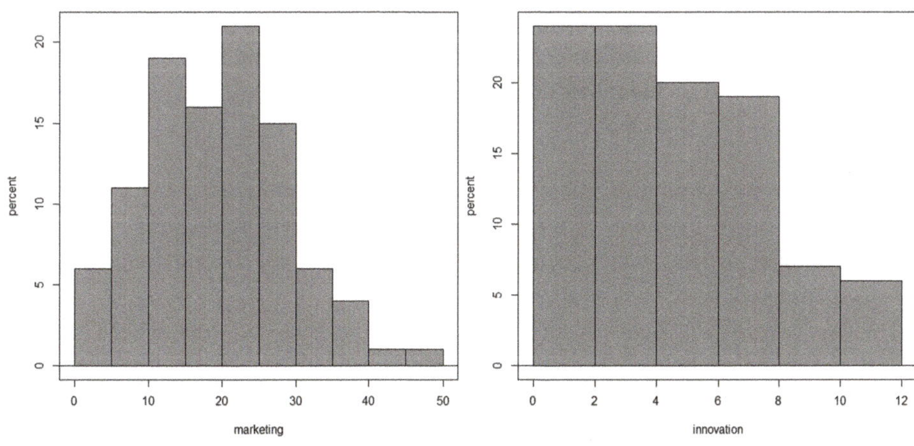

5.7 The illustrations from Solution 5.6 show that neither the marketing variable nor the Product innovation variable is truly symmetrical. Symmetry would exist if we could fold the respective chart in the middle and both sides would lie on top of each other.

5.8 Frequency table:

Class Growth Rate	Number	Relative Share	Frequency Density
from over … to …	n_i	f_i	f_i^*
−10 to −5	3	0.03	0.006

Class Growth Rate	Number	Relative Share	Frequency Density
−5 to 0	8	0.08	0.016
0 to 5	27	0.27	0.054
5 to 25	62	0.62	0.031

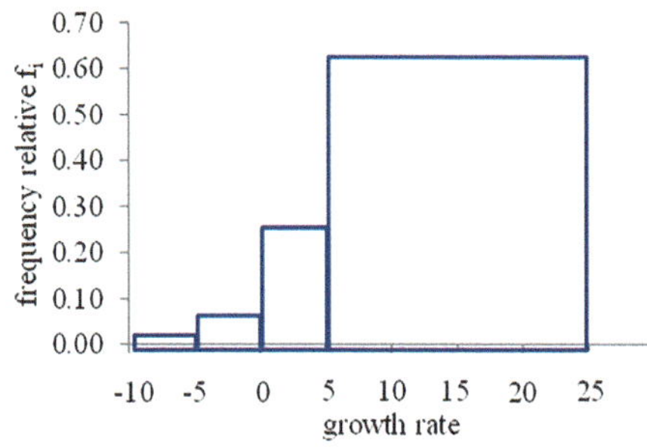

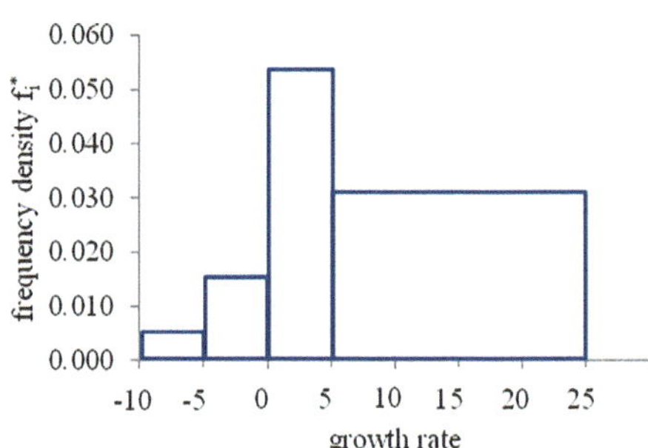

Relative frequency chart and histogram:
In the relative frequency chart, the right bar is much more dominant. This is due to the larger class width compared to the other classes. The histogram corrects the different class widths and is therefore preferable.

5.9–12 Here, everyone is challenged to properly prepare the data and try to create the charts. A small note: Creating publication-ready graphics with R is not entirely easy and requires a deep engagement with R's command language. If that's too tedious, you should rather draw the graphs with Excel.

Chapter 6

6.1 For nominal data with two characteristics, we use the phi coefficient. For nominal data with more than two characteristics, the contingency coefficient is used. For ordinal data, the correlation coefficient of Spearman is used. For metric data, we use the correlation coefficient of Bravais-Pearson.

6.2 Many correlations between two variables are spurious correlations. Therefore, whether there is actually a relationship between two variables requires theoretical considerations.

6.3 A correlation coefficient merely shows how two variables relate to each other. It does not reveal whether one variable influences the other or vice versa. Statistical tools cannot accomplish this. Causality is a matter of theory.

6.4 In the left image, we can approximately draw a straight line. Here, the calculation of the Bravais-Pearson correlation coefficient is sensible. The middle image shows, except for one outlier, a perfect positive relationship. Before we calculate the Bravais-Pearson correlation coefficient, we need to deal with the outlier. The right image also shows a perfect, but non-linear relationship. The simple calculation of the Bravais-Pearson correlation coefficient (without transforming the relationship into a linear one) leads to incorrect results.

6.5 Scatterplot for the variables growth and marketing

growth rate
20
18
16
14
12
10
8
6
4
2
0
0
5
10
15
20
25
30
35
marketing

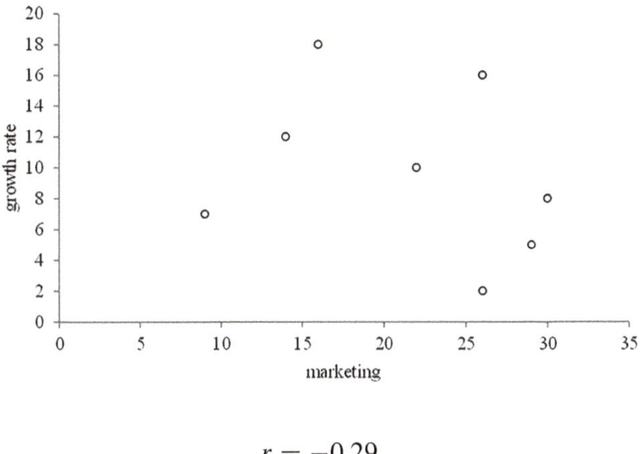

$$r = -0.29$$

6.6 Consider the respective scatterplots before calculation.

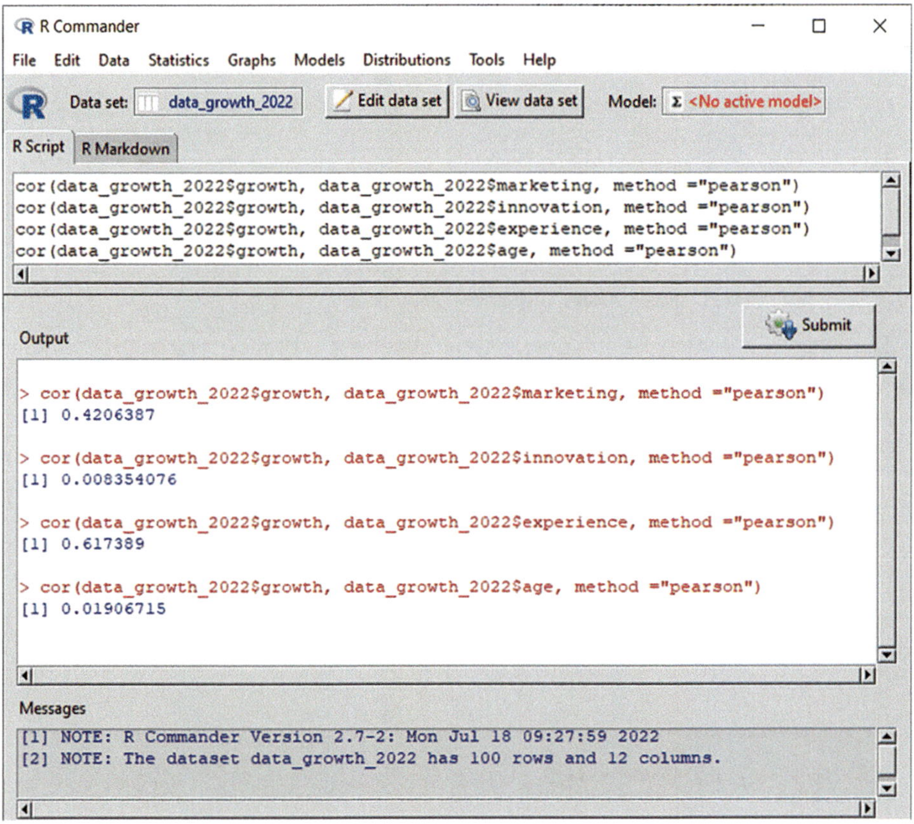

R Commander
File Edit Data Statistics Graphs Models Distributions Tools Help
Data set: data_growth_2022
Edit data set
View data set
Model: Σ <No active model>
R Script
R Markdown
cor(data_growth_2022$growth, data_growth_2022$marketing, method ="pearson")
cor(data_growth_2022$growth, data_growth_2022$innovation, method ="pearson")
cor(data_growth_2022$growth, data_growth_2022$experience, method ="pearson")
cor(data_growth_2022$growth, data_growth_2022$age, method ="pearson")
Submit
Output
> cor(data_growth_2022$growth, data_growth_2022$marketing, method ="pearson")
[1] 0.4206387
> cor(data_growth_2022$growth, data_growth_2022$innovation, method ="pearson")
[1] 0.008354076
> cor(data_growth_2022$growth, data_growth_2022$experience, method ="pearson")
[1] 0.617389
> cor(data_growth_2022$growth, data_growth_2022$age, method ="pearson")
[1] 0.01906715
Messages
[1] NOTE: R Commander Version 2.7-2: Mon Jul 18 09:27:59 2022
[2] NOTE: The dataset data_growth_2022 has 100 rows and 12 columns.

6.7 $r_{Sp} = 0.60$

6.8 Result:

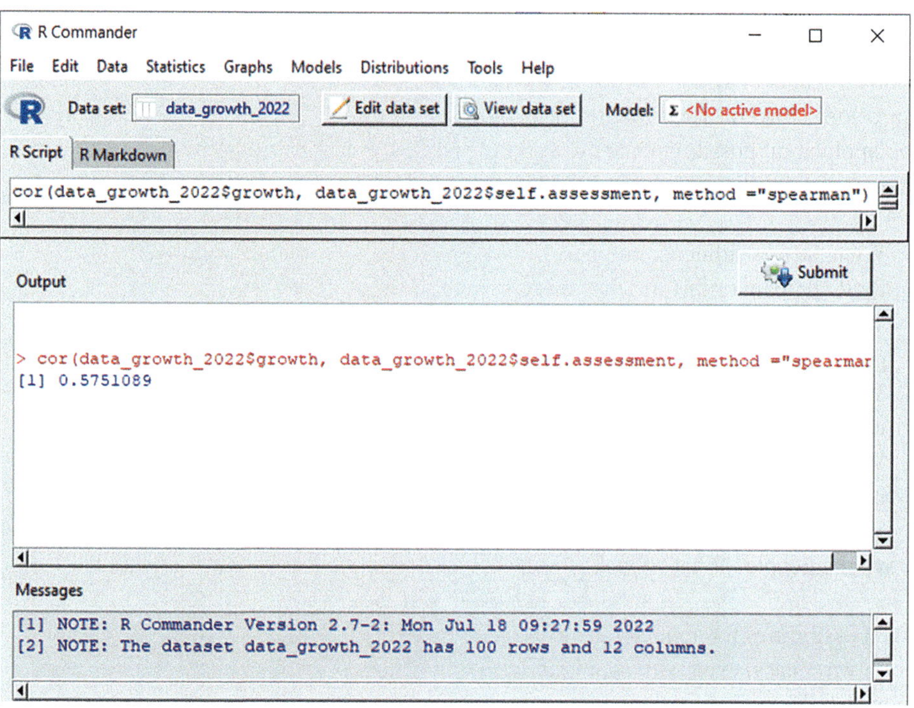

6.9 Result:

R Commander

File Edit Data Statistics Graphs Models Distributions Tools Help

Data set: data_growth_2022 Edit data set View data set Model: Σ <No active model>

R Script R Markdown

```
# Phi coefficient:
(25*26-40*9)/sqrt(65*35*34*66)
```

Output Submit

```
Frequency table:
      f_sector
f_sex   0  1
     0 25 40
     1  9 26


> # Phi coefficient:

> (25*26-40*9)/sqrt(65*35*34*66)
[1] 0.12835
```

Messages

```
[4] NOTE: The dataset data_growth_2022 has 100 rows and 14 columns.
```

6.10 Result:

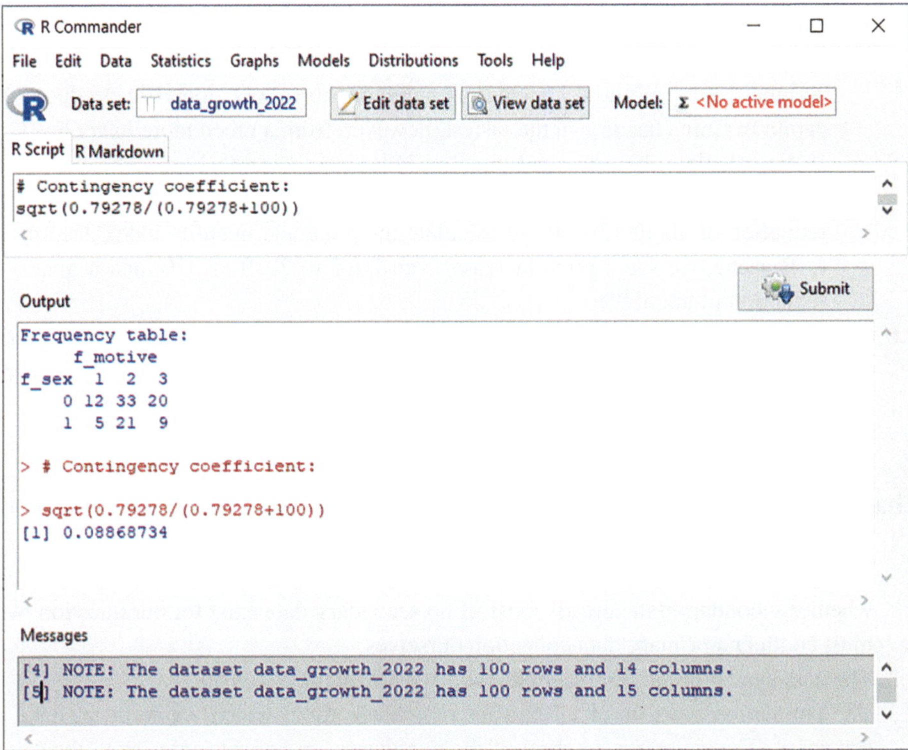

6.11 Consider the scatterplot before calculation. The calculation of the correlation coefficient according to Bravais-Pearson results in a value of $r = -0.32$. However, there is an outlier, without this the value is $r = -0.79$. Depending on whether we calculate with or without the outlier, we either get a weak negative or a strong negative relationship between GDP per capita and CO_2 emissions per capita. Generally, wealthier countries emit less CO_2 per capita than poorer countries. Whether we calculate with or without the outlier depends on whether the outlier is relevant for our analysis.

6.12 The correlation coefficient between the shares of VW and Daimler is $r = 0.99$. We thus have an almost perfect positive relationship, i.e., the value is moving in exactly the same direction. The correlation coefficient between the shares of VW and SAP is $r = 0.56$ and that between Daimler and SAP $r = 0.48$, so we have a weaker positive relationship. The relationship interests us, for example, when we aim for risk diversification in our investment portfolio. Then the values of our shares should not move exactly the same. However, there is also an outlier in the data here.

Chapter 7

7.1–3 See the examples in the book.

7.4 Measured at current prices, the gross domestic product per capita in China in the year 2018 is 363% of the value of 2007, in Germany it is 114% and in Switzerland 130%. Without considering inflation, the gross domestic product per capita in China has grown the fastest, however, from a much more lower level.

7.5 If we calculate the price and quantity index according to Laspeyres, we see a price increase from 2015 to 2019 of 25%. This is partially offset by a quantity reduction of about 13%. If we calculate the price and quantity index according to Paasche, we see a price increase from 2015 to 2019 of 31% and a quantity reduction of about 9%.

7.6 From the base date of 10.06.2020, the index has developed as follows until 24.06.2020: 100.00 (10.6), 96.62, 93.73, 96.18, 98.94, 98.98, 99.20, 98.95, 98.18, 99.02, 95.54 (24.6). The average value of the shares is 4.46% (95.54 − 100 = −4.46) lower on 24.6 than on 10.6.

Chapter 8

8.1 First, the problem and the question must be clearly defined. Second, it is necessary to evaluate which data are needed to answer the question. Third, it must be checked whether secondary data already exist. If no secondary data exist for our question, we must conduct a primary data collection ourselves.

8.2 The data can be found, for example, on the World Bank's website.

8.3 The population consists of all persons or objects about which we want to make a statement. The sample is a subset of the population. In order to infer from the sample to the population, we need an unbiased representative sample. The prerequesite to obtain such a sample is to draw the sample at random.

8.4 $\bar{x}_{\text{growth}} = 7.1$, $\text{CI}_{90\,\%} = [6.19; 8.01]$, $\text{CI}_{95\,\%} = [6.01; 8.18]$, $\text{CI}_{99\,\%} = [5.67; 8.53]$

8.5
$$n = \frac{2^2 \times 1.96^2 \times sd^2}{\text{CIW}^2_{95\,\%}} = \frac{2^2 \times 1.96^2 \times 5^2}{2^2} = 96.04$$

8.6 $p_{\text{industry}} = 0.34$, $\text{CI}_{90\,\%} = [0.263; 0.417]$, $\text{CI}_{95\,\%} = [0.248; 0.432]$, $\text{CI}_{99\,\%} = [0.219; 0.461]$

8.7
$$n = \frac{2^2 \times 1.64^2 \times p(1-p)}{\text{CIW}^2_{90\,\%}} = \frac{2^2 \times 1.64^2 \times 0.25(1-0.25)}{0.08^2} = 315.19$$

8.8 Validity means that our variables actually measure the subject of the investigation. Reliability means that we measure reliably.

8.9 Statement one is correct, we can reliably measure the wrong thing. Statement two is not correct, if we do not measure reliably, then the variable is also not a good proxy for the subject of investigation.

Chapter 9

9.1 H_0: There is no relationship between turnover growth and the proportion of turnover spent on innovation.
H_A: There is a relationship between turnover growth and the proportion of revenue spent on innovation.
H_0: There is no difference between male and female founders in terms of the industry experience they have at the time of founding.
H_A: There is a difference between male and female founders in terms of the industry experience they have at the time of founding.
H_0: There is no relationship between smoking and life expectancy.
H_A: There is a relationship between smoking and life expectancy.

9.2 H_0: There is no or a negative correlation between turnover growth and the proportion of revenue spent on innovation.
H_A: There is a positive correlation between turnover growth and the proportion of revenue spent on innovation.
H_0: Female founders have less or equal industry experience than male founders at the time of founding.
H_A: Female founders have more industry experience than male founders at the time of founding.
H_0: The level of cigarette consumption has no or a positive influence on life expectancy.
H_A: The level of cigarette consumption has a negative influence on life expectancy.

9.5 No relationship and no difference are clearly defined statements and thus testable.

Chapter 10

10.1 Normal distributions:

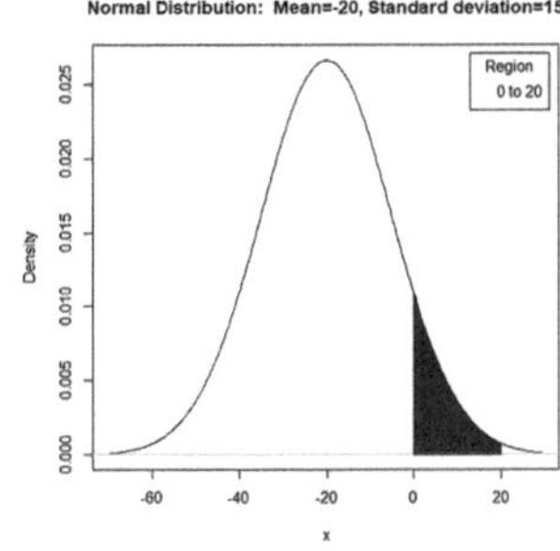

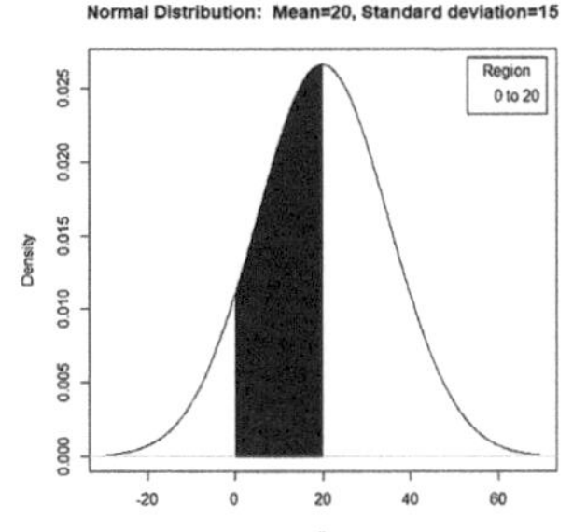

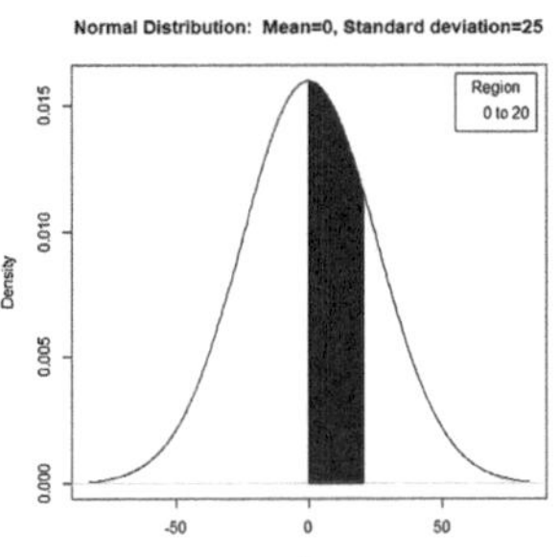

10.2 $z_{(x=-25)} = -3$, $z_{(x=-20)} = -2.67$, $z_{(x=-10)} = -2$, $z_{(x=-7)} = -1.8$, $z_{(x=5)} = -1$, $z_{(x=20)} = 0$, $z_{(x=35)} = 1$, $z_{(x=47)} = 1.8$, $z_{(x=50)} = 2$, $z_{(x=60)} = 2.67$, $z_{(x=65)} = 3$

10.3 The area to the right of $z = 0.5$ is 30.85%, the area to the right of $z = 0.75$ is 22.66%, the area to the right of $z = 1.0$ is 15.87%, the area to the right of $z = 1.26$ is 10.38%.

10.4 The area to the left of $z = -0.5$ is 30.85%, the area to the left of $z = -0.75$ is 22.66%, the area to the left of $z = -1.0$ is 15.87%, the area to the left of $z = -1.26$ is 10.38%.

10.5 The areas to the right and left of the respective positive and negative z-values are equal in size due to the symmetry property of the standard normal distribution.

10.6 The probability of discovering a company founder older than 42 years is 15.87%. The probability of discovering a company founder younger than 21 years is 2.28%. The probability of discovering a company founder in the age range of 42 to 49 years is 13.59%.

10.7 The probability is 6.58%.

10.8 To belong to the top 5% of students, the score must exceed 96.5.

Chapter 11

11.1 The significance level α is on the one hand the probability with which we reject the null hypothesis, on the other hand it is the error probability of incorrectly rejecting the null hypothesis.

11.2 Statements:

- The statement that exceeding the critical value leads to the rejection of H_0 is correct.
- Even if one wanted to, one could not set the α-error to zero. The axes of the distribution function approach the x-axis asymptotically, but never reach it.
- The statement that the smaller the α-error, the better the result, is not correct. Firstly, with a small α-error, we reject the null hypothesis based on our sample findings less often, even if this might be the correct decision. Secondly, there is not only a α-error but also a β-error, which increases with decreasing α.
- The statement that the choice of the α-error depends on the consequences of a false rejection of the null hypothesis is correct.

11.3 A two-sided test is used when we have an undirectional null and alternative hypothesis. A left-sided or right-sided test is used when directional null and alternative hypotheses are present.

11.4 Two-sided test:

H_0: Company founders are on average 40 years old.
H_A: Company founders are on average not 40 years old.

Left-sided test:

H_0: Company founders are on average 40 years old or older.
H_A: Company founders are on average younger than 40 years old.

Right-sided test:

H_0: Company founders are on average 40 years old or younger.
H_A: Company founders are on average older than 40 years old.

11.5 The α-error is the probability of rejecting H_0 even though H_0 is correct. The β-error is the probability of not rejecting H_0 even though H_0 is incorrect. If you reduce the α-error, the β-error becomes larger and vice versa.

Chapter 12

12.1 Step 1: Null hypothesis, alternative hypothesis, significance level

- H_0: $\mu = 10$
- H_A: $\mu \neq 10$
- $\alpha = 0.1$

Step 2: Test distribution and test statistic

- Test distribution is the t-distribution with $df = n - 1 = 99$ degrees of freedom.
- Test statistic is $t = \frac{\bar{x}-\mu}{\hat{\sigma}_{\bar{x}}}$.

Step 3: Rejection area and critical value

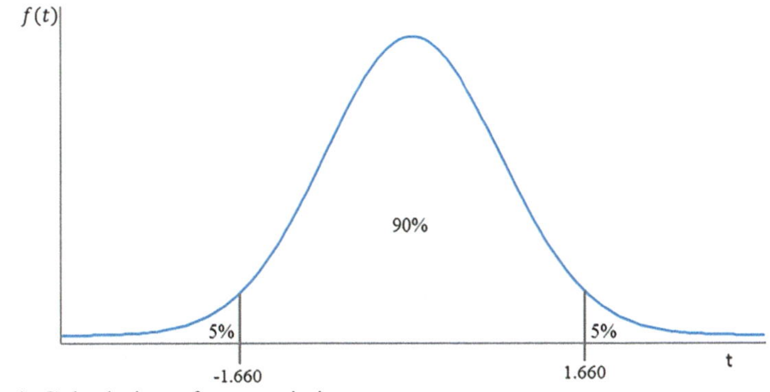

Step 4: Calculation of test statistic

- $t = \frac{\bar{x}-\mu}{\hat{\sigma}_{\bar{x}}} = \frac{7.42-10}{\frac{3.51}{\sqrt{100}}} = -7.35$
- Cohen's $d = \frac{\bar{x}-\mu}{s} = \frac{7.42-10}{3.51} = -0.74$

Step 5: Decision and interpretation

- The calculated t-value is smaller than the critical t-value, we reject H_0.
- The brought industry professional experience is not ten years, but is less. Cohen's d indicates a medium-sized effect.

12.2 Step 1: Null hypothesis, alternative hypothesis, significance level

- H_0: $\mu = 5$
- H_A: $\mu \neq 5$
- $\alpha = 0.05$

Step 2: Test distribution and test statistic

- Test distribution is the t -distribution with $df = n-1 = 99$ degrees of freedom.
- Test statistic is $t = \frac{\bar{x}-\mu}{\hat{\sigma}_{\bar{x}}}$.

Step 3: Rejection area and critical value

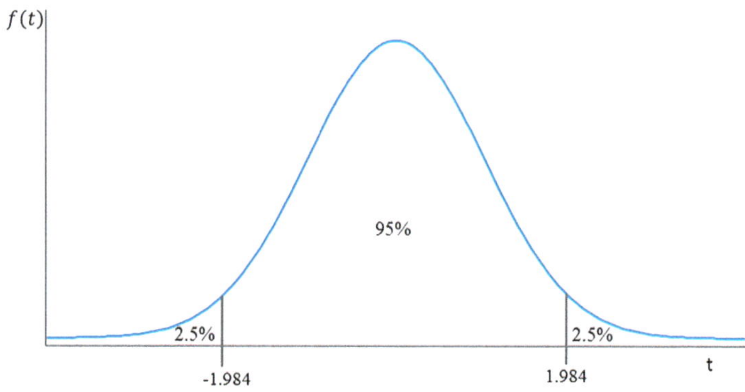

Step 4: Calculation of test statistic

- $t = \frac{\bar{x}-\mu}{\hat{\sigma}_{\bar{x}}} = \frac{7.1-5}{\frac{5.53}{\sqrt{100}}} = 3.80$
- Cohen's $d = \frac{\bar{x}-\mu}{s} = \frac{7.1-5}{5.53} = 0.38$

Step 5: Decision and interpretation

- The calculated t -value is larger than the critical t -value, we reject H_0.
- The average company growth is greater than 5%, so our company has grown below average. However, according to Cohen's d, the effect is rather small or only 0.38 standard deviations.

12.3 Step 1: Null hypothesis, alternative hypothesis, significance level

- H_0: $\mu = 5$
- H_A: $\mu \neq 5$
- $\alpha = 0.05$

Step 2: Test distribution and test statistic

- Test distribution is the t -distribution with $df = n - 1 = 24$ degrees of freedom.
- Test statistic is $t = \frac{\bar{x}-\mu}{\hat{\sigma}_{\bar{x}}}$.

Step 3: Rejection area and critical value

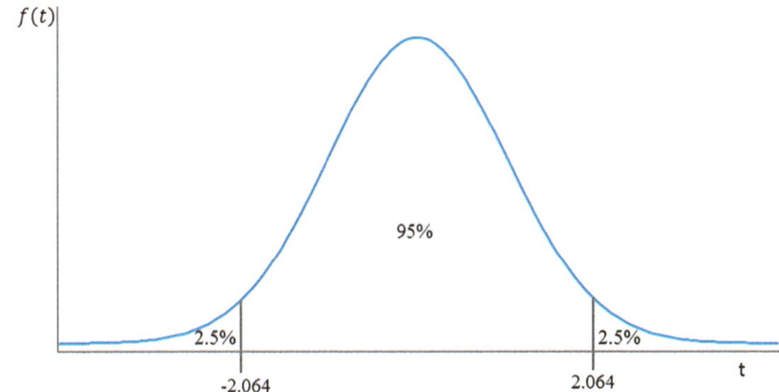

Step 4: Calculation of test statistic

- $t = \frac{\bar{x} - \mu}{\hat{\sigma}_{\bar{x}}} = \frac{8 - 5}{\frac{4.16}{\sqrt{25}}} = 3.61$
- Cohen's $d = \frac{\bar{x} - \mu}{s} = \frac{8 - 5}{4.16} = 0.72$

Step 5: Decision and interpretation

- The calculated t -value is larger than the critical t -value, we reject H_0.
- The average company growth is greater than 5%, thus our company has grown below average. The effect is medium according to Cohen's d.

12.4 Step 1: Null hypothesis, alternative hypothesis, significance level

- H_0: $\mu \geq 15$
- H_A: $\mu < 15$
- $\alpha = 0.1$

Step 2, 3 and 4: R Commander

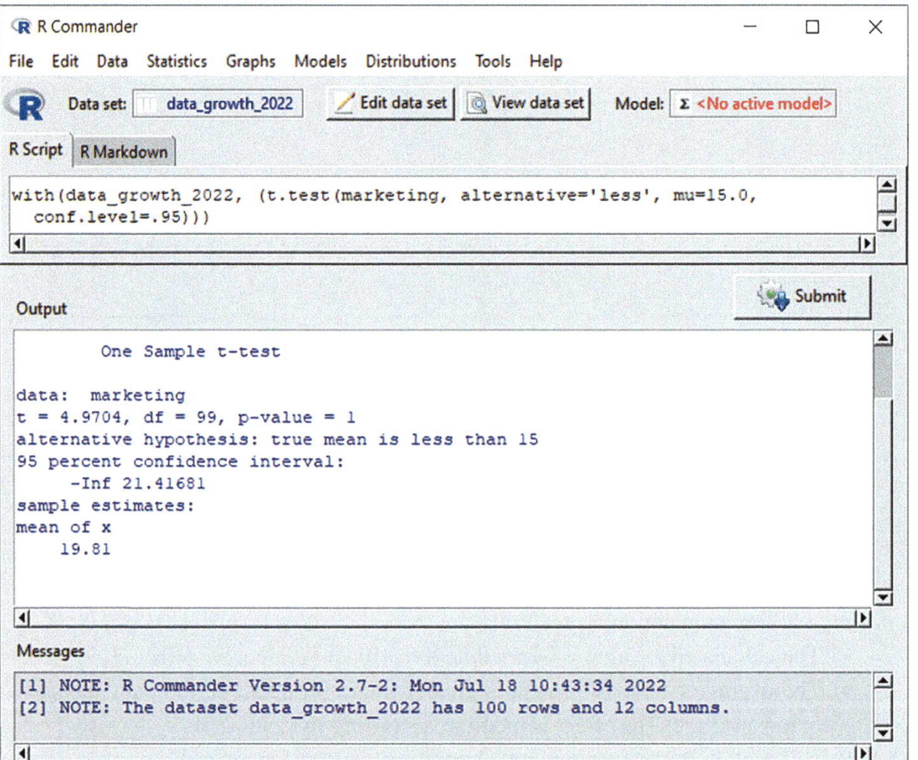

Since we do not reject H_0, we do not necessarily have to calculate the effect size.

Step 5: Decision and interpretation

- The calculated p-value is larger than 10%, we do not reject H_0.
- The proportion allocated for marketing has not decreased.

12.5 Step 1: Null hypothesis, alternative hypothesis, significance level

- H_0: $\mu \leq 5$
- H_A: $\mu > 5$
- $\alpha = 0.05$

Step 2, 3 and 4: R Commander

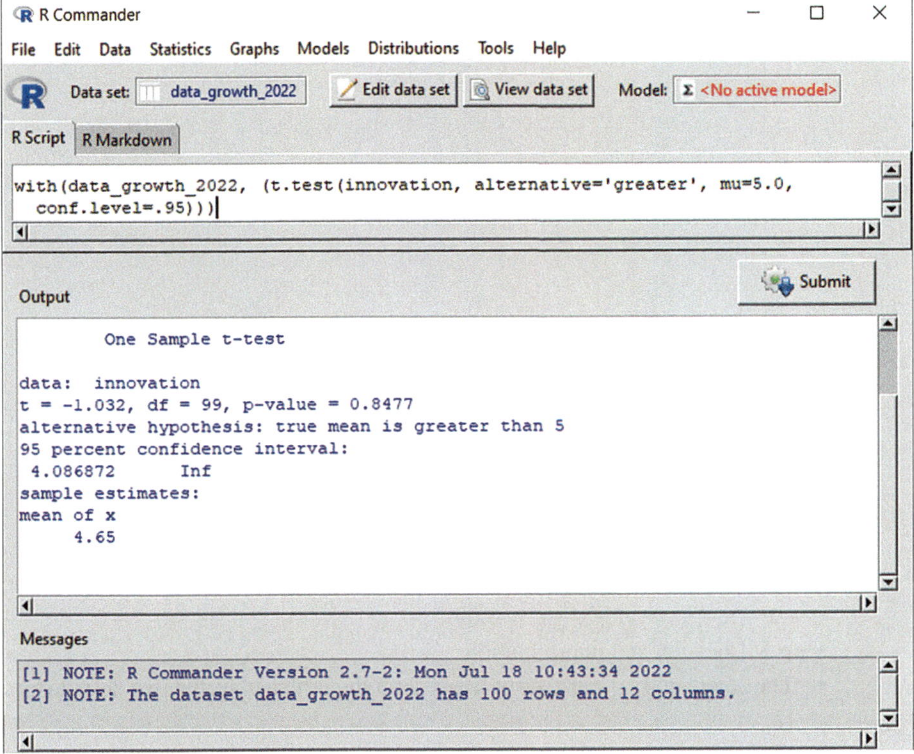

- Since we do not reject H_0, we do not necessarily have to calculate the effect size.

Step 5: Decision and interpretation

- The calculated p-value is larger than 5%, we do not reject H_0.
- The proportion allocated for innovation has not increased.

Chapter 13

13.1 Step 1: Null hypothesis, alternative hypothesis, significance level

- H_0: $\mu_m - \mu_f \leq 0$
- H_A: $\mu_m - \mu_f > 0$
- $\alpha = 0.1$

Step 2: Test distribution and test statistic

- Test distribution is the t -distribution with $df = n_m + n_f - 2 = 98$ degrees of freedom.
- Test statistic is $t = \frac{\bar{x}_1 - \bar{x}_2}{\hat{\sigma}_{\bar{x}_1, \bar{x}_2}}$.

Step 3: Rejection area and critical value

Step 4: Calculation of test statistic

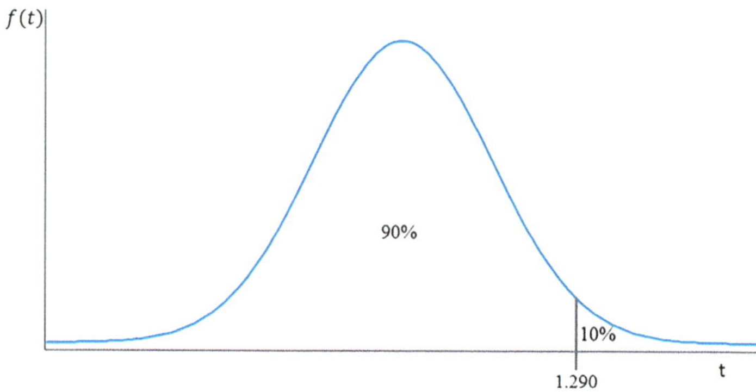

- $t = \frac{\bar{x}_1 - \bar{x}_2}{\hat{\sigma}_{\bar{x}_1,\bar{x}_2}} = \frac{34.77 - 33.29}{\sqrt{\left[\frac{(65-1)\times 7.55^2+(35-1)\times 7.76^2}{65+35-2}\right]\left[\frac{65+35}{65\times 35}\right]}} = 0.93$

- The effect size does not necessarily have to be calculated, as we do not reject H_0. A calculation would yield the following value: $r = \sqrt{\frac{t^2}{t^2+df}} = \sqrt{\frac{0.93^2}{0.93^2+98}} = 0.09.$

Step 5: Decision and interpretation

- The calculated t -value is not larger than the critical t -value, we do not reject H_0.
- Men are not older than women when starting a business, or there is no effect.

13.2–3 Step 1: Null hypothesis, alternative hypothesis, significance level

- H_0: $\mu_i - \mu_s = 0$
- H_A: $\mu_i - \mu_s \neq 0$
- $\alpha = 0.05$

Step 2, 3 and 4: R Commander (we test two-sided with equal variances, the variances are slightly different, we should consider performing the test also with unequal variances)

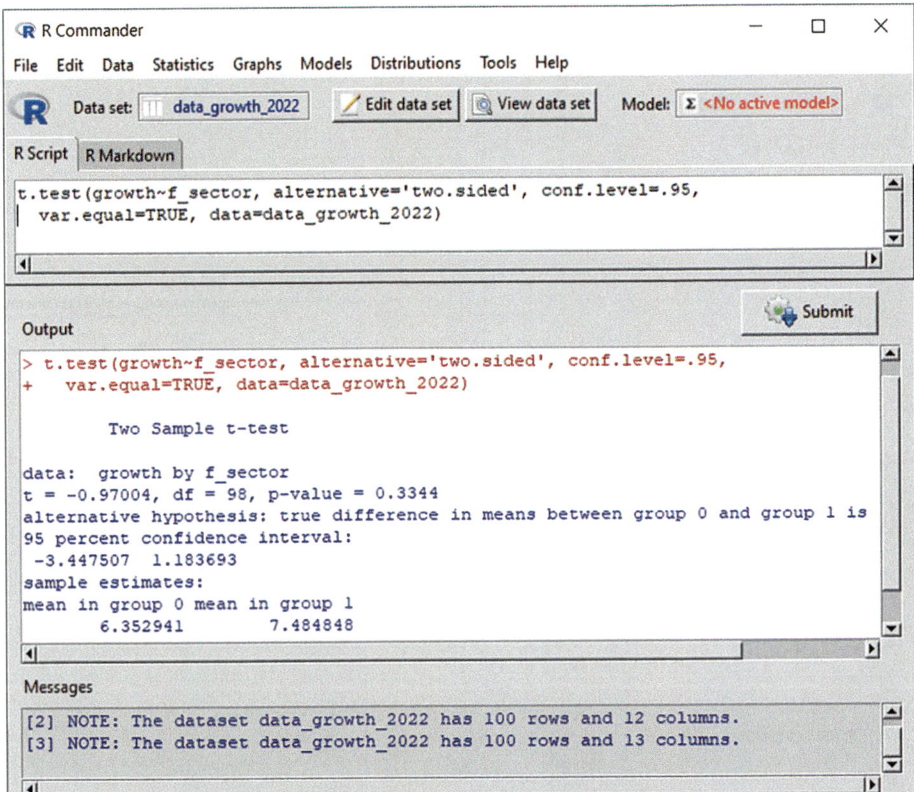

- Since we do not reject H_0, we do not necessarily have to calculate the effect size.

Step 5: Decision and interpretation

- The calculated p-value is larger than 5%, we do not reject H_0.
- There is no difference in the growth of industrial and service companies.

13.4–5 Step 1: Null hypothesis, alternative hypothesis, significance level

- H_0: $\mu_i - \mu_s \leq 0$
- H_A: $\mu_i - \mu_s > 0$
- $\alpha = 0.01$

Step 2, 3 and 4: R Commander (we test one-sided with equal variances, the variances are approximately equal)

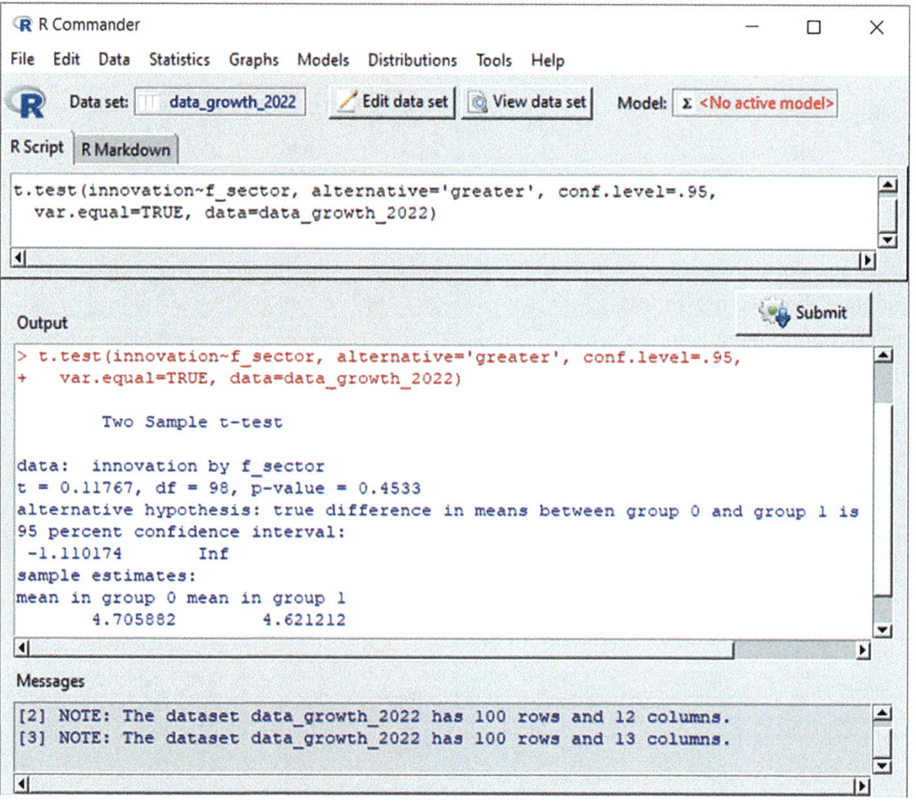

- Since we do not reject H_0, we do not necessarily have to calculate the effect size. Step 5: Decision and interpretation
- The calculated p-value is larger than 1%, we do not reject H_0.
- Industrial companies do not spend more on innovation than service companies.

13.6–7 Step 1: Null hypothesis, alternative hypothesis, significance level

- H_0: $\mu_m - \mu_f = 0$
- H_A: $\mu_m - \mu_f \neq 0$
- $\alpha = 0.1$

Step 2, 3 and 4: Commander (we test two-sided with equal variances, the variances are approximately equal)

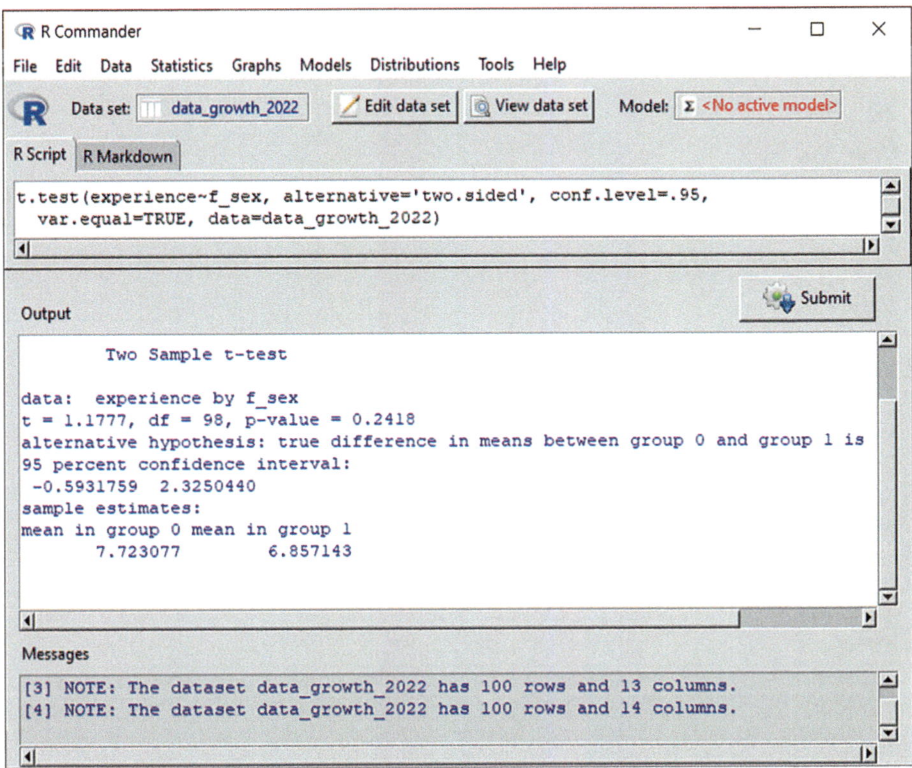

- The effect size does not need to be calculated, as we do not reject H_0.

Step 5: Decision and interpretation

- The calculated p-value is larger than 10%, we do not reject H_0.
- Women and men are not different in terms of industry professional experience.

13.8 Step 1: Null hypothesis, alternative hypothesis, significance level

- H_0: $\mu_m - \mu_f \leq 0$
- H_A: $\mu_m - \mu_f > 0$
- $\alpha = 0.1$ (is not specified in the task, can also be set differently)
  Step 2, 3 and 4: R Commander (we test one-sided with equal variances, the variances are approximately equal)

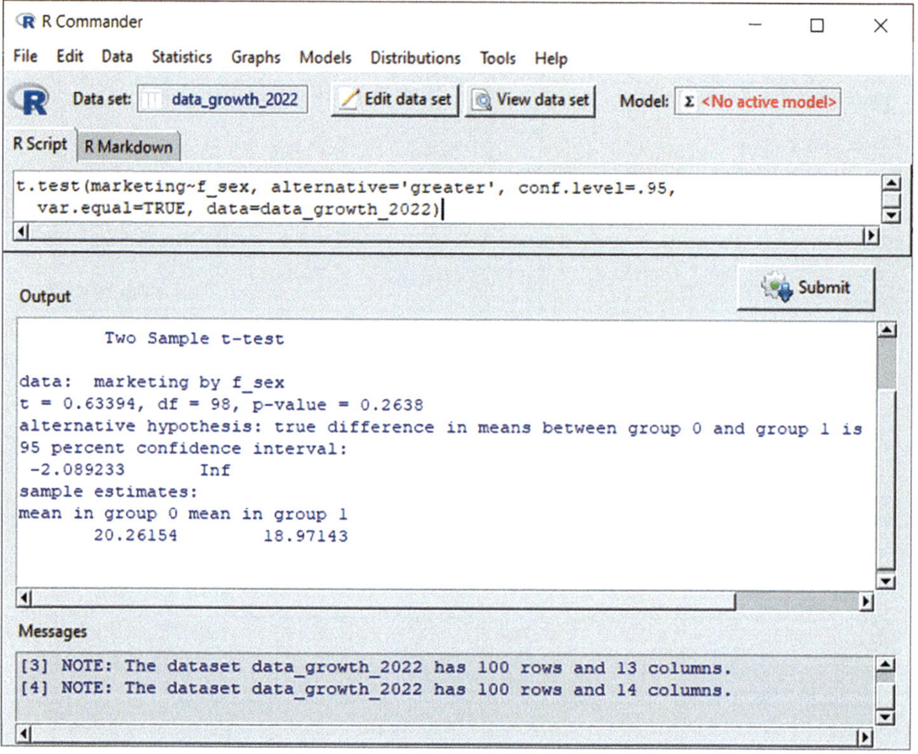

- We do not reject H_0, i.e., the effect size is not relevant.

Step 5: Decision and interpretation

- The calculated p-value is larger than 10%, we do not reject H_0.
- Men do not spend more on marketing than women.

13.9 The arithmetic mean, the standard deviation, and the sample size for both groups are required.

13.10 The Mann-Whitney test is available to us.

Chapter 14

14.1–2 Step 1: Null hypothesis, alternative hypothesis, significance level

- H_0: $\mu_{\text{pre}} - \mu_{\text{post}} \leq 0$
- H_A: $\mu_{\text{pre}} - \mu_{\text{post}} > 0$
- $\alpha = 0.1$

Step 2: Test distribution and test statistic

- Test distribution is the t-distribution with $df = n - 1 = 14$ degrees of freedom.
- Test statistic is $t = \frac{\sum d_i}{\sqrt{\frac{n\sum d_i^2 - (\sum d_i)^2}{n-1}}}$.

Step 3: Rejection area and critical value

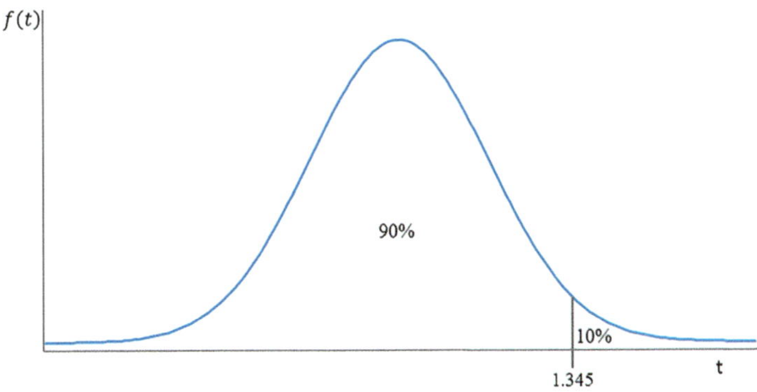

Step 4: Calculation of test statistic

- $t = \frac{\sum d_i}{\sqrt{\frac{n\sum d_i^2 - (\sum d_i)^2}{n-1}}} = \frac{17}{\sqrt{\frac{15 \times 161 - 289}{15-1}}} = 1.38$
- Effect size $r = \sqrt{\frac{t^2}{t^2 + df}} = \sqrt{\frac{1.38^2}{1.38^2 + 14}} = 0.35$

Step 5: Decision and interpretation

- The calculated t-value is larger than the critical t-value, we reject H_0.
- The training measure works and achieves a medium effect.

The following output shows the results with the R Commander:

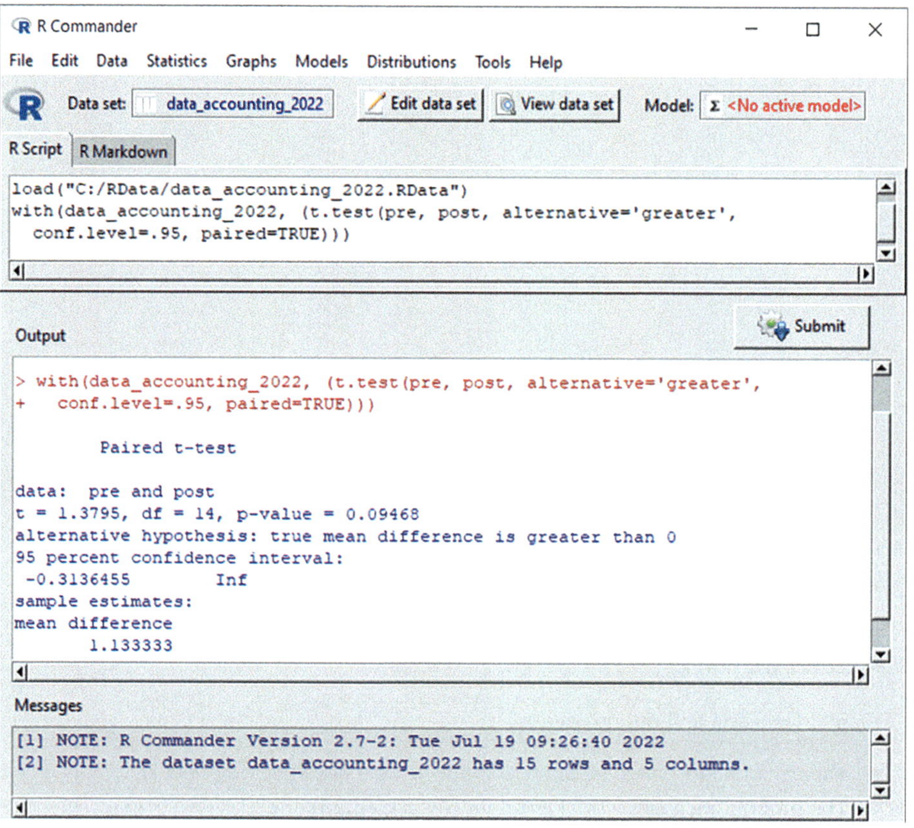

14.3–4 Step 1: Null hypothesis, alternative hypothesis, significance level

- H_0: $\mu_{\text{without}} - \mu_{\text{with}} \geq 0$
- H_A: $\mu_{\text{without}} - \mu_{\text{with}} < 0$
- $\alpha = 0.01$ (14.3) and $\alpha = 0.1$ (14.4)

Step 2, 3 and 4: R Commander (the manual calculation is omitted)

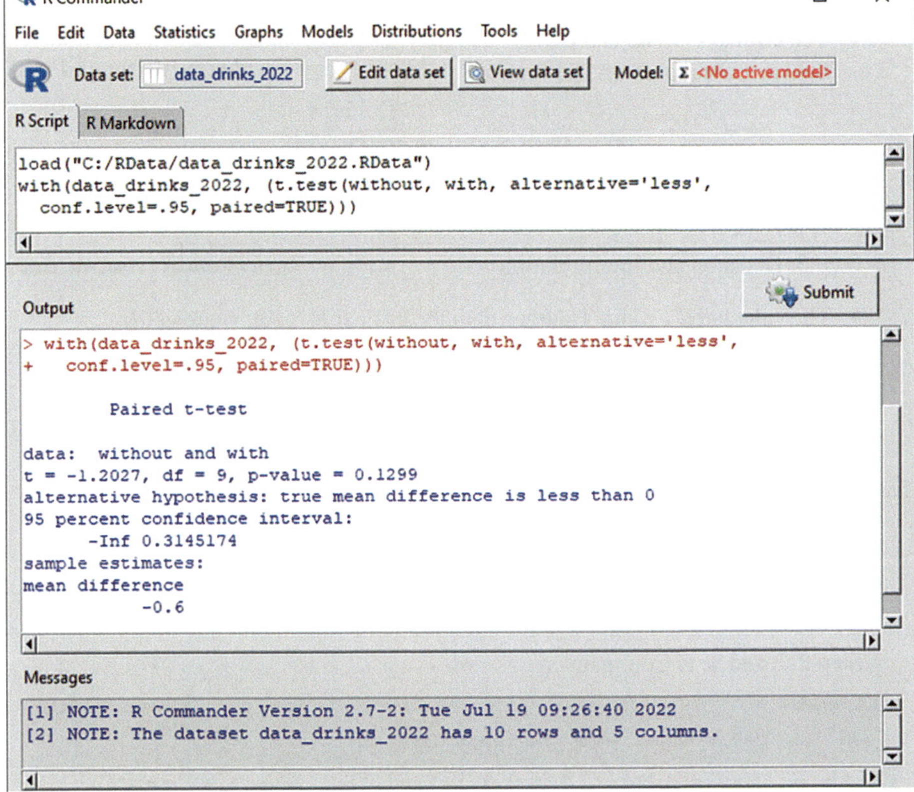

- The effect size does not need to be calculated, as H_0 is not rejected.

Step 5: Decision and interpretation

- The calculated p-value is larger than 10%, we do not reject H_0.
- According to our study design, energy drinks do not increase performance.

Chapter 15

15.1 Step 1: Null hypothesis, alternative hypothesis, significance level

- H_0: $\mu_1 = \mu_2 = \mu_3$
- H_A: $\mu_i = \mu_j$ for at least one pair of ij
- $\alpha = 0.1$

Step 2: Test distribution and test statistic

- The test distribution is the F distribution with 2 and 9 degrees of freedom.
- The test statistic is

Step 3: Rejection area and critical value

$$F = \frac{SS_{\text{explained}}/(G-1)}{SS_{\text{not explained}}/(G \times (K-1))}$$

- The test distribution is the F-distribution with $\alpha = 10\%$ and $df_1 = 2$ and $df_2 = 9$.
- The critical F-value is 3.01.

  Step 4: Calculation of Test Statistic
- $$F = \frac{SS_{\text{explained}}/(G-1)}{SS_{\text{not explained}}/(G \times (K-1))} = \frac{721.17/(3-1)}{88.50/(3 \times (4-1))} = 36.67$$
- Effect size $r = \sqrt{\frac{SS_{\text{explained}}}{SS_{\text{total}}}} = \sqrt{\frac{721.17}{809.67}} = 0.94$

Step 5: Decision and Interpretation

- The calculated F value is larger than the critical F value, we reject H_0.
- The effect size indicates a large effect, this is also reflected in the absolute numbers.
- There is at least a difference in working time between two groups, to determine betwenn which ones, the posthoc test should still be performed.

15.2 Step 1: Null hypothesis, alternative hypothesis, significance level

- H_0: $\mu_1 = \mu_2 = \mu_3 = \mu_4$
- H_A: $\mu_i = \mu_j$ for at least one pair of ij
- $\alpha = 0.05$

Step 2, 3 and 4: R Commander

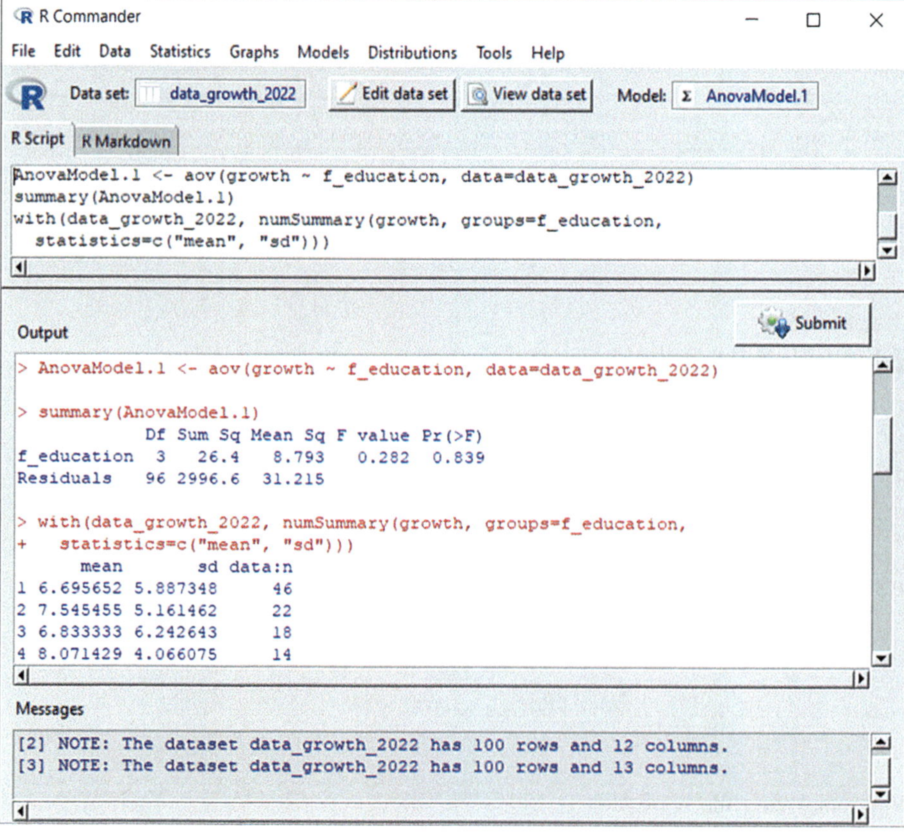

- We do not need to calculate the effect size and the post hoc test, as we do not reject H_0.

Step 5: Decision and interpretation

- The calculated probability is larger than 5%, we do not reject H_0.
- The result suggests that there is no difference in company growth between founders with different levels of education.

15.3 Step 1: Null hypothesis, alternative hypothesis, significance level

- H_0: $\mu_1 = \mu_2 = \mu_3 = \mu_4$
- H_A: $\mu_i = \mu_j$ for at least one pair of ij
- $\alpha = 0.1$

Step 2, 3 and 4: R Commander

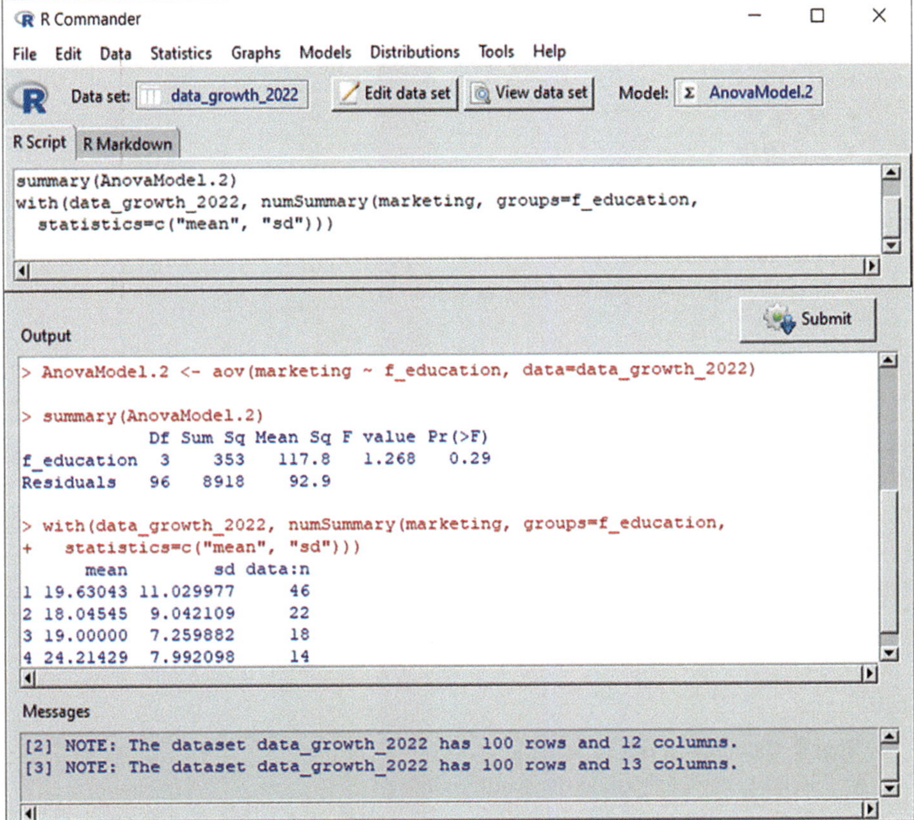

- We do not need to calculate the effect size and the posthoc test, as we do not reject H_0.

Step 5: Decision and interpretation

- The calculated probability is larger than 10%, we do not reject H_0.
- The result suggests that there is no difference in marketing expenditures between founders with different educational backgrounds.

15.4 Step 1: Null hypothesis, alternative hypothesis, significance level

- H_0: $\mu_1 = \mu_2$
- H_A: $\mu_1 \neq \mu_2$
- $\alpha = 0.05$

Step 2, 3, 4 and 5: test for a difference between means with independent samples and analysis of variance with the R Commander

- We do not exceed the critical value in both cases and do not reject H_0.
- There is no difference in the weekly working hours between female and male founders.
- Both with the analysis of variance and with the independent samples t-test, we come to the same conclusion.

Chapter 16

16.1 Step 1: Null hypothesis, alternative hypothesis, significance level

- H_0: $r = 0$
- H_A: $r \neq 0$
- $\alpha = 0.05$

Step 2: Test distribution and test statistic

- Test distribution is the t -distribution with $df = 8 - 2 = 6$ degrees of freedom.
- Test statistic is $t = \frac{r \times \sqrt{n-2}}{\sqrt{1-r^2}}$.

Step 3: Rejection area and critical value

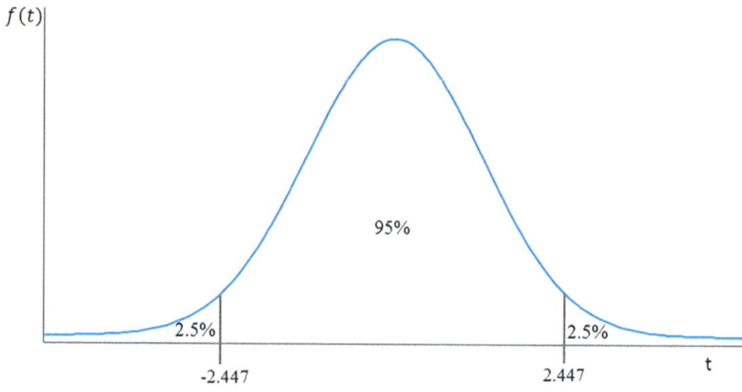

Step 4: Calculation of test statistic

- First, we need to look at the scatterplot.
- Then we can calculate the Bravais-Pearson correlation coefficient. This is $r = -0.29$.
- $$t = \frac{-0.29 \times \sqrt{8-2}}{\sqrt{1-(-0.29)^2}} = -0.74$$

Step 5: Decision and interpretation

- The calculated t -value is not larger or smaller than the critical t -values, we do not reject H_0.
- There is no relationship between the two variables.

16.2 Step 1: Null hypothesis, alternative hypothesis, significance level

- $H_0: r = 0$
- $H_A: r \neq 0$
- $\alpha = 0.05$

Step 2, 3 and 4: R Commander (we examine the scatterplot in advance)

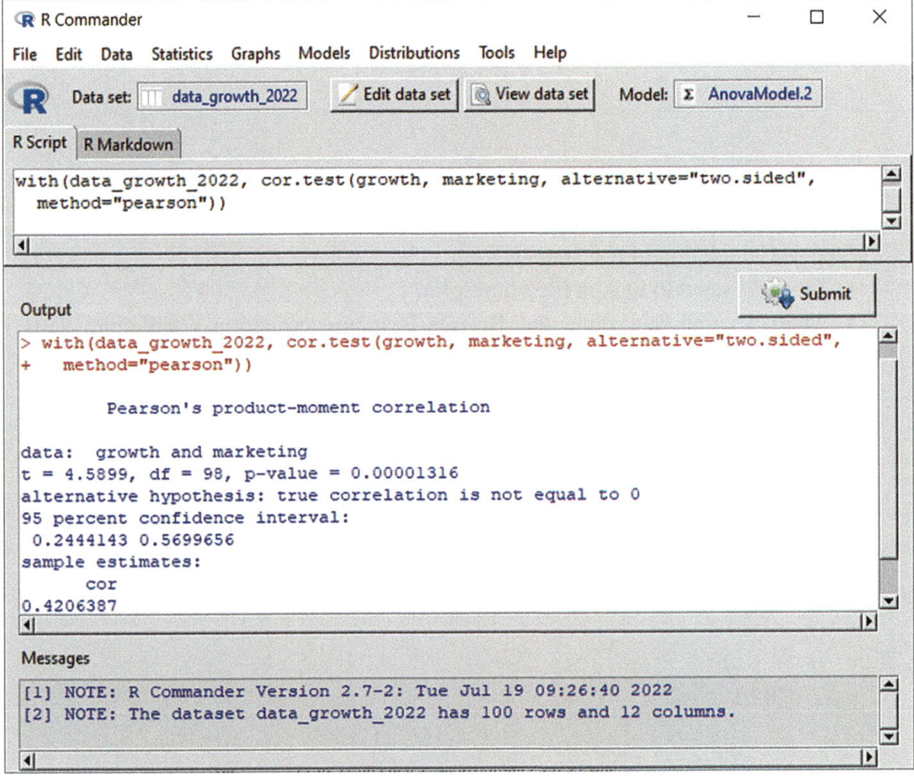

Step 5: Decision and interpretation

- The calculated p-value is less than 5%, we reject H_0.
- There is a moderate relationship between the two variables.

16.3 Step 1: Null hypothesis, alternative hypothesis, significance level

- $H_0: r \leq 0$
- $H_A: r > 0$
- $\alpha = 0.01$

Step 2: Test distribution and test statistic

- Test distribution is the t -distribution with $df = n - 2 = 8 - 6 = 2$ degrees of freedom.
- Test statistic is $t = \frac{r \times \sqrt{n-2}}{\sqrt{1-r^2}}$.

Step 3: Rejection area and critical value

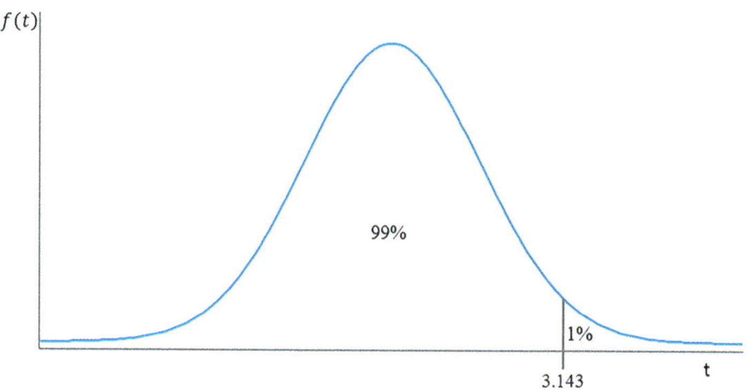

Step 4: Calculation of test statistic

- First, we need to look at the scatterplot.
- Then we can determine the Bravais-Pearson correlation coefficient. This is $r = 0.90$.
- $t = \dfrac{0.90 \times \sqrt{8-2}}{\sqrt{1-0.90^2}} = 5.06$

Step 5: Decision and interpretation

- The calculated t -value is greater than the critical t -value, we reject H_0.
- There is a strong positive relationship between the two variables.

16.4 Step 1: Null hypothesis, alternative hypothesis, significance level

- H_0: $r \leq 0$
- H_A: $r > 0$
- $\alpha = 0.01$

Step 2, 3 and 4: R Commander (we examine the scatterplot in advance)

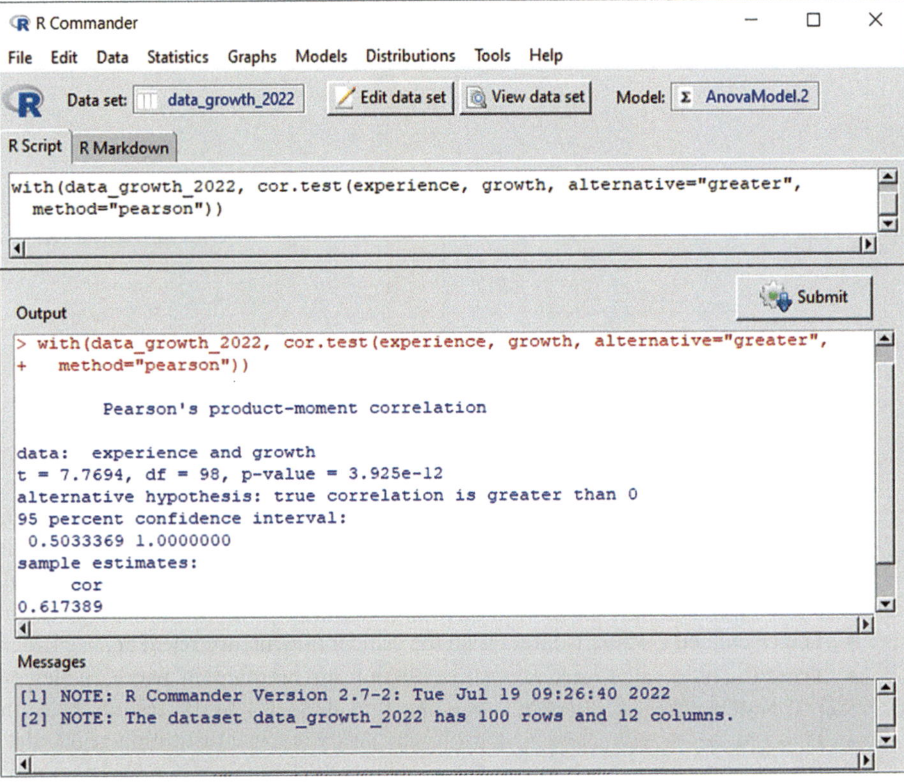

Step 5: Decision and interpretation

- The calculated p-value is less than 1%, we reject H_0.
- There is a strong positive relationship between the two variables.

16.5 Step 1: Null hypothesis, alternative hypothesis, significance level

- H_0: $r_{\text{Sp}} \leq 0$
- H_A: $r_{\text{Sp}} > 0$
- $\alpha = 0.1$

Step 2: Test distribution and test statistic

- Test distribution is the t-distribution with $df = n - 2 = 8 - 2 = 6$ degrees of freedom.
- Test statistic is $t = \frac{r_{\text{Sp}}}{\sqrt{\frac{1-r_{\text{Sp}}^2}{n-2}}}$.

Step 3: Rejection area and critical value

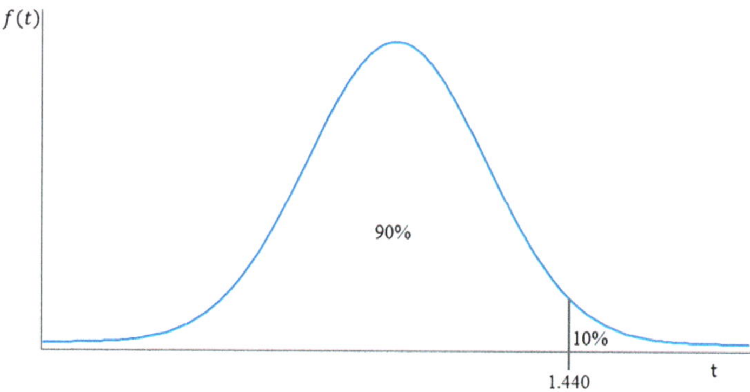

Step 4: Calculation of the test statistic

- We calculate the Spearman correlation coefficient. This is $r = 0.57$.
- $t = \frac{0.57}{\sqrt{\frac{1-0.57^2}{8-2}}} = 1.70$

Step 5: Decision and interpretation

- The calculated t -value is larger than the critical t -value, we reject H_0.
- There is a moderate to strong positive relationship between the two variables.

16.6 Step 1: Null hypothesis, alternative hypothesis, significance level

- H_0: $r_{Sp} \leq 0$
- H_A: $r_{Sp} > 0$
- $\alpha = 0.1$

Step 2, 3 and 4: R Commander

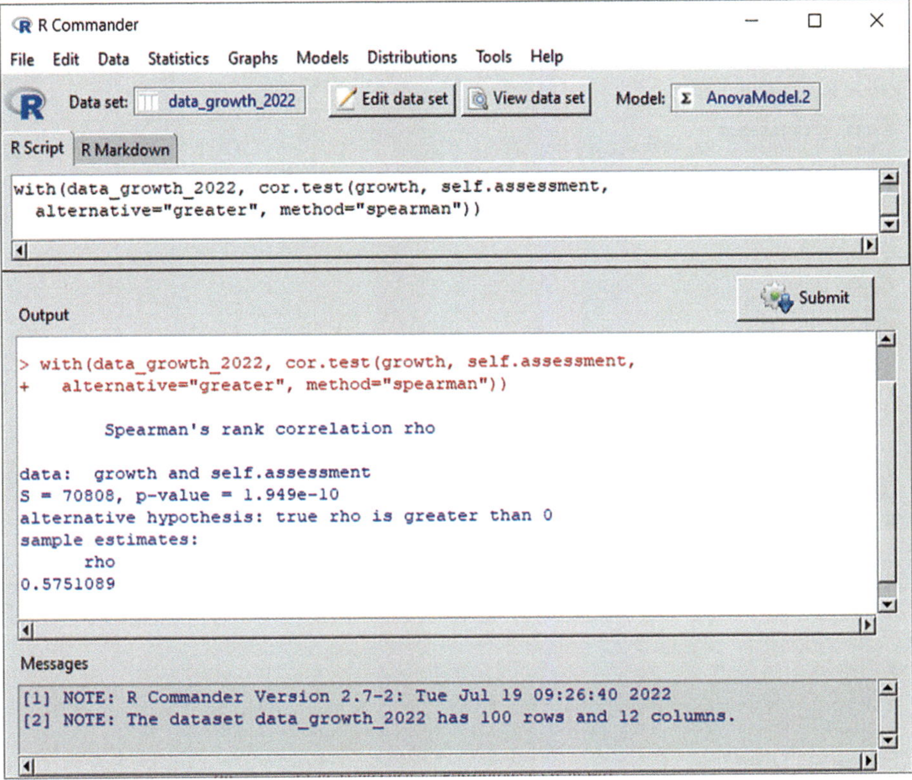

Step 5: Decision and interpretation

- The calculated p-value is less than 10%, we reject H_0.
- There is a moderate to strong positive relationship between both variables.

16.7 Step 1: Null hypothesis, alternative hypothesis, significance level

- H_0: $r_\Phi = 0$
- H_A: $r_\Phi \neq 0$
- $\alpha = 0.05$

Step 2, 3 and 4: R Commander (we first need to convert the required variables into factors)

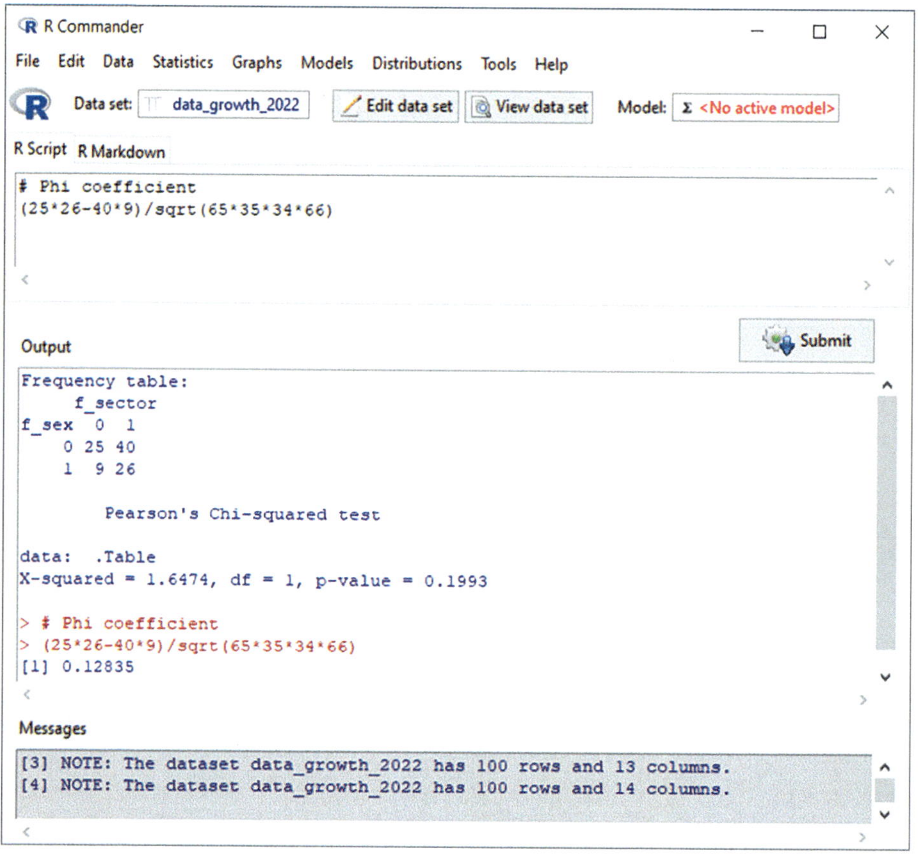

Step 5: Decision and interpretation

- The calculated p-value is larger than 5%, we do not reject H_0.
- There is no relationship between the two variables.

16.8 Step 1: Null hypothesis, alternative hypothesis, significance level

- H_0: $C = 0$
- H_A: $C > 0$
- $\alpha = 0.1$

Step 2, 3 and 4: R Commander (we first need to convert the required variables into factors)

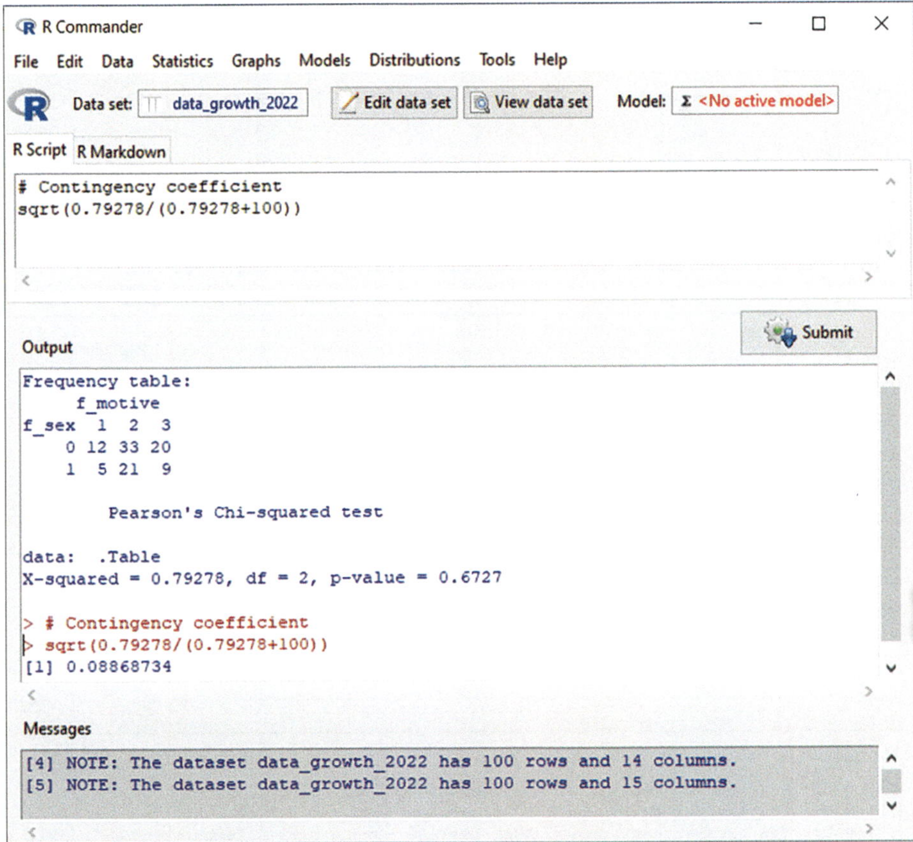

Step 5: Decision and interpretation

- The calculated p-value is larger than 10%, we do not reject H_0.
- There is no relationship between the two variables.

16.9 There is only a significant relationship between the shares of Daimler and VW, i.e., based on the data, we can assume that these two stock values move in the same direction in the population.

Chapter 17

17.1 Step 1: Null hypothesis, alternative hypothesis, significance level

- $H_0: \pi_1 = \pi_2 = 0.5$
- $H_A: \pi_1 = \pi_2 \neq 0.5$
- $\alpha = 0.05$

Step 2, 3 and 4: R Commander (we first need to convert the required variable into factors)

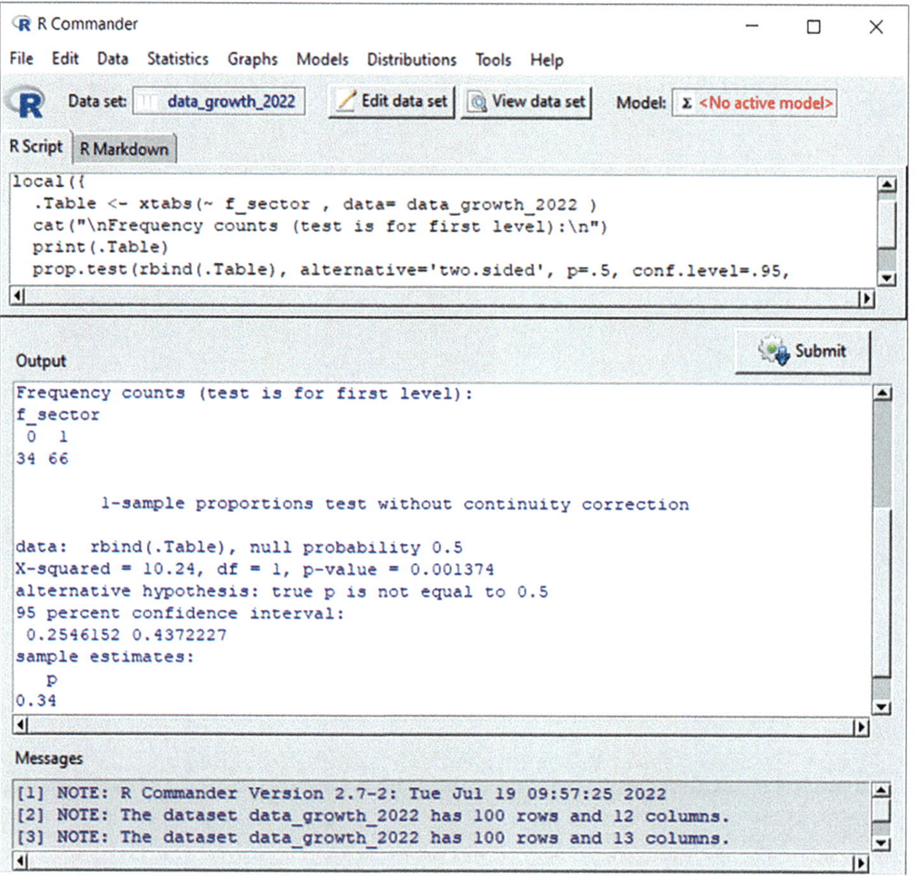

Step 5: Decision and interpretation

- The calculated p-value is less than 5%, we reject H_0.
- Industrial foundations and service foundations are not equally frequent.

17.2 Step 1: Null hypothesis, alternative hypothesis, significance level

- H_0: $\pi_{\text{women}} = \pi_{\text{men}}$
- H_A: $\pi_{\text{women}} \neq \pi_{\text{men}}$
- $\alpha = 0.05$

Step 2, 3 and 4: R Commander (we first need to convert the required variables into factors)

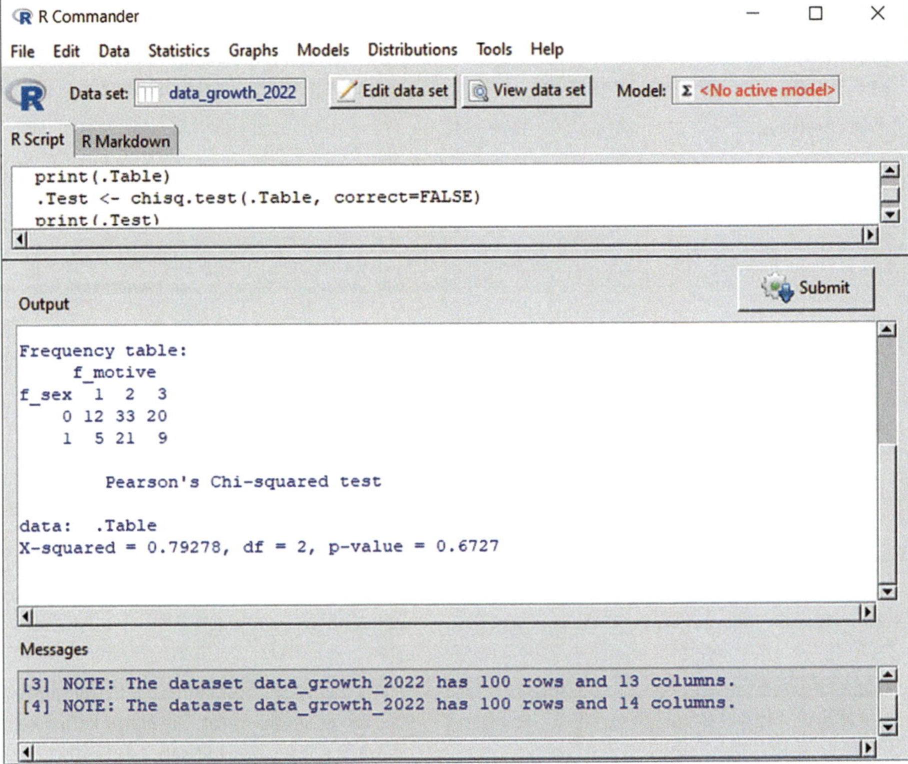

Step 5: Decision and interpretation

- The calculated *p*-value is larger than 5%, we do not reject H_0.
- There are no differences in the motives for founding between men and women.

17.3 Step 1: Null hypothesis, alternative hypothesis, significance level

- H_0: $\pi_{\text{pre}} = \pi_{\text{post}}$
- H_A: $\pi_{\text{pre}} \neq \pi_{\text{post}}$
- $\alpha = 0.1$

Step 2, 3 and 4: R Commander (we first need to convert the required variables of the dataset data_awareness_campaign into factors)

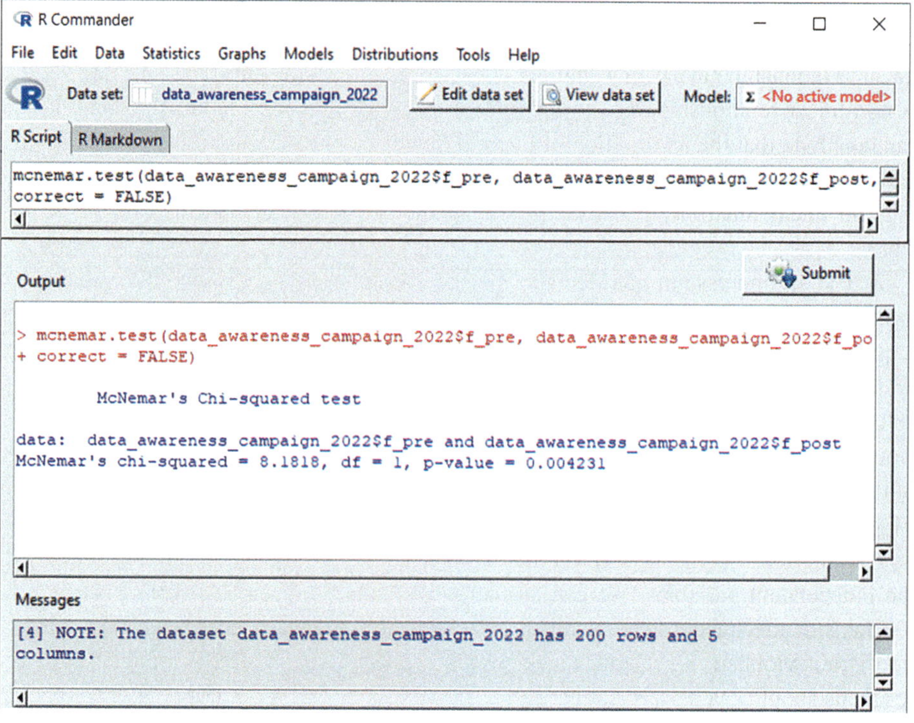

Step 5: Decision and interpretation

- The calculated p-value is less than 10%, we reject H_0.
- The awareness campaign has an impact on dietary behavior.

Chapter 19

19.1 Without a theory about the relationship, the result of the regression analysis is worthless. Only with the help of theory can we make a statement about whether one variable actually influences another variable. If we do not have this statement, we do not know what happens to Y when we change X, as we no not know anything about causality.

19.2. Result:

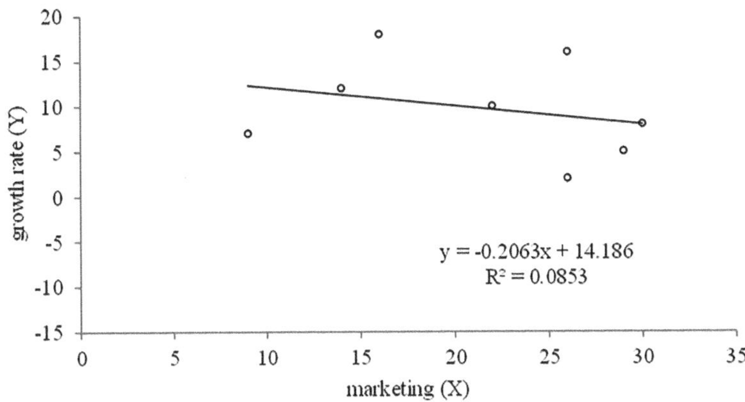

- The slope of the regression line is slightly negative, the growth rate tends to decrease as the effort for marketing increases.
- The R^2 is only 8.5%, i.e. the explanatory power of the regression line is very low.

19.3 The predicted growth rate is 10.06%. The forecast is very unreliable, as the coefficient of determination R^2 is very low (see task 19.2).

19.4 Result:

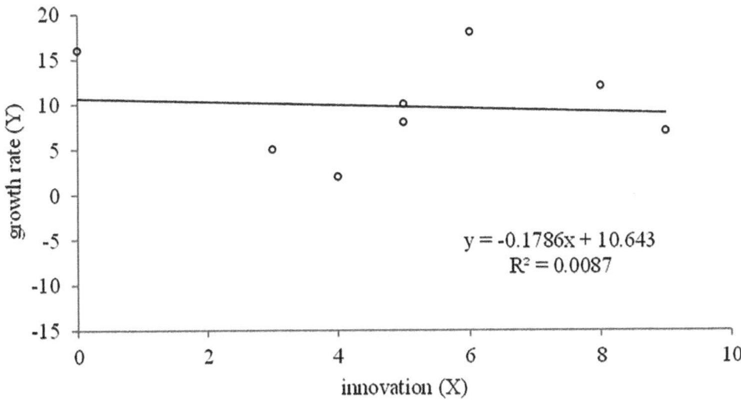

- The regression line is $\hat{y} = 10.643 - 0.1786x$. The coefficient of determination is 0.87%.
- The R^2 is only 0.87%, i.e., the explanatory power approaches zero.

19.5 The predicted growth rate is 7.07%. We have two problems with the forecast: 1) it is an out-of-sample forecast, 2) the coefficient of determination R^2 is very low (see task 19.4).

19.6 Result (only the regression calculation for the variables growth rate and marketing is shown below):

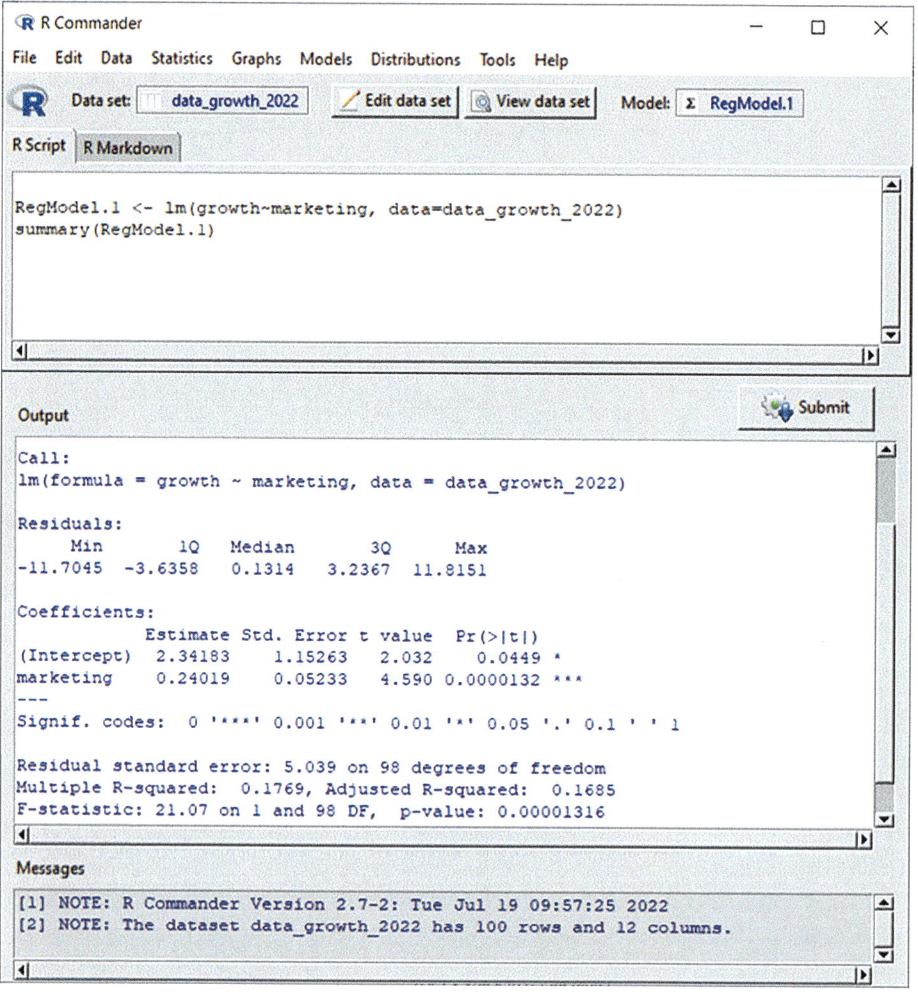

- There seems to be a relationship between growth and marketing and growth and experience.
- There seems to be no relationship between growth and innovation and growth and age.

19.7 Both theoretically and in view of task 19.6, it is very likely that not only one variable but several have an influence on the growth rate.

Chapter 20

20.1. Result:

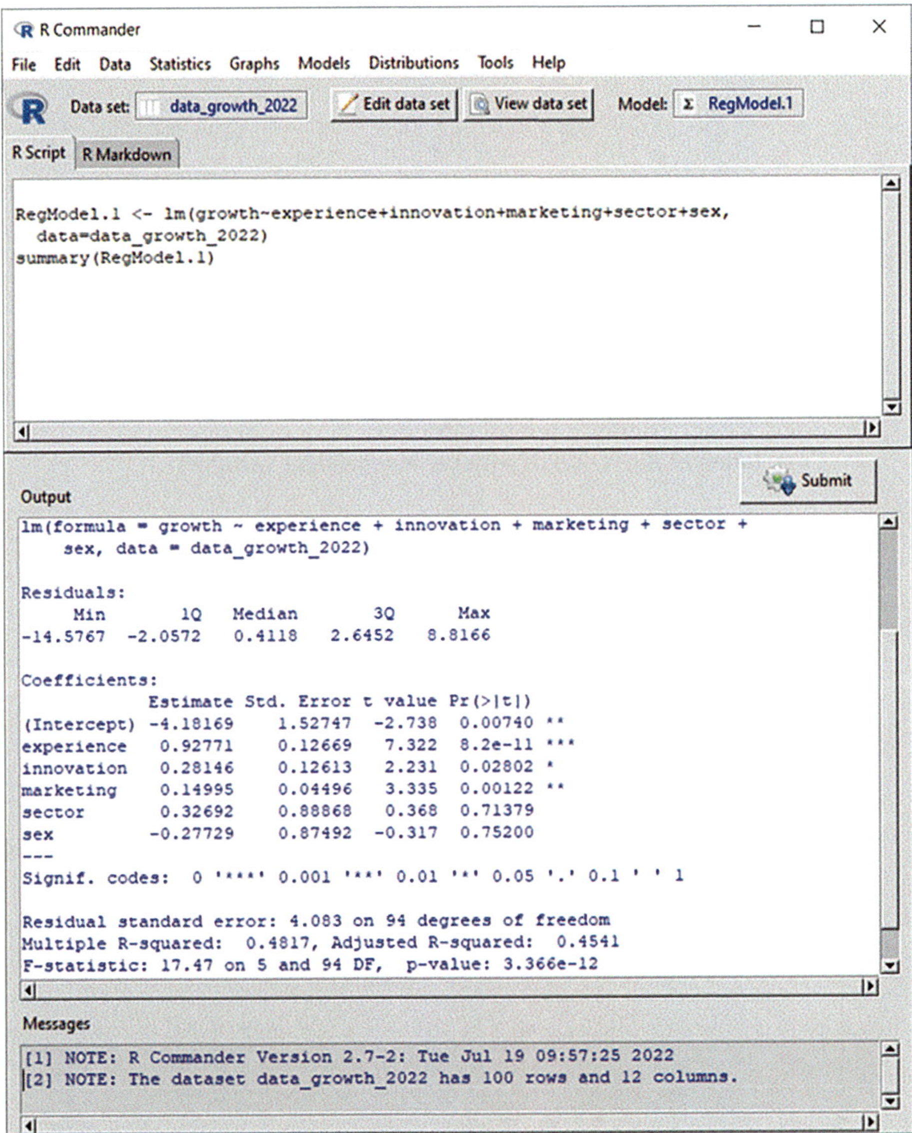

For interpretation, compare Chap. 20.

20.2 Result

Linearity:

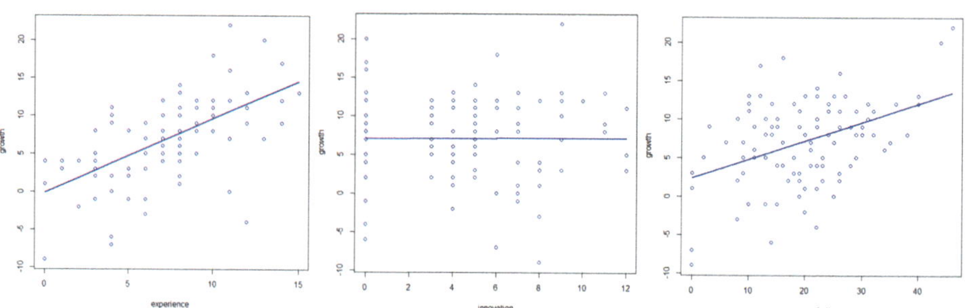

- The scatterplots show either a linear or no relationship between the dependent and independent variables.
- A non-linear relationship is not discernible.

Correlation of the residuals with the independent variables:

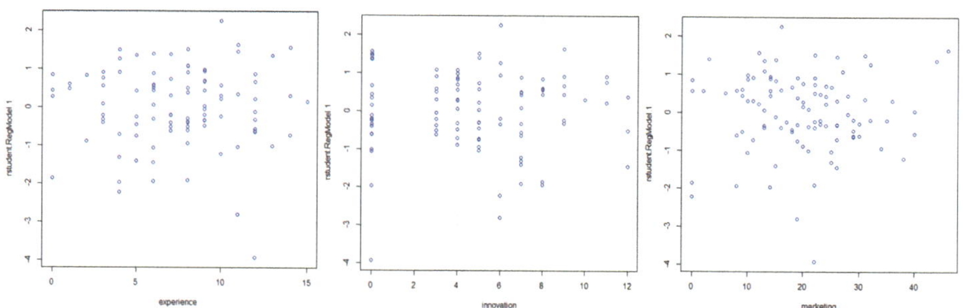

- No relationship can be identified between the independent variables and the residuals.

Constancy of the variance of deviations:

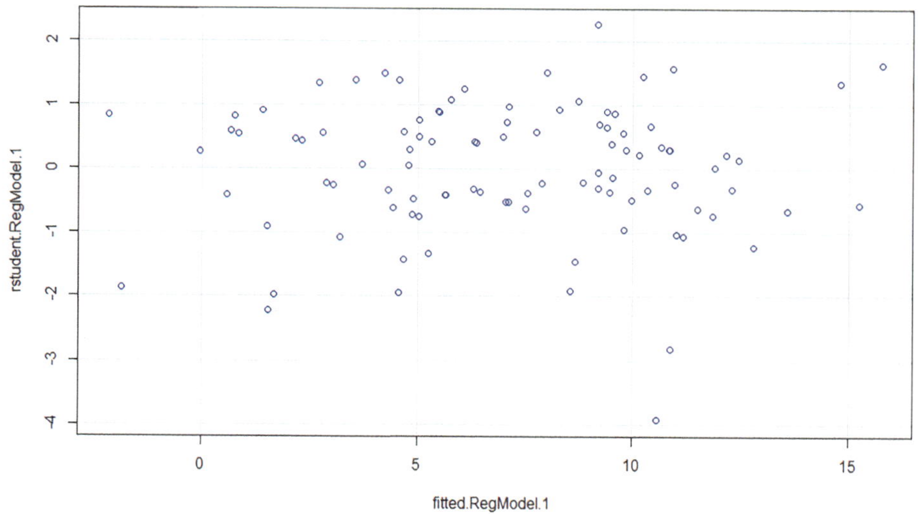

- The variation of the residuals does not change with the increase or decrease of the estimated $\hat{Y}$ values (the residuals scatter largely within the same bandwidth).

No correlation between two or more independent variables:

	Marketing	Product Improvement Experience	Experience
Marketing	1		
Product Improvement	−0.03	1	
Experience	0.26	−0.27	1

- There is no correlation between the independent variables.

No correlation of deviations to each other:

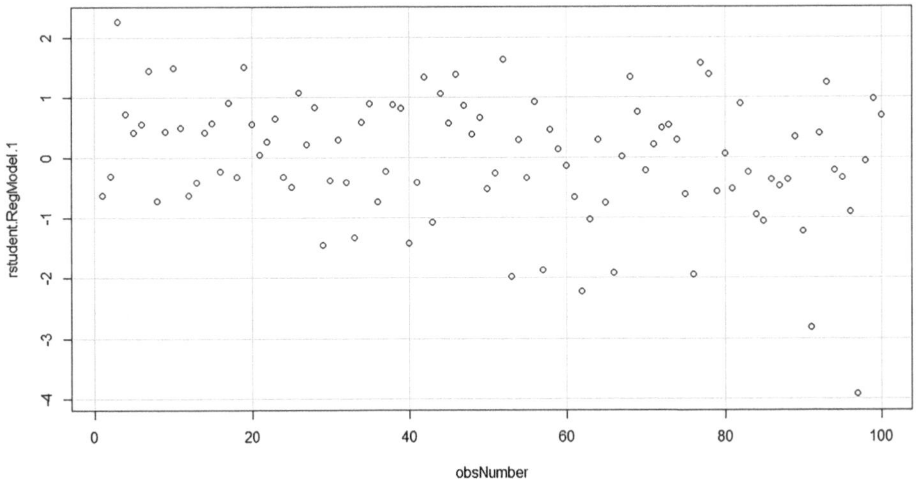

- In the figure, we discover no pattern.

Normal distribution of deviations:

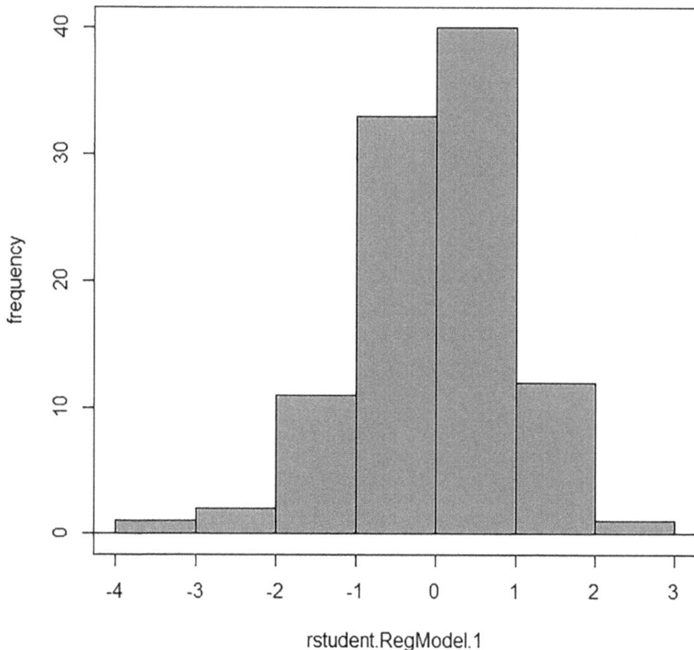

- The histogram shows a slightly left-skewed, but largely normally distributed distribution of deviations.

20.3 The calculation of the correlation shows that there is no longer any correlation between the independent variables. This is an indication that there is no problem with the assumption of multicollinearity anymore.

Appendix 2: The Standard Normal Distribution N (0,1)

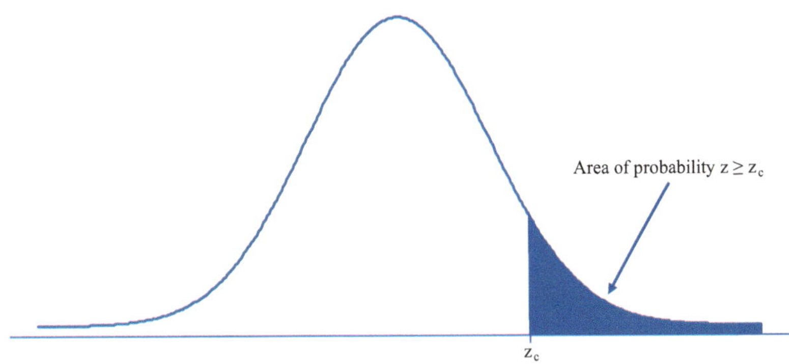

z_k → ↓	second decimal place									
	.00	.01	.02	.03	.04	.05	.06	.07	.08	.09
0.0	0.5000	0.4960	0.4920	0.4880	0.4840	0.4801	0.4761	0.4721	0.4681	0.4641
0.1	0.4602	0.4562	0.4522	0.4483	0.4443	0.4404	0.4364	0.4325	0.4286	0.4247
0.2	0.4207	0.4168	0.4129	0.4090	0.4052	0.4013	0.3974	0.3936	0.3897	0.3859
0.3	0.3821	0.3783	0.3745	0.3707	0.3669	0.3632	0.3594	0.3557	0.3520	0.3483
0.4	0.3446	0.3409	0.3372	0.3336	0.3300	0.3264	0.3228	0.3192	0.3156	0.3121
0.5	0.3085	0.3050	0.3015	0.2981	0.2946	0.2912	0.2877	0.2843	0.2810	0.2776
0.6	0.2743	0.2709	0.2676	0.2643	0.2611	0.2578	0.2546	0.2514	0.2483	0.2451
0.7	0.2420	0.2389	0.2358	0.2327	0.2296	0.2266	0.2236	0.2206	0.2177	0.2148
0.8	0.2119	0.2090	0.2061	0.2033	0.2005	0.1977	0.1949	0.1922	0.1894	0.1867
0.9	0.1841	0.1814	0.1788	0.1762	0.1736	0.1711	0.1685	0.1660	0.1635	0.1611
1.0	0.1587	0.1562	0.1539	0.1515	0.1492	0.1469	0.1446	0.1423	0.1401	0.1379
1.1	0.1357	0.1335	0.1314	0.1292	0.1271	0.1251	0.1230	0.1210	0.1190	0.1170
1.2	0.1151	0.1131	0.1112	0.1093	0.1075	0.1056	0.1038	0.1020	0.1003	0.0985
1.3	0.0968	0.0951	0.0934	0.0918	0.0901	0.0885	0.0869	0.0853	0.0838	0.0823
1.4	0.0808	0.0793	0.0778	0.0764	0.0749	0.0735	0.0721	0.0708	0.0694	0.0681
1.5	0.0668	0.0655	0.0643	0.0630	0.0618	0.0606	0.0594	0.0582	0.0571	0.0559
1.6	0.0548	0.0537	0.0526	0.0516	0.0505	0.0495	0.0485	0.0475	0.0465	0.0455
1.7	0.0446	0.0436	0.0427	0.0418	0.0409	0.0401	0.0392	0.0384	0.0375	0.0367
1.8	0.0359	0.0351	0.0344	0.0336	0.0329	0.0322	0.0314	0.0307	0.0301	0.0294
1.9	0.0287	0.0281	0.0274	0.0268	0.0262	0.0256	0.0250	0.0244	0.0239	0.0233
2.0	0.0228	0.0222	0.0217	0.0212	0.0207	0.0202	0.0197	0.0192	0.0188	0.0183
2.1	0.0179	0.0174	0.0170	0.0166	0.0162	0.0158	0.0154	0.0150	0.0146	0.0143
2.2	0.0139	0.0136	0.0132	0.0129	0.0125	0.0122	0.0119	0.0116	0.0113	0.0110
2.3	0.0107	0.0104	0.0102	0.0099	0.0096	0.0094	0.0091	0.0089	0.0087	0.0084
2.4	0.0082	0.0080	0.0078	0.0075	0.0073	0.0071	0.0069	0.0068	0.0066	0.0064
2.5	0.0062	0.0060	0.0059	0.0057	0.0055	0.0054	0.0052	0.0051	0.0049	0.0048
2.6	0.0047	0.0045	0.0044	0.0043	0.0041	0.0040	0.0039	0.0038	0.0037	0.0036
2.7	0.0035	0.0034	0.0033	0.0032	0.0031	0.0030	0.0029	0.0028	0.0027	0.0026
2.8	0.0026	0.0025	0.0024	0.0023	0.0023	0.0022	0.0021	0.0021	0.0020	0.0019
2.9	0.0019	0.0018	0.0018	0.0017	0.0016	0.0016	0.0015	0.0015	0.0014	0.0014
3.0	0.001350									
3.5	0.000233									
4.0	0.000032									
5.0	0.0000003									

Appendix 3: The t-Distribution

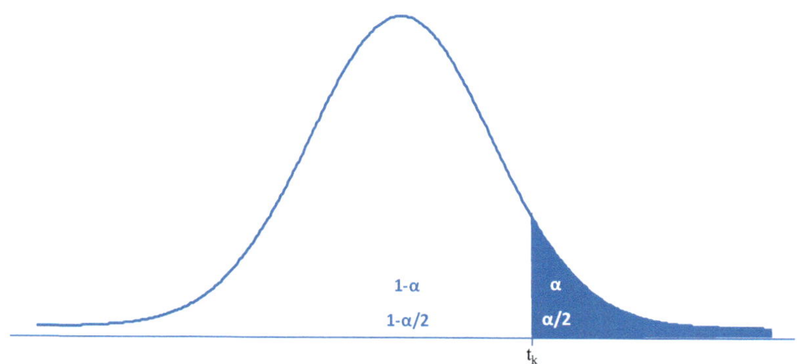

					α or. α/2				
df	**0.200**	**0.150**	**0.125**	**0.100**	**0.075**	**0.050**	**0.025**	**0.010**	**0.005**
1	1.376	1.963	2.414	3.078	4.165	6.314	12.706	31.821	63.657
2	1.061	1.386	1.604	1.886	2.282	2.920	4.303	6.965	9.925
3	0.978	1.250	1.423	1.638	1.924	2.353	3.182	4.541	5.841
4	0.941	1.190	1.344	1.533	1.778	2.132	2.776	3.747	4.604
5	0.920	1.156	1.301	1.476	1.699	2.015	2.571	3.365	4.032
6	0.906	1.134	1.273	1.440	1.650	1.943	2.447	3.143	3.707
7	0.896	1.119	1.254	1.415	1.617	1.895	2.365	2.998	3.499
8	0.889	1.108	1.240	1.397	1.592	1.860	2.306	2.896	3.355
9	0.883	1.100	1.230	1.383	1.574	1.833	2.262	2.821	3.250
10	0.879	1.093	1.221	1.372	1.559	1.812	2.228	2.764	3.169
11	0.876	1.088	1.214	1.363	1.548	1.796	2.201	2.718	3.106
12	0.873	1.083	1.209	1.356	1.538	1.782	2.179	2.681	3.055
13	0.870	1.079	1.204	1.350	1.530	1.771	2.160	2.650	3.012
14	0.868	1.076	1.200	1.345	1.523	1.761	2.145	2.624	2.977
15	0.866	1.074	1.197	1.341	1.517	1.753	2.131	2.602	2.947
16	0.865	1.071	1.194	1.337	1.512	1.746	2.120	2.583	2.921
17	0.863	1.069	1.191	1.333	1.508	1.740	2.110	2.567	2.898
18	0.862	1.067	1.189	1.330	1.504	1.734	2.101	2.552	2.878
19	0.861	1.066	1.187	1.328	1.500	1.729	2.093	2.539	2.861
20	0.860	1.064	1.185	1.325	1.497	1.725	2.086	2.528	2.845
21	0.859	1.063	1.183	1.323	1.494	1.721	2.080	2.518	2.831
22	0.858	1.061	1.182	1.321	1.492	1.717	2.074	2.508	2.819
23	0.858	1.060	1.180	1.319	1.489	1.714	2.069	2.500	2.807
24	0.857	1.059	1.179	1.318	1.487	1.711	2.064	2.492	2.797
25	0.856	1.058	1.178	1.316	1.485	1.708	2.060	2.485	2.787
26	0.856	1.058	1.177	1.315	1.483	1.706	2.056	2.479	2.779
27	0.855	1.057	1.176	1.314	1.482	1.703	2.052	2.473	2.771
28	0.855	1.056	1.175	1.313	1.480	1.701	2.048	2.467	2.763
29	0.854	1.055	1.174	1.311	1.479	1.699	2.045	2.462	2.756
30	0.854	1.055	1.173	1.310	1.477	1.697	2.042	2.457	2.750
40	0.851	1.050	1.167	1.303	1.468	1.684	2.021	2.423	2.704
50	0.849	1.047	1.164	1.299	1.462	1.676	2.009	2.403	2.678
100	0.845	1.042	1.157	1.290	1.451	1.660	1.984	2.364	2.626
200	0.843	1.039	1.154	1.286	1.445	1.653	1.972	2.345	2.601

Appendix 4: The χ 2-Distribution

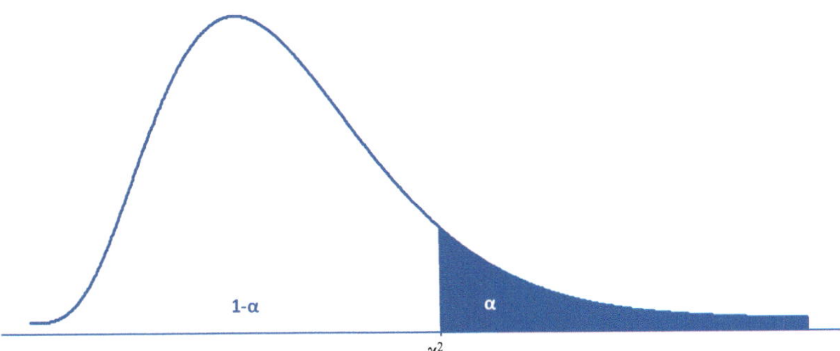

	α								
df	0.200	0.150	0.125	0.100	0.075	0.050	0.025	0.010	0.005
1	1.642	2.072	2.354	2.706	3.170	3.841	5.024	6.635	7.879
2	3.219	3.794	4.159	4.605	5.181	5.991	7.378	9.210	10.597
3	4.642	5.317	5.739	6.251	6.905	7.815	9.348	11.345	12.838
4	5.989	6.745	7.214	7.779	8.496	9.488	11.143	13.277	14.860
5	7.289	8.115	8.625	9.236	10.008	11.070	12.833	15.086	16.750
6	8.558	9.446	9.992	10.645	11.466	12.592	14.449	16.812	18.548
7	9.803	10.748	11.326	12.017	12.883	14.067	16.013	18.475	20.278
8	11.030	12.027	12.636	13.362	14.270	15.507	17.535	20.090	21.955
9	12.242	13.288	13.926	14.684	15.631	16.919	19.023	21.666	23.589
10	13.442	14.534	15.198	15.987	16.971	18.307	20.483	23.209	25.188
11	14.631	15.767	16.457	17.275	18.294	19.675	21.920	24.725	26.757
12	15.812	16.989	17.703	18.549	19.602	21.026	23.337	26.217	28.300
13	16.985	18.202	18.939	19.812	20.897	22.362	24.736	27.688	29.819
14	18.151	19.406	20.166	21.064	22.180	23.685	26.119	29.141	31.319
15	19.311	20.603	21.384	22.307	23.452	24.996	27.488	30.578	32.801
16	20.465	21.793	22.595	23.542	24.716	26.296	28.845	32.000	34.267
17	21.615	22.977	23.799	24.769	25.970	27.587	30.191	33.409	35.718
18	22.760	24.155	24.997	25.989	27.218	28.869	31.526	34.805	37.156
19	23.900	25.329	26.189	27.204	28.458	30.144	32.852	36.191	38.582
20	25.038	26.498	27.376	28.412	29.692	31.410	34.170	37.566	39.997
21	26.171	27.662	28.559	29.615	30.920	32.671	35.479	38.932	41.401
22	27.301	28.822	29.737	30.813	32.142	33.924	36.781	40.289	42.796
23	28.429	29.979	30.911	32.007	33.360	35.172	38.076	41.638	44.181
24	29.553	31.132	32.081	33.196	34.572	36.415	39.364	42.980	45.559
25	30.675	32.282	33.247	34.382	35.780	37.652	40.646	44.314	46.928
26	31.795	33.429	34.410	35.563	36.984	38.885	41.923	45.642	48.290
27	32.912	34.574	35.570	36.741	38.184	40.113	43.195	46.963	49.645
28	34.027	35.715	36.727	37.916	39.380	41.337	44.461	48.278	50.993
29	35.139	36.854	37.881	39.087	40.573	42.557	45.722	49.588	52.336
30	36.250	37.990	39.033	40.256	41.762	43.773	46.979	50.892	53.672
40	47.269	49.244	50.424	51.805	53.501	55.758	59.342	63.691	66.766
50	58.164	60.346	61.647	63.167	65.030	67.505	71.420	76.154	79.490
100	111.667	114.659	116.433	118.498	121.017	124.342	129.561	135.807	140.169
200	216.609	220.744	223.186	226.021	229.466	233.994	241.058	249.445	255.264

Appendix 5.1: The F-Distribution ($\alpha = 10\ \%$)

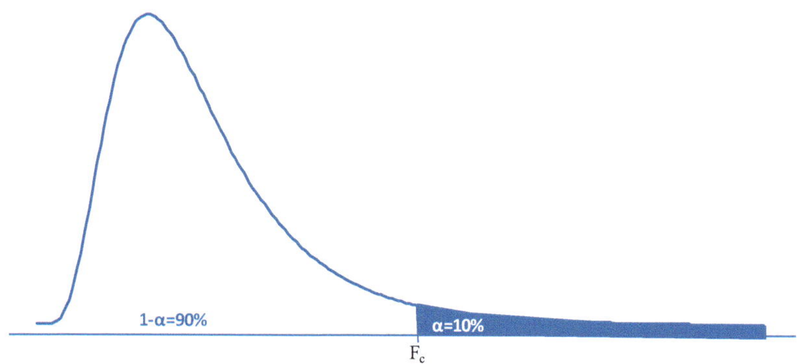

df_2	df_1: 1	2	3	4	5	6	7	8	9	10	15	20	30	50
1	39.86	49.50	53.59	55.83	57.24	58.20	58.91	59.44	59.86	60.19	61.22	61.74	62.26	62.69
2	8.53	9.00	9.16	9.24	9.29	9.33	9.35	9.37	9.38	9.39	9.42	9.44	9.46	9.47
3	5.54	5.46	5.39	5.34	5.31	5.28	5.27	5.25	5.24	5.23	5.20	5.18	5.17	5.15
4	4.54	4.32	4.19	4.11	4.05	4.01	3.98	3.95	3.94	3.92	3.87	3.84	3.82	3.80
5	4.06	3.78	3.62	3.52	3.45	3.40	3.37	3.34	3.32	3.30	3.24	3.21	3.17	3.15
6	3.78	3.46	3.29	3.18	3.11	3.05	3.01	2.98	2.96	2.94	2.87	2.84	2.80	2.77
7	3.59	3.26	3.07	2.96	2.88	2.83	2.78	2.75	2.72	2.70	2.63	2.59	2.56	2.52
8	3.46	3.11	2.92	2.81	2.73	2.67	2.62	2.59	2.56	2.54	2.46	2.42	2.38	2.35
9	3.36	3.01	2.81	2.69	2.61	2.55	2.51	2.47	2.44	2.42	2.34	2.30	2.25	2.22
10	3.29	2.92	2.73	2.61	2.52	2.46	2.41	2.38	2.35	2.32	2.24	2.20	2.16	2.12
11	3.23	2.86	2.66	2.54	2.45	2.39	2.34	2.30	2.27	2.25	2.17	2.12	2.08	2.04
12	3.18	2.81	2.61	2.48	2.39	2.33	2.28	2.24	2.21	2.19	2.10	2.06	2.01	1.97
13	3.14	2.76	2.56	2.43	2.35	2.28	2.23	2.20	2.16	2.14	2.05	2.01	1.96	1.92
14	3.10	2.73	2.52	2.39	2.31	2.24	2.19	2.15	2.12	2.10	2.01	1.96	1.91	1.87
15	3.07	2.70	2.49	2.36	2.27	2.21	2.16	2.12	2.09	2.06	1.97	1.92	1.87	1.83
16	3.05	2.67	2.46	2.33	2.24	2.18	2.13	2.09	2.06	2.03	1.94	1.89	1.84	1.79
17	3.03	2.64	2.44	2.31	2.22	2.15	2.10	2.06	2.03	2.00	1.91	1.86	1.81	1.76
18	3.01	2.62	2.42	2.29	2.20	2.13	2.08	2.04	2.00	1.98	1.89	1.84	1.78	1.74
19	2.99	2.61	2.40	2.27	2.18	2.11	2.06	2.02	1.98	1.96	1.86	1.81	1.76	1.71
20	2.97	2.59	2.38	2.25	2.16	2.09	2.04	2.00	1.96	1.94	1.84	1.79	1.74	1.69
21	2.96	2.57	2.36	2.23	2.14	2.08	2.02	1.98	1.95	1.92	1.83	1.78	1.72	1.67
22	2.95	2.56	2.35	2.22	2.13	2.06	2.01	1.97	1.93	1.90	1.81	1.76	1.70	1.65
23	2.94	2.55	2.34	2.21	2.11	2.05	1.99	1.95	1.92	1.89	1.80	1.74	1.69	1.64
24	2.93	2.54	2.33	2.19	2.10	2.04	1.98	1.94	1.91	1.88	1.78	1.73	1.67	1.62
25	2.92	2.53	2.32	2.18	2.09	2.02	1.97	1.93	1.89	1.87	1.77	1.72	1.66	1.61
26	2.91	2.52	2.31	2.17	2.08	2.01	1.96	1.92	1.88	1.86	1.76	1.71	1.65	1.59
27	2.90	2.51	2.30	2.17	2.07	2.00	1.95	1.91	1.87	1.85	1.75	1.70	1.64	1.58
28	2.89	2.50	2.29	2.16	2.06	2.00	1.94	1.90	1.87	1.84	1.74	1.69	1.63	1.57
29	2.89	2.50	2.28	2.15	2.06	1.99	1.93	1.89	1.86	1.83	1.73	1.68	1.62	1.56
30	2.88	2.49	2.28	2.14	2.05	1.98	1.93	1.88	1.85	1.82	1.72	1.67	1.61	1.55
40	2.84	2.44	2.23	2.09	2.00	1.93	1.87	1.83	1.79	1.76	1.66	1.61	1.54	1.48
50	2.81	2.41	2.20	2.06	1.97	1.90	1.84	1.80	1.76	1.73	1.63	1.57	1.50	1.44
100	2.76	2.36	2.14	2.00	1.91	1.83	1.78	1.73	1.69	1.66	1.56	1.49	1.42	1.35
200	2.73	2.33	2.11	1.97	1.88	1.80	1.75	1.70	1.66	1.63	1.52	1.46	1.38	1.31

Appendix 5.2: The F-Distribution ($\alpha = 5\%$)

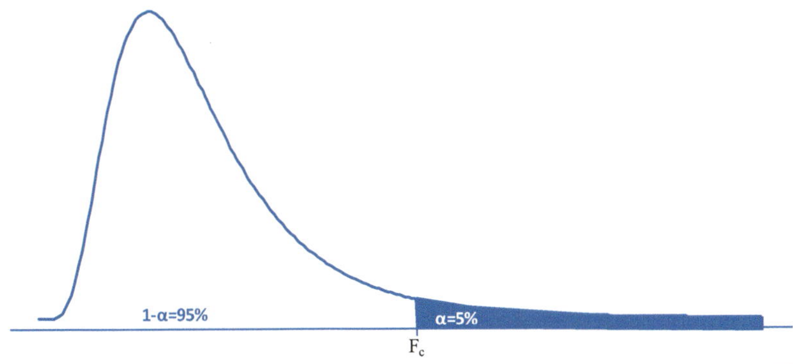

df_2 \ df_1	1	2	3	4	5	6	7	8	9	10	15	20	30	50
1	161.45	199.50	215.71	224.58	230.16	233.99	236.77	238.88	240.54	241.88	245.95	248.01	250.10	251.77
2	18.51	19.00	19.16	19.25	19.30	19.33	19.35	19.37	19.38	19.40	19.43	19.45	19.46	19.48
3	10.13	9.55	9.28	9.12	9.01	8.94	8.89	8.85	8.81	8.79	8.70	8.66	8.62	8.58
4	7.71	6.94	6.59	6.39	6.26	6.16	6.09	6.04	6.00	5.96	5.86	5.80	5.75	5.70
5	6.61	5.79	5.41	5.19	5.05	4.95	4.88	4.82	4.77	4.74	4.62	4.56	4.50	4.44
6	5.99	5.14	4.76	4.53	4.39	4.28	4.21	4.15	4.10	4.06	3.94	3.87	3.81	3.75
7	5.59	4.74	4.35	4.12	3.97	3.87	3.79	3.73	3.68	3.64	3.51	3.44	3.38	3.32
8	5.32	4.46	4.07	3.84	3.69	3.58	3.50	3.44	3.39	3.35	3.22	3.15	3.08	3.02
9	5.12	4.26	3.86	3.63	3.48	3.37	3.29	3.23	3.18	3.14	3.01	2.94	2.86	2.80
10	4.96	4.10	3.71	3.48	3.33	3.22	3.14	3.07	3.02	2.98	2.85	2.77	2.70	2.64
11	4.84	3.98	3.59	3.36	3.20	3.09	3.01	2.95	2.90	2.85	2.72	2.65	2.57	2.51
12	4.75	3.89	3.49	3.26	3.11	3.00	2.91	2.85	2.80	2.75	2.62	2.54	2.47	2.40
13	4.67	3.81	3.41	3.18	3.03	2.92	2.83	2.77	2.71	2.67	2.53	2.46	2.38	2.31
14	4.60	3.74	3.34	3.11	2.96	2.85	2.76	2.70	2.65	2.60	2.46	2.39	2.31	2.24
15	4.54	3.68	3.29	3.06	2.90	2.79	2.71	2.64	2.59	2.54	2.40	2.33	2.25	2.18
16	4.49	3.63	3.24	3.01	2.85	2.74	2.66	2.59	2.54	2.49	2.35	2.28	2.19	2.12
17	4.45	3.59	3.20	2.96	2.81	2.70	2.61	2.55	2.49	2.45	2.31	2.23	2.15	2.08
18	4.41	3.55	3.16	2.93	2.77	2.66	2.58	2.51	2.46	2.41	2.27	2.19	2.11	2.04
19	4.38	3.52	3.13	2.90	2.74	2.63	2.54	2.48	2.42	2.38	2.23	2.16	2.07	2.00
20	4.35	3.49	3.10	2.87	2.71	2.60	2.51	2.45	2.39	2.35	2.20	2.12	2.04	1.97
21	4.32	3.47	3.07	2.84	2.68	2.57	2.49	2.42	2.37	2.32	2.18	2.10	2.01	1.94
22	4.30	3.44	3.05	2.82	2.66	2.55	2.46	2.40	2.34	2.30	2.15	2.07	1.98	1.91
23	4.28	3.42	3.03	2.80	2.64	2.53	2.44	2.37	2.32	2.27	2.13	2.05	1.96	1.88
24	4.26	3.40	3.01	2.78	2.62	2.51	2.42	2.36	2.30	2.25	2.11	2.03	1.94	1.86
25	4.24	3.39	2.99	2.76	2.60	2.49	2.40	2.34	2.28	2.24	2.09	2.01	1.92	1.84
26	4.23	3.37	2.98	2.74	2.59	2.47	2.39	2.32	2.27	2.22	2.07	1.99	1.90	1.82
27	4.21	3.35	2.96	2.73	2.57	2.46	2.37	2.31	2.25	2.20	2.06	1.97	1.88	1.81
28	4.20	3.34	2.95	2.71	2.56	2.45	2.36	2.29	2.24	2.19	2.04	1.96	1.87	1.79
29	4.18	3.33	2.93	2.70	2.55	2.43	2.35	2.28	2.22	2.18	2.03	1.94	1.85	1.77
30	4.17	3.32	2.92	2.69	2.53	2.42	2.33	2.27	2.21	2.16	2.01	1.93	1.84	1.76
40	4.08	3.23	2.84	2.61	2.45	2.34	2.25	2.18	2.12	2.08	1.92	1.84	1.74	1.66
50	4.03	3.18	2.79	2.56	2.40	2.29	2.20	2.13	2.07	2.03	1.87	1.78	1.69	1.60
100	3.94	3.09	2.70	2.46	2.31	2.19	2.10	2.03	1.97	1.93	1.77	1.68	1.57	1.48
200	3.89	3.04	2.65	2.42	2.26	2.14	2.06	1.98	1.93	1.88	1.72	1.62	1.52	1.41

Appendix 5.3: The F-Distribution ($\alpha = 1\ \%$)

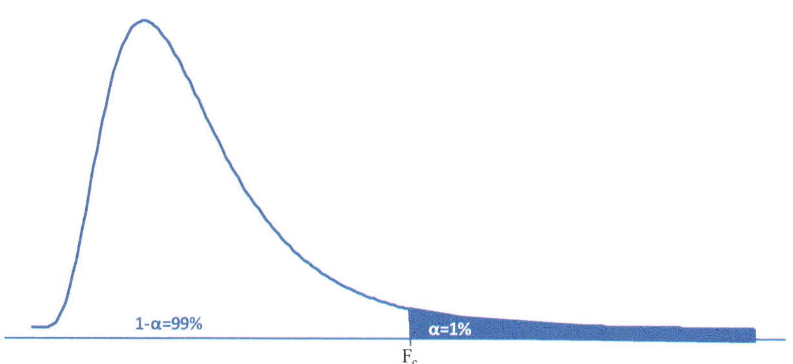

	df1													
df2	1	2	3	4	5	6	7	8	9	10	15	20	30	50
1	4052.2	4999.5	5403.4	5624.6	5763.6	5859.0	5928.4	5981.1	6022.5	6055.8	6157.3	6208.7	6260.6	6302.5
2	98.50	99.00	99.17	99.25	99.30	99.33	99.36	99.37	99.39	99.40	99.43	99.45	99.47	99.48
3	34.12	30.82	29.46	28.71	28.24	27.91	27.67	27.49	27.35	27.23	26.87	26.69	26.50	26.35
4	21.20	18.00	16.69	15.98	15.52	15.21	14.98	14.80	14.66	14.55	14.20	14.02	13.84	13.69
5	16.26	13.27	12.06	11.39	10.97	10.67	10.46	10.29	10.16	10.05	9.72	9.55	9.38	9.24
6	13.75	10.92	9.78	9.15	8.75	8.47	8.26	8.10	7.98	7.87	7.56	7.40	7.23	7.09
7	12.25	9.55	8.45	7.85	7.46	7.19	6.99	6.84	6.72	6.62	6.31	6.16	5.99	5.86
8	11.26	8.65	7.59	7.01	6.63	6.37	6.18	6.03	5.91	5.81	5.52	5.36	5.20	5.07
9	10.56	8.02	6.99	6.42	6.06	5.80	5.61	5.47	5.35	5.26	4.96	4.81	4.65	4.52
10	10.04	7.56	6.55	5.99	5.64	5.39	5.20	5.06	4.94	4.85	4.56	4.41	4.25	4.12
11	9.65	7.21	6.22	5.67	5.32	5.07	4.89	4.74	4.63	4.54	4.25	4.10	3.94	3.81
12	9.33	6.93	5.95	5.41	5.06	4.82	4.64	4.50	4.39	4.30	4.01	3.86	3.70	3.57
13	9.07	6.70	5.74	5.21	4.86	4.62	4.44	4.30	4.19	4.10	3.82	3.66	3.51	3.38
14	8.86	6.51	5.56	5.04	4.69	4.46	4.28	4.14	4.03	3.94	3.66	3.51	3.35	3.22
15	8.68	6.36	5.42	4.89	4.56	4.32	4.14	4.00	3.89	3.80	3.52	3.37	3.21	3.08
16	8.53	6.23	5.29	4.77	4.44	4.20	4.03	3.89	3.78	3.69	3.41	3.26	3.10	2.97
17	8.40	6.11	5.18	4.67	4.34	4.10	3.93	3.79	3.68	3.59	3.31	3.16	3.00	2.87
18	8.29	6.01	5.09	4.58	4.25	4.01	3.84	3.71	3.60	3.51	3.23	3.08	2.92	2.78
19	8.18	5.93	5.01	4.50	4.17	3.94	3.77	3.63	3.52	3.43	3.15	3.00	2.84	2.71
20	8.10	5.85	4.94	4.43	4.10	3.87	3.70	3.56	3.46	3.37	3.09	2.94	2.78	2.64
21	8.02	5.78	4.87	4.37	4.04	3.81	3.64	3.51	3.40	3.31	3.03	2.88	2.72	2.58
22	7.95	5.72	4.82	4.31	3.99	3.76	3.59	3.45	3.35	3.26	2.98	2.83	2.67	2.53
23	7.88	5.66	4.76	4.26	3.94	3.71	3.54	3.41	3.30	3.21	2.93	2.78	2.62	2.48
24	7.82	5.61	4.72	4.22	3.90	3.67	3.50	3.36	3.26	3.17	2.89	2.74	2.58	2.44
25	7.77	5.57	4.68	4.18	3.85	3.63	3.46	3.32	3.22	3.13	2.85	2.70	2.54	2.40
26	7.72	5.53	4.64	4.14	3.82	3.59	3.42	3.29	3.18	3.09	2.81	2.66	2.50	2.36
27	7.68	5.49	4.60	4.11	3.78	3.56	3.39	3.26	3.15	3.06	2.78	2.63	2.47	2.33
28	7.64	5.45	4.57	4.07	3.75	3.53	3.36	3.23	3.12	3.03	2.75	2.60	2.44	2.30
29	7.60	5.42	4.54	4.04	3.73	3.50	3.33	3.20	3.09	3.00	2.73	2.57	2.41	2.27
30	7.56	5.39	4.51	4.02	3.70	3.47	3.30	3.17	3.07	2.98	2.70	2.55	2.39	2.25
40	7.31	5.18	4.31	3.83	3.51	3.29	3.12	2.99	2.89	2.80	2.52	2.37	2.20	2.06
50	7.17	5.06	4.20	3.72	3.41	3.19	3.02	2.89	2.78	2.70	2.42	2.27	2.10	1.95
100	6.90	4.82	3.98	3.51	3.21	2.99	2.82	2.69	2.59	2.50	2.22	2.07	1.89	1.74
200	6.76	4.71	3.88	3.41	3.11	2.89	2.73	2.60	2.50	2.41	2.13	1.97	1.79	1.63

GPSR Compliance

The European Union's (EU) General Product Safety Regulation (GPSR) is a set of rules that requires consumer products to be safe and our obligations to ensure this.

If you have any concerns about our products, you can contact us on ProductSafety@springernature.com

In case Publisher is established outside the EU, the EU authorized representative is:

Springer Nature Customer Service Center GmbH
Europaplatz 3
69115 Heidelberg, Germany

Batch number: 10382077

Printed by Printforce, the Netherlands